Preliminary Edition, Revised

WAYNE GIESAU

Beginning Algebra
Once and For All

Jane A. Jamsen
Gwendolyn K. Hetler

Mathematics and Computer Science Department
Northern Michigan University

KENDALL/HUNT PUBLISHING COMPANY
4050 Westmark Drive   Dubuque, Iowa 52002

Cover image © 2002 Dynamic Graphics

2008 Printing

Copyright © 2000 by Kendall/Hunt Publishing Company

ISBN 978-0-7872-9146-4

All rights reserved. No part of this publication may be reproduced,
stored in a retrieval system, or transmitted, in any form or by any means,
electronic, mechanical, photocopying, recording, or otherwise,
without the prior written permission of the copyright owner.

Printed in the United States of America
10 9 8 7 6 5 4

Contents

Chapter One – Building the Foundation, 1

1.1 Integers: Concepts, Models, and Operations, 1
—order and absolute value
—introduction to the coordinate system & graphing points
—models and rules for integer operations
—introductory calculator usage
—introduction to the use of variables
—*problem set 1.1*, 15
  • *exploration problems*, 18
  • *solutions to problem set 1.1*, 19

1.2 Development and Use of the Order of Operations, 21
—exponential notation
—evaluating mathematical expressions
—translation from words to symbols
—informal equation solving
  • "diagraming" and solving an equation, the pan-balance method
  • cover-up method
  • trial and error
—*problem set 1.2*, 32
  • *exploration problems*, 33
  • *solutions to problem set 1.2*, 34

1.3 Properties of Integers and Algebraic Expressions, 36
—simplifying expressions using the properties of integers
—properties of equality
—introduction to polynomials
  • simplifying and factoring
  • multiplying using algebra tiles
—*problem set 1.3*, 46
  • *exploration problems*, 48
  • *solutions to problem set 1.3*, 49

1.4 Integers in Problem Solving, 51
—finding the midpoint of a line segment
  • on the number line and coordinate system
—number sequences
—problem solving strategies
—*problem set 1.4*, 59
  • *exploration problems*, 62
  • *solutions to problem set 1.4*, 63

1.5 Connecting Integers and Geometry, 65
—graphing rectangles and triangles and transformations in the coordinate plane
—exploring perimeter and area of rectangles and triangles
  • on the coordinate system
  • using tiles
  • on the geoboard
—perimeter and area in applications
  • developing and interpreting graphical representations
—*problem set 1.5*, 77
  • *exploration problems*, 80
  • *solutions to problem set 1.5*, 81

—*chapter 1 review*, 83
- *exploration problems*, 88
- *solutions to chapter 1 review*, 89

Chapter Two – The Role of Rational Numbers in Algebra, 91

2.1 Rational Numbers: Concepts, Models, Addition, Subtraction, Comparison, 91
—definition and models
- area, number line, and sets
- part-to-whole and whole-to-part

—mental arithmetic and estimation
—models and rules for rational number addition and subtraction
- equivalencies
- ordering
- rational expressions

—*problem set 2.1*, 104
- *exploration problems*, 105
- *solutions to problem set 2.1*, 107

2.2 Rational Numbers: Multiplication, Division, Applications, 108
—models and rules for rational number multiplication and division
- complex fractions

—variable work extended
—*problem set 2.2*, 115
- *exploration problems*, 116
- *solutions to problem set 2.2*, 117

2.3 Development and Use of the Properties of Rational Numbers, 119
—polynomials with rational number coeffficients
—*problem set 2.3*, 123
- *exploration problems*, 124
- *solutions to problem set 2.3*, 125

2.4 Informal Equation Solving, 126
—cover-up method extended
—*problem set 2.4*, 130
- *exploration problems*, 131
- *solutions to problem set 2.4*, 132

—*chapter 2 review*, 133
- *exploration problems*, 135
- *solutions to chapter 1 review*, 137

Chapter Three – Dimensional Analysis and Other Applications of Fractions, Decimals, and Percent, 138

3.1 Ratios & Unit Rates, 138
—part-to-part and part-to-whole
—proportions
- solving word problems
- similarity

—*problem set 3.1*, 146
- *exploration problems*, 148
- *solutions to problem set 3.1*, 150

3.2 Relationships Among Fractions, Decimals, and Percents, 151
—area models
—terminating and repeating decimals
—equivalencies and order
—size changes
—estimation and mental arithmetic in application problems
—*problem set 3.2*, 159
- *exploration problems*, 160
- *solutions to problem set 3.2*, 162

3.3 Rational Numbers in Problem Solving, 164
—mark–up, discount, commission
—applications in geometry
—interpreting graphs
- concept of functions
—*problem set 3.3*, 172
- *exploration problems*, 1
- *solutions to problem set 3.3*, 176
—*chapter 3 review*, 177
- *exploration problems*, 179
- *solutions to chapter 3 review*, 180

Chapter Four – Probability and Statistics, 182

4.1 Concepts of Probability, 182
—concrete representations
—calculating simple probabilities
—geometric probability
—*problem set 4.1*, 189
- *exploration problems*, 192
- *solutions to problem set 4.1*, 194

4.2 Tree Diagrams and Compound Events, 195
—development of addition and multiplication rules
—sampling with and without replacement
—*problem set 4.2*, 201
- *exploration problems*, 203
- *solutions to problem set 4.2*, 204

4.3 Statistics, 206
—summarizing data
- line plots, stem and leaf plots, bar graphs, circle graphs
—*problem set 4.3*, 212
- *exploration problems*, 214
- *solutions to problem set 4.3*, 215

4.4 Measures of Clustering and Box Plots, 217
—mean, median, mode
- meanings, calculations and comparisons
—range
—box and whisker plots
—*problem set 4.4*, 224
- *exploration problems*, 225
- *solutions to problem set 4.4*, 227
—*chapter 4 review*, 228
- *exploration problems*, 231
- *solutions to chapter 4 review*, 232

Chapter Five – Linear Equations and Inequalities, 235

5.1 Formal Equation Solving, 235
—connecting the formal and informal methods
—formal equation solving
- additive and multiplicative inverses
- writing equations from problems

—*problem set 5.1*, 244
- *exploration problems*, 246
- *solutions to problem set 5.1*, 247

5.2 Development of Algebraic Equations in Two Unknowns, 249
—mathematical relations
—mapping
—generating equations from charts, graphs and tables
—graphing in two variables
- meaning of "solution"

—graphing systems of equations
—*problem set 5.2*, 257
- *exploration problems*, 260
- *solutions to problem set 5.2*, 261

5.3 Literal Equations, 264
—use of formulas
—*problem set 5.3*, 269
- *exploration problems*, 270
- *solutions to problem set 5.3*, 271

5.4 Using Variables in Problem Solving, 275
—using tables and lists to generate equations
—writing and solving equations in problem contexts, extended
—*problem set 5.4*, 283
- *exploration problems*, 285
- *solutions to problem set 5.4*, 286

5.5 Inequalities in One and Two Variables, 287
—solving inequalities in one variable
- graphing the solution

—graphing inequalities in two variables (optional)
—applications of inequalities
- at most, at least, interpreting the solution

—*problem set 5.5*, 298
- *exploration problems*, 299
- *solutions to problem set 5.5*, 300

— *chapter 5 review*, 304
- *exploration problems*, 307
- *solutions to chapter 5 review*, 309

Chapter Six – Exponents and Radicals, 313

6.1 Integral Exponents, 313
—development of rules
—zero & negative exponents
—exploration with the calculator
—using exponent rules to solve equations
—*problem set 6.1*, 321
- *exploration problems*, 322
- *solutions to problem set 6.1*, 323

6.2 Exponents in Applications, 325
—scientific notation
—inflation & depreciation
—population growth & decline
  • graphical representations
—simple and compound interest
—approximating solutions of exponential equations
—*problem set 6.2*, 336
  • *exploration problems*, 338
  • *solutions to problem set 6.2*, 339

6.3 Radicals and Rational Exponents, 340
—geometric model using the Pythagorean Theorem
—relationship of radicals and exponents
—exploration with the calculator
—rational & irrational numbers
  • decimal equivalents and approximations
—introduction to simplying radicals
—*problem set 6.3*, 349
  • *exploration problems*, 350
  • *solutions to problem set 6.3*, 351

6.4 Applications of Rational Exponents and Radicals, 353
—the distance formula
—finding the midpoint of any line segment
—perimeter and area of geometric figures revisited
  • on the coordinate system
  • using the geoboard model
—*problem set 6.4*, 362
  • *exploration problems*, 365
  • *solutions to problem set 6.4*, 366
— *chapter 6 review*, 367
  • *exploration problems*, 370
  • *solutions to chapter 6 review*, 371

Chapter Seven – Functions, 374

7.1 The Function Concept, 374
—function diagrams
—equations, ordered pairs and graphs
—applications
—*problem set 7.1*, 382
  • *exploration problems*, 386
  • *solutions to problem set 7.1*, 387

7.2 Linear Functions, 389
—linear equations and rate of change
—slope-intercept form of linear equations
—applications of linear functions
—*problem set 7.2*, 400
  • *exploration problems*, 402
  • *solutions to problem set 7.2*, 403

7.3 **Quadratic and Exponential Equations and Functions, 406**
—quadratics using charts and graphs
- applications
—analyzing and graphing exponential functions
—*problem set 7.3, 415*
- *exploration problems, 418*
- *solutions to problem set 7.3, 419*

7.4 **Functional Relationships, 423**
—distance as related to time and rate
—business
- interest amounts as related to length of mortgage
—geometric applications
- surface area
- surface area and volume
—*problem set 7.4, 437*
- *exploration problems, 439*
- *solutions to problem set 7.4, 440*

7.5 **Explorations with the Graphing Calculator, 442**
—reexamining linear and quadratic equations
- solutions
- application problems
- introduction to polynomial equations
- graphical solutions to polynomial equations
—median-median line and line of best fit
—*problem set 7.5, 451*
- *exploration problems, 452*
- *solutions to problem set 7.5, 453*
— *chapter 7 review, 454*
- *exploration problems, 458*
- *solutions to chapter 7 review, 459*

CHAPTER ONE - Building the Foundation

Section 1.1 - Integers: Concepts, Models, and Operations

As children, we learn to count the number of digits on one hand by reciting "one, two, three, four, five." These are examples of the **counting** or **natural numbers**. In set notation, the counting numbers are represented:

$$N = \{1, 2, 3, 4, 5, ...\}$$

To represent the number of cookies on a plate after someone has eaten them all, the set of counting numbers is not sufficient. The set of **whole numbers,** which includes *zero*, is represented as

$$W = \{0, 1, 2, 3, 4, 5, ...\}$$

We also require representations for losses as well as gains, for debits as well as credits, for temperatures above and below zero. In these cases, the "direction" of the number is as important as the value of the number itself. The set of **integers** can be represented:

$$I = \{...,-3, -2, -1, 0, 1, 2, 3, 4, ...\}$$

and can be depicted graphically on the temperature scale as shown below

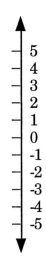

or on the number line

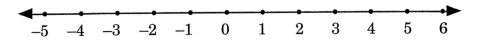

Note that numbers down from zero on the temperature scale and to the left of zero on the number line are negative in value while numbers up and to the right of zero are positive.

In general, positive integers represent gain and negative integers represent loss.

Also, we can compare integers, that is, order integers, by using the "less than" sign (<) or the "greater than" sign (>). For any pair of integers, an integer to the left on the number line is smaller in value than an integer to its right, and, conversely, an integer on the right is greater in value that an integer to its left. For example, 2 < 3, 3 > 2, but –2 > –3.

This concept may be easier to understand if you look at the temperature scale and think about "warmth." Two degrees is *less warm* than 3 degrees, 3° is *warmer* than 2°, but 2° below zero is *warmer* than 3° below zero.

Consider the *relative positions* of 3 and –3 on the number line above. –3 can be thought of as the opposite of +3 (or in the opposite direction from zero) and +3 can be thought of as the opposite of –3, but both 3 and –3 are 3 units from zero. Therefore, we say that –3 and +3 have the same **absolute value**.

Mathematically, the distance a point is from zero on a number line, regardless of the direction, is called its **absolute value**. Absolute value of a quantity is symbolized by placing the quantity between two vertical lines. For example, both positive two and negative two are two units from zero on the number line, and this fact is stated symbolically by: $|2| = 2$ and $|-2| = 2$.

Sample Problems

1. Describe a few contexts like temperature that would require both positive and negative integers.

   Possible answers: Gains and losses in football yardage, gains and losses on the stock market, gains and losses in student enrollment in schools, assets and debits in budget considerations.

2. If n is an integer >2, find all points on the number line that make the statement true. (Notice that we are using a letter, n, to ask this question. The use of a letter, called a *variable* in mathematics, is a distinguishing feature of algebra and allows us to make some useful general statements.)

   Solution:

   ```
   ├─┼─┼─●─●─●─●─●─→
   0 1 2 3 4 5
   ```
   which we can also write as {3, 4, 5, ...}

3. If $|n| = 4$, where n is an integer, what value(s) of n would make the statement true?

   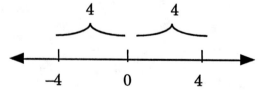

   Solution: 4 or –4.

Challenge Problem

Let $-a$ represent the opposite of a, for any integer a. Is $-a$ always negative? Choose values for a to justify your answer.

Solution: Case 1:   If $a = 2$, then $-a = -2$;
             Case 2:   If $a = -2$, then $-a =$ the opposite of (-2), written symbolically as $-(-2), = 2$
Therefore, $-a$ is not always negative.

Extending the set of integers to a two-dimensional graph, the **Cartesian coordinate system** is labeled as shown.

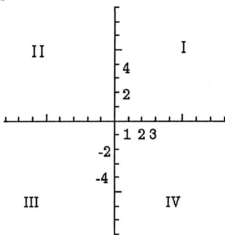

The two perpendicular lines intersect in a point called the **origin**. The lines themselves, called **axes**, divide the plane into 4 quadrants numbered I, II, III, and IV in a counterclockwise direction as shown above.

To depict a point in any quadrant, we use an **ordered pair** of integers which are two values given *in the order they are used*. The **origin** is named by the ordered pair **(0,0)**.

Let's consider the point (2,3). The first value, 2, gives the horizontal distance and direction from the *origin* and the second value, 3, gives the vertical distance and direction from *the x-value*, in this case, 2. In mathematical terms, the two values are called the ***x*- and *y*-coordinates**, respectively. When plotting a point, the directions indicated by the signs of the integers are critical.

Next consider the point (–3,5). Here the x-coordinate is –3 and the y-coordinate is 5. The horizontal distance is on the "negative" side of the origin but the vertical distance from $x = -3$ is up 5 units (a positive direction). These points and a few more points are plotted and labeled for you on the Cartesian coordinate system below.

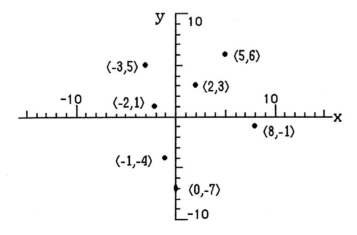

Sample Problems

1. Find two points in each quadrant and discuss how the points in each quadrant can be identified in general.

   Some examples:
   - Quadrant I: (2,3), (5,9)
   - Quadrant II: (–2,3), (–5,10)
   - Quadrant III: (–3,–4), (–2,–5)
   - Quadrant IV: (3,–4), (1,–5)

   For points in quadrant I, both coordinates are positive; for points in quadrant II, only the *x*-coordinate is negative; for points in quadrant III, both the *x*-coordinate and *y*-coordinate are negative; and for points in quadrant IV, only the *y*-coordinate is negative.

2. Find the distance between the two points indicated on the number line below.

   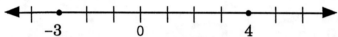

   Solution: Since a unit interval is the distance between any two consecutive points, count the number of unit intervals that separate the two points, 4 and –3. There are 7 intervals between the points so 4 and –3 are 7 units apart.

Integer Addition

Work with integers can be modeled on the temperature scale using directed line segments.

Example 1: If the temperature in Marquette, MI at 11:00 p.m. on January 30 was 5° C, and the temperature dropped 6° C during the night, what was the temperature in the early morning of January 31?

Drawing an arrow from 0 to +5 and then drawing an arrow in the opposite direction to represent a drop of 6°, results in a temperature reading of –1° C the morning of January 31.

Mathematically, 5 + (–6) = –1

Example 2: During the morning of January 31, if the temperature in Marquette, MI rose from –1° to 9° C, what was the change in temperature from the early morning?

Mathematically, this change is symbolized by

– 1 + □ = 9   where □ represents the temperature change.

To find the missing value, we can count, using the vertical number line, the directed distance from −1 to +9.

The temperature change between early morning and noon on that day in Marquette was a 10° rise.

In other words, − 1 + ⬚10 = 9   10° increase

Note: We were not initially given the amount of temperature change in this example − we only knew the direction of the change and the result.

<u>Sample Problems</u>

1. Write the meaning of the following in terms of temperature changes and then find the answer.

   $$-2 + (-8)$$

   Possible solution: The temperature fell 2° between midnight and 6:00 A.M. and then fell another 8 degrees by noon.
   The resultant temperature change from midnight to 12 noon on that day was a drop of 10°, mathematically shown as: −2 + (−8) = −10.

2. Write the meaning of −15 + 45 in terms of deposits and withdrawals from a checking account and find the answer.

   Possible solution: A withdrawal of $15 was made and then a deposit of $45 was made. The net result is an additional $30 in the account.
   Mathematically,   −15 + 45 = 30

Rules for Addition of Integers

A vertical or horizontal number line can always be used to help add integers, but it will be more efficient to use the example problems to generalize the situations and develop rules for the addition of integers.

In this text, the term "digit(s)" will be used to indicate a number itself, without regard to sign. In other words, "digit" is treated as being synonymous with the absolute value of the number.

In a sample problem above, we showed that −2 + (−8) = −10; from previous experience we know that 2 + 8 = 10. In general, when adding integers:

> 1) **If the signs of the integers are the same, add the digits and use the same sign.**

Other examples: 2 + 13 = 15 and −2 + (−13) = −15

We also illustrated these problems: 5 + (–6) = –1 and –15 + 45 = 30. In general:

> 2) **If the signs of the integers are different, take the difference of the digits and use the sign of the larger digit.**

Other examples: 2 + (–8) = –6 and –2 + 8 = 6

Note that the numbers added are called **addends**; the result is the **sum**.

Integer Subtraction

Subtraction of integers is sometimes difficult to illustrate using a temperature scale or number line, so we'll use charged particles instead. First we need some definitions. A positive particle (ion) is denoted by "+", a negative particle (ion) is denoted by "–" and a neutral particle is denoted by "$\pm$".

To depict the difference of –6 and –2, start with a set of 6 negatively charged particles as shown in fig.1. Two negative particles (representing the amount to be subtracted) are then circled and removed, or taken out (fig. 2), leaving behind 4 negative particles (fig 3).

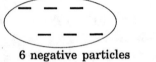
6 negative particles
fig. 1

2 particles taken out
fig. 2

4 negative particles remaining
fig. 3

Thus, – 6 – (– 2) = –4. Note also that – 6 + 2 = –4.

Before we develop a rule, we need to examine another subtraction problem, 5 – (–2).

Using the charged particle model, we begin with five *positively* charged particles but we need to subtract two *negatively* charged particles. Starting with the original set of 5 positive particles, we add some neutral particles, combinations of positive and negative particles, and thus a possible source of the negative particles we wish to take out. <u>Remember that adding and/or having extra neutral charges in any set does not affect the overall charge on the set.</u> We then remove two negative particles.

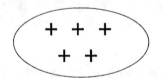
5 positive particles

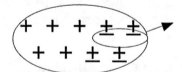
5 positive & 4 neutral particles,
2 negative particles taken out

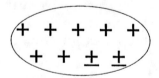
7 positive particles

Therefore, the difference of 5 and (–2) is +7. In other words, 5 – (–2) = 7. Recall again from our addition work that 5 + 2 = 7.

Even though the problems were different, the two examples above have something in common. We first started with a subtraction problem, found the answer using the charged particle model, and then compared the results to a similar addition problem. In each case, the subtraction problem and the related addition problem generated the same answer. The distinction, however, is that the second number in each addition problem was the <u>opposite</u> of that in its subtraction counterpart. Examining our results, it would then seem that the following statements are true:

$$-6 - (-2) = -6 + 2 = -4 \quad \text{and} \quad 5 - (-2) = 5 + 2 = 7$$

In fact, we can change any subtraction problem to an addition problem as long as we add the opposite of (or **additive inverse** of) the number we are subtracting. In general, for all integers a and b,

$$\mathbf{a - b = a + (-b)}$$

This general statement is extremely useful for several reasons:

1) We can use the addition rules in these problems (and that saves memorizing *more* rules); and

2) When we are faced with problems such as $-14 + 23 - (-12) - 34 + (-18)$ we can change all the subtraction operations to addition of the additive inverses and then add the numbers in any order we wish (more on this later when we discuss properties of integers).

In the example given in 2) above, we will change the "subtractions" to "additions of the additive inverses" where appropriate, and we have:

$$-14 + 23 - (-12) - 34 + (-18)$$
$$= -14 + 23 + (12) + (-34) + (-18)$$

We can then add all the negative numbers together, all the positive numbers together and finally take the difference only once. In other words, we have partial sums

$$-14 + (-34) + (-18) = -66 \quad \text{and} \quad 23 + 12 = 35$$
$$\text{and finally,} \quad -66 + 35 = -31$$

> For all integers a and b, use the statement, $\mathbf{a - b = a + (-b)}$, and the rules for the addition of integers will apply.

Note: a is called the **minuend**, b is the **subtrahend**; the result is the **difference**.

A useful application of integer subtraction is finding the length of a line segment separating any two integers on a number line (or temperature scale). Remember that we did this type of problem earlier by simply counting the number of intervals between the two integers.

If the endpoints of the line segment are -2 and 10, for example, subtract the integer on the left (the smaller integer) from the integer on the right (the larger integer) and we get

$$10 - (-2) = 10 + (2) = 12 \text{ intervals}$$

which can be verified on the number line.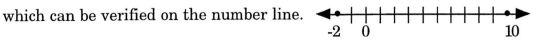

Sample Problems

1. Model –6 – 2 using charged particles.

    Solution: Since this problem is really –6 – (+2), begin with 6 negatively charged particles, rename the set by adding two (or more) neutral charges and then remove 2 positive charges from the set.

    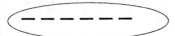  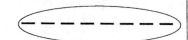

    There are 8 negative charges in the final set, so

    $$-6 - 2 = -8$$

2. Perform the indicated operations using the rules of integers.

    a) –6 + 4   b) –3 – (–2)   c) 6 – (–4)   d) 8 + (–14) – 2 – (–12)

    Solutions:
    a) –2
    b) –3 – (–2)
    = –3 + 2
    = –1
    c) 6 – (–4)
    = 6 + (+4)
    = 10
    d) 8 + (–14) – 2 – (–12)
    = 8 + 12 + (–14) + (–2)
    = 20 + (–16)
    = 4

3. On some scientific calculators, –6 is keyed in by pressing 6 and then the +/– key. Others, including the graphing calculator, have a "(–)" key. By pressing that key and then the number, the display will show a negative number. On scientific calculators, the subtraction key is used for the operation of subtraction.

    To key in 8 + (–14) – 2 – (–12) , press

    8   +   14   +/–   –   2   –   12   +/–   =

    The result is 4.

    On the graphing calculator, press

    8   +   (–)   14   –   2   –   (–)   12   Enter

    and the result again should be 4.

    Now use your calculator to evaluate: –8 – (–2) + 4 + (–20)

    Solution: –22

4. The sum of two integers is 33 and the difference between the larger and the smaller is 9. Find the two integers.

    Solution: One approach is to consider two integers that add up to 33; then subtract the smaller from the larger to check their difference. For example, if we guess that one number is 30 and the other is 3, the sum is 33 but the difference is 27 and we want a difference of 9. We could keep trying sums and taking differences, but a more organized approach is to make a table and try various sets of integers that add to 33.

Larger Integer	Smaller Integer	Sum	Difference
33	0	33	33
28	5	33	23
23	10	33	13 (too big!)
18	15	33	3 (too small!)
22	11	33	11
21	12	33	9 (perfect!)

Note here that as we choose a larger value for the "smaller integer" the difference comes closer to our "goal" difference of 9. In the fourth line of the table, however, the difference is too small and that means we need to backtrack and try a value smaller than 15 in column two. Continuing in this manner and trying other combinations of integers eventually leads to the solution: the smaller integer is 12 and the larger is 21.

Integer Multiplication

$3(-2)$ can be thought of as 3 groups of (-2). On the number line,

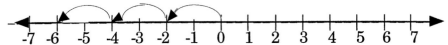

and it's easy to see that $3(-2) = -6$.

[$3(-2)$ means $3 \times (-2)$. The parentheses can be used immediately next to a quantity to indicate the operation of multiplication. A raised dot also indicates multiplication, as in $-2 \cdot 3$. However, the dot is _not_ used before a negative number, as in $2 \cdot -3$.]

(Note that what we are doing here is $-2 + (-2) + (-2) = -6$. It should be apparent that <u>multiplication is really repeated addition</u>.)

Since the meaning of "-3 groups of -2" is fairly unclear, to multiply -3 by -2 consider the pattern exemplified on <u>each side below</u>. Based on a fact such as $3(-2)$ that we can illustrate on the number line, we can do the left and right sides independently.

$$\begin{array}{r}3(-2) = -6\\2(-2) = -4\\1(-2) = -2\\0(-2) = 0\\-1(-2) = 2\\-2(-2) = 4\\-3(-2) = 6\end{array}$$

Decrease each multiplier by 1.   Add 2 to each previous result to generate the next answer.

Finally: $-3(-2) = 6$

Another way to view the product of -3 and -2 is as follows:

Since -3 can be thought of as the <u>opposite</u> of 3, think of $-3(-2)$ as the <u>opposite</u> of $3(-2)$.

We know that 3(–2) = – 6, so the opposite of 3(–2) must be +6 and, thus,
–[3(–2)] = –(–6) = –3(–2) = 6.

Also, to find the opposite of any number, x, we can think of it as $(-1)x$ which = $-x$.

Recalling from our examples, 3(–2) = –6 and (–3)(–2) = +6, we can write the rules for multiplication of any integers in general as:

> 1) **If two integers having the same sign are multiplied, the result will be positive.**
> 2) **If two integers having different signs are multiplied, the result will be negative.**

Numbers that are multiplied are called **factors**; the result is called the **product**.

Examples: 3(9) = 27;   (–3)(–9) = 27;   –6(5) = –30;   6(–5) = –30.

Note: The sign of the larger digit *alone* does not determine the sign of the answer as it does in the addition of integers. For example, 6(–5) = (–6)(5) = –30, but 6 + (–5) = +1 and –6 + 5 = –1.

Integer Division

Several models can be used to illustrate integer division. Let's consider the problem: –6 ÷ 3. If we think of this as 6 negatively charged particles divided into 3 groups of equal size, we can picture two negatively charged particles in each group, as shown in the diagram below.

Therefore, –6 ÷ 3 = –2.

Also, because division is defined in terms of multiplication by:

$$a \div b = c \text{ if and only if } cb = a \text{ for all integers } a, b \text{ and } c,$$

then –6 ÷ 3 = ☐ if and only if ☐ • 3 = –6.   [–6 ÷ 3 can also be written as $-6/3$.]

Using the multiplication rules for integers,   ☐ = –2.
Since every division problem can thus be converted to a multiplication problem, the general rules for division of integers are the same as for multiplication of integers.

> **In division:**
> 1) If the integers have the same sign, the result will be positive.
> 2) If the integers have different signs, the result will be negative.

In $a \div b = c$, a is called the **dividend**, b is the **divisor**, and c is the **quotient**.

Sample Problems

1. Use a pattern to show that $(-1)(-1) = 1$.

   Solution: $3(-1) = -3$
   $2(-1) = -2$
   $1(-1) = -1$ Add 1 to each preceding value on the right
   $0(-1) = 0$ and use a smaller number (by 1) each time
   and $(-1)(-1) = 1$ as a multiplier on the left.

2. Evaluate $\dfrac{(-2)(-3)(6)}{(-1)(9)}$

   Solution: $\dfrac{(-2)(-3)(6)}{(-1)(9)} = \dfrac{+36}{-9} = -4$

3. A sequence of integers is described by: $-6, -4, -2, 0, \ldots$ Use the relationship between consecutive terms in the sequence to find the next two terms.

   Solution:

   $-6 + \boxed{2} = -4$

   $-4 + \boxed{2} = -2$

   $-2 + \boxed{2} = 0$

   (or take the difference in any two consecutive terms, such as $-2 - (-4) = 2$)

   2 is added to each term to get the next term in the sequence. Thus, we have $-6, -4, -2, 0,$ **2, 4.**

The Meaning of Variables

Early in this section we used a variable, n, to represent elements in the set of integers. Later, we stated rules for addition, subtraction, multiplication, and division of integers by using variables to represent integers in general, rather than by referencing specific integer values. Variables are also used to assist us in problem solving by expressing numerical statements in symbolic form. For example, in a class of 30 students, if each student calculated what his/her age would be in 5 years, we might have statements from students like

" In 5 years, I'll be $18 + 5 = 23$ years old" or
"In 5 years, I'll be $32 + 5 = 37$ years old."

How could we generalize the students' responses so that they could apply to any student in the class? If we let n represent the age of *any* student, then

$n + 5$ would represent *any* student's age in 5 years.

Similarly, $n - 5$ would represent *any* student's age 5 years ago.

Next, consider the following tables. Compare the number of circles on the left and right sides of each table.

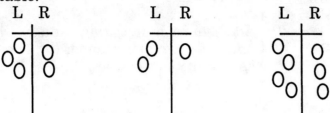

Even though there are different numbers of circles in each of the tables, the tables do have something in common. In each case, the left side has one more circle than the right side. To compare and describe the relationship between the number of circles on the left and right sides we could write the specific relationships as:

$$3 > 2, \qquad 2 > 1, \qquad \text{and} \qquad 4 > 3;$$

or we could write the relationships in terms of equality, as

$$3 - 1 = 2, \qquad 2 - 1 = 1, \qquad \text{and} \qquad 4 - 1 = 3.$$

These last statements help to identify the pattern, or commonality, of the three tables because in each case 1 was subtracted from the value on the left side to generate the corresponding number of circles on the right. In general, for *any* table indicating this specific relationship between left and right sides,

if L represents the number of circles on the left, then

$L - 1$ could represent the number of circles on the right.

At the same time, if R represents the number of circles on the right, then

$R + 1$ would represent the number of circles on the left!

Consider another situation in which the relationship between two quantities is summarized in a table as follows:

L	R
−1	3
0	0
2	−6
3	−9

Since the numbers in the right-hand column can be calculated by multiplying the numbers on the left by −3, we could say that values on the right can be represented by −3L, or as a statement of equality, $R = -3L$; and since numbers on the left can be represented by $R/{-3}$, we can write the equation $L = R/{-3}$.

Instead of using the designations of Left and Right when identifying the columns in the table, recall our work with ordered pairs, and change the labels to use x and y instead. Then we have the following:

12

x	y
−1	3
0	0
2	−6
3	−9

If x represents a number on the left, then $-3x$ would represent the corresponding number on the right, or, in equation form, $y = -3x$.

At the same time, if y represents a number on the right, then $y/_{-3}$ would represent the corresponding number on the left, and this can be written as $x = y/_{-3}$.

Finally, consider the following array of dots with a box that partially covers the full array as pictured.

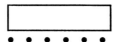

The problem is to find the total number of dots in the entire array, given that the rows under the box also contain 6 dots each. What is unknown here is the exact number of rows. If we were to assume that there are 3 rows of dots *hidden*, then there would then be a *total* of 4 rows with 6 dots each or 4 (6) = 24 dots. But since the number of rows is not given, it would be reasonable to write:

Number of dots = ☐ • 6 where ☐ represents the unknown number of rows.

Since the box may represent *any* number, we can use another symbol, such as a letter, to designate the total number of rows. Using *n* to represent the total number of rows, we then have:

Number of dots = n • 6.

More work with variables will be done throughout the text, but it is important to understand why letters can be used and how they can be used.

Sample Problems

1. One classroom in the science building contains *d* desks per row and 8 rows of desks. How many desks are in that classroom?

   Solution: 8 • d = 8d desks (When using variables, it is common to omit the multiplication sign when numbers and variables, or several variables, are multiplied.)

2. There are 38 coins in a coin box, nickels and dimes only. Using a variable, express the number of nickels and the number of dimes in this box.

Solution:
> If there are 10 nickels in the coin box, there must be
> $$38 - 10 = 28 \text{ dimes}$$
> and if there are 20 nickels, then there are
> $$38 - 20 = 18 \text{ dimes}.$$
> So letting n = the number of nickels, the <u>remaining</u> coins must be dimes and we can write the number of dimes as $38 - n$.

Each of the two sample problems just presented are examples of simple "word problems" or application problems. Frequently throughout the text you will have opportunities to design and write word problems as well as to solve them. Since problem-solving is a primary focus of this book, analyzing word problems in order to translate words into symbols is an important skill to acquire. Each application problem must provide the potential problem-solver with essential information and it must also ask a question. In addition, non-essential information may be included. It is the responsibility of the problem-solver to distinguish between important and unimportant information in order to analyze the problem.

Problem Set 1.1

1. Use the model specified to depict each of the following.

   a) $-4 + 2$ (number line)
   b) $-10 + (-5)$ (temperature scale)
   c) $3(-4)$ (number line)
   d) $8 - (-2)$ (charged particles)
   e) $-2 - (-6)$ (charged particles)

2. Write the subtraction problem and a related addition problem depicted in the diagram below.

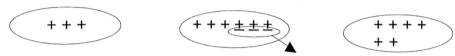

3. Show how you could use the charged particle model to illustrate $-4 + 9$.

4. Write a word problem in real-life context that uses the operation of addition on the integers -30 and 5 to find the solution. Remember to give enough information and then to ask a question. You need not answer the question you ask.

5. Give an example of a number that is a whole number but not a natural number.

6. Give an example of a number that is an integer but not a whole number.

7. Evaluate each of the following without a calculator; verify with a calculator:

   a) $-2 + (-3) - (4 \cdot 3 - 2)$
   b) $4 - (-5) + (-18) - 32$
   c) $\dfrac{(6)(0)(-5)}{(-10)}$
   d) $\dfrac{(-6)(4)(2)(-1)}{-3}$
   e) $\dfrac{(12)(-4) + 18}{(-3)(-2)}$

8. Two cars, one traveling north and the other south, leave Denver, CO on U.S. 25. In two hours, the northbound car has traveled 125 miles and the southbound car has traveled 110 miles. After two hours, how far apart are these cars?

9. Use your calculator to evaluate each of the following.

   a) $-3 + 4 - (-2) + 6 - 18$
   b) $\dfrac{(-3)(-2)(14)}{-4 + (-2)}$
   c) $3(4 - 5) - 2 + 8(-12)$

10. Find *all* values that could be put into each window to make each statement true.

    a) $|-6| = \square$
    b) $8 - \square = 5$
    c) $\dfrac{\square}{-2} = -10$
    d) $\square + 8 = 3$
    e) $|\square| = -2$
    f) $|\square + 3| = 1$
    g) $|3 - \square| = 5$

11. Identify any of the following that represent negative integers.

    a) $-|3|$
    b) $|-3|$
    c) $-(-3)$
    d) $-|-2|$

12. Find the next 3 terms in the following sequence: 2, –4, 8, –16, __ , __ , __ .

13. Put a ">" or "<" sign between each pair of integers below to make the expression true.

    a)　　4　　–6　　　　　　b)　　–2　　–5　　　　　　c)　　–4　　0

    d)　|–5|　　– 5　　　　　e)　4　　|–5|

14. Plot each of the following ordered pairs on a coordinate system.

    a) (1,–5)　　　b) (–2,–4)　　　c) (0,4)　　　d) (4,0)　　　e) (3,–4)

15. The temperature in Fargo, ND at 11:00 P.M. on Feb. 2 was 8° F. It dropped 10° during the night, then rose 20° during the daytime of Feb. 3.

    a) What was the temperature at the end of daytime hours on Feb. 3 in Fargo?

    b) What was the net change in temperature in Fargo from 11:00 P.M. on Feb. 2 until evening Feb. 3?

16. Find the distance between points A and B on the integer number line,

    a) by counting unit intervals

    b) by taking the difference

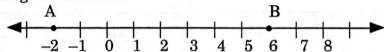

17. The altitude of Mt. McKinley's peak is approximately 20,270 feet above sea level and the altitude of the deepest point of the Marianas Trench in the Pacific Ocean is approximately 35,800 feet below sea level. Find the difference in altitudes between these two points, showing the expression which you evaluate.

18. The perimeter of a geometric figure is the distance around the figure, starting at one point and "walking" around the figure until returning to the starting point. A rectangle is positioned on the coordinate system below. Find the coordinates of each vertex and then find the perimeter of the rectangle.

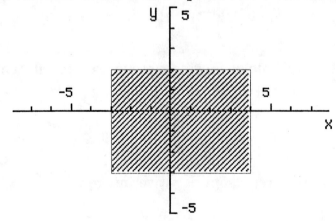

19. A company gained 2 points on the stock exchange in each of 3 weeks and then lost 1 point in each of 2 weeks. Find the net change (gain or loss) in this company's stocks in these 5 weeks.

20. The graph below shows a company's sales rounded to the nearest million dollars for the years 1987 to 1992, inclusive. Answer the following questions with respect to this graph.

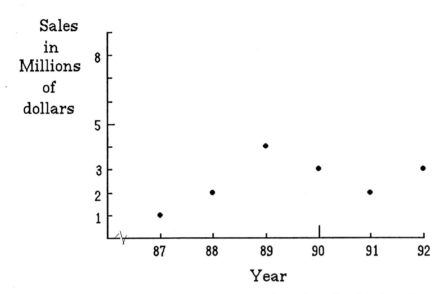

   a) In what year were the company's sales the highest?

   b) Use an integer to describe the difference in sales between 1989 and 1990.

   c) Calculate the average sales for the years 1990, 1991, and 1992.

21. The sum of two integers is 65 and the difference between the larger and smaller is 37. Find the integers. Use a chart to help you.

22. To raise money for charity, the Pi Theta Gamma sorority sold a total of 236 tickets to an air-band competition. The price for each student ticket was $4 and for each adult ticket was $6. If n represents the number of student tickets sold,

   a) find the number of adult tickets sold with respect to n;

   b) find an expression that represents the total value of the ticket sales.

23. The smallest angle of a triangle is x degrees. Another angle is twice the size of the smallest and the third angle is 20° larger than the smallest. Write the measures of the second and third angles of this triangle in terms of the size of the smallest.

24. A westbound car leaves Duluth, MN on U.S. 2 traveling at an average rate of 50 mph while an eastbound car leaves Duluth on U.S. 2 traveling at an average rate of 48 mph. (Remember that d = rt.)

   a) Find the distance each has traveled in 3 hours.

b) How far apart are the cars in 3 hours?

c) Find the distance each has traveled in n hours.

d) Find their distance apart after n hours.

Exploration problems

1. a) Given the rectangle below, find the missing coordinates.

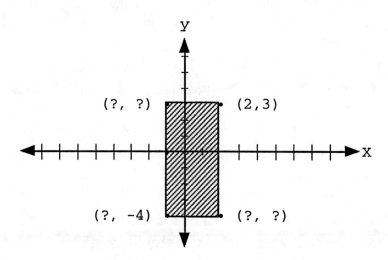

b) On a coordinate system, draw at least two rectangles, each with a perimeter of 20 units, and give the coordinates of the vertices.

c) Find the midpoints of each of the sides of this rectangle. Give each midpoint as a set of x and y coordinates.

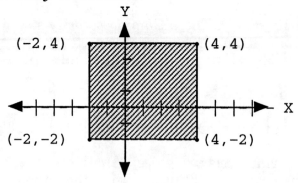

2. Write an example problem in which the *difference* between two *negative* integers gives:
    a) a positive integer as a result.

    b) a negative integer as a result.

Solutions - Problem Set 1.1

1. a)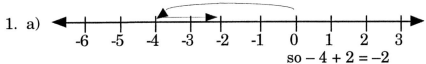
   so $-4 + 2 = -2$

   b)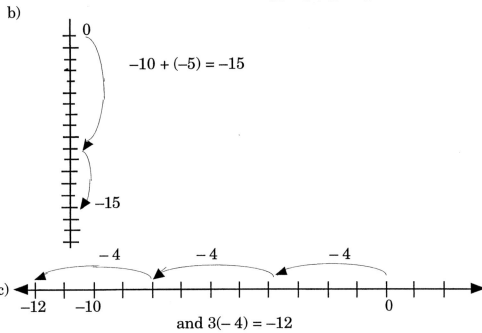

   c) and $3(-4) = -12$

   d)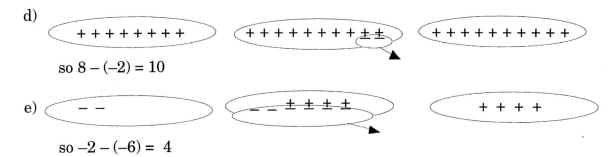
   so $8 - (-2) = 10$

   e) so $-2 - (-6) = 4$

2. $3 - (-3) = 6$ and $3 + 3 = 6$

3.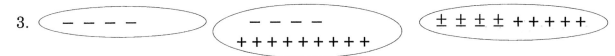

   To 4 negative charges add 9 positive charges. Four of the positive charges will combine with the 4 negative charges to form 4 neutral particles, leaving 5 positive charges as the result.

4. Example: A team lost 30 yards and then gained 5 yards in two consecutive plays. What was the net gain or loss in yardage?

5. The only number is 0.

6. Example: -4

7. a) -15 b) -41 c) 0 d) -16 e) -5

8. 235 miles apart

9. a) –9 b) –14   c) –101

10. a) 6   b) 3   c) 20   d) –5   e) no solution   f) – 4 or –2   g) –2 or 8

11. Negative integers: –|3| and –|– 2| [that is, a) and d)]

12. 32, –64, 128

13. a) >   b) >   c) <   d) >   e) <

14.

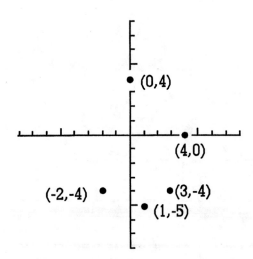

15. a) 8 + (–10) + 20 = +18, so the temperature was 18°.
    b) 18 – 8 = 10. There was a 10° rise in temperature.

16. a) there are 8 intervals   b) $|6 - (- 2)| = 8$

17. 56,070 feet

18. The coordinates are (4,2), (4,–3), (–3,2) and (–3,–3) and P = 24 graph units.

19. 4 point gain

20. a) 1989   b) loss of 1 million, or –1 million   c) $2,666,666.67 or $2\frac{2}{3}$ million

21. The integers are 14 and 51.

22. a) The number of adult tickets is $236 - n$.   b) Sales = $4n + 6(236 - n)$

23. The second angle measures $2x$ degrees and the third measures $x + 20$ degrees.

24. a) Westbound car: 50(3) = 150 miles;   Eastbound car: 48(3) = 144 miles
    b) Distance apart after 3 hours = 294 miles
    c) $50n$ and $48n$, respectively
    d) $50n + 48n$

Section 1.2 - Development and Use of the Order of Operations

In grade school, after we learned how to do addition, we found out that there was a shortcut for repeated addition: multiplication. As noted in the last section, it is quicker and easier to write 7 • 6 than it is to write 6 + 6 + 6 + 6 + 6 + 6 + 6.

Similarly, when we multiply one quantity by itself two or more times, there is an efficient way of expressing the product: exponential notation.

Instead of writing 3 • 3 we can write 3^2. Instead of writing a•a•a we can write a^3. We can also write 5^4 instead of 5 • 5 • 5 • 5. The raised number, called an **exponent**, tells how many times the other number, the **base**, is used as a factor. Clearly, when the exponent is large, its use represents a savings of space, time, and energy, and we can tell how many times a number (or variable) is to be multiplied by itself without having to count the number of times it was written.

Scientific calculators usually have two keys, $\boxed{x^2}$ and $\boxed{y^x}$, which are used to evaluate expressions containing exponents. If we wish to square a number, that is, multiply it by itself, we press $\boxed{x^2}$. Thus, 4 squared is keyed 4 x^2. For *any* exponent, we can use $\boxed{y^x}$. To evaluate 5^4, for example, press 5 y^x 4 = . On a *graphing* calculator, 5^4 would be keyed 5 ^ 4 =, using the caret symbol before the exponent.

For the correct key sequence to be used with *your* calculator, look in the direction manual which came with the calculator.

When we were introduced to the operations of arithmetic, we began with addition and then learned subtraction, multiplication, and division, probably in that order. In evaluating a mathematical expression that includes two or more of these operations, however, we may perform the operations in a different order from the way we first learned them, and we need to know what that order is. Your scientific calculator is programmed with the correct order, and you can use that fact to discover the order for yourself.

Look at each example, *guess* what you think the answer should be, and write it down. Then, *using your calculator*, key in the problem from left to right and compare that answer with your guess. If the calculator answer is different from your guess, then note carefully the order you think the CALCULATOR executed the operations. Two operations are used in each example. [You'll need to press "=" at the end of each problem in order to get the calculator answer, and you'll need to locate the parentheses keys because they appear in some of the examples.]

1. 2 + 4 − 3
2. 5 − 6 + 1
3. 5 − (6 + 1)
4. −6 • 4 ÷ 2
5. 6 ÷ 2 • 3
6. 6 ÷ (2 • 3)

Correct answers:
1. 3
2. 0
3. −2
4. −12
5. 9
6. 1

In what order does the calculator execute addition and subtraction?

In what order does the calculator execute multiplication and division?

What is the effect of including parentheses?

Conclusions:

> Additions and subtractions are executed IN THE ORDER THEY ARE ENCOUNTERED from left to right.
> Multiplications and divisions are executed IN THE ORDER THEY ARE ENCOUNTERED from left to right.
> Operations inside parentheses are done first.

Continue to guess the answers and verify them with your calculator.

7.	$-3 + 7 \cdot 2$		12.	$4 \cdot 5 - 1$
8.	$2 \cdot 7 + 3$		13.	$-15 - 5 \cdot 2$
9.	$(-3 + 7) \cdot 2$		14.	$(-15 - 5) \cdot 2$
10.	$4 + 6 \div 2$		15.	$8 - 4 \div 2$
11.	$(4 + 6) \div 2$		16.	$(8 - 4) \div 2$

Correct answers:
- 7. 11
- 8. 17
- 9. 8
- 10. 7
- 11. 5
- 12. 19
- 13. −25
- 14. −40
- 15. 6
- 16. 2

In what order does the calculator execute addition and multiplication? Addition and division? In what order does the calculator execute subtraction and multiplication? Subtraction and division?

Conclusion:

> Multiplication and division are done before addition and subtraction.

What is the effect of including parentheses?

Conclusion:

> The operation in parentheses is done first.

Let's continue by guessing and verifying results as before.

17.	$5 + 3^2$	19.	$10 - 3^2$	21.	$2 \cdot 5^2$	23.	$20 \div 2^2$
18.	$(5 + 3)^2$	20.	$(10 - 3)^2$	22.	$(2 \cdot 5)^2$	24.	$(20 \div 2)^2$

Correct answers:
- 17. 14
- 18. 64
- 19. 1
- 20. 49
- 21. 50
- 22. 100
- 23. 5
- 24. 100

What conclusion can be drawn about using exponents and other operations?

Conclusion:

> Exponents are applied before any other operations are executed.

What is the effect of including parentheses?

Conclusion:

> The operation in parentheses is done first.

To summarize our conclusions, we can describe what is known as

---

The Fundamental Order of Operations or **Algebraic Logic with Hierarchy**

1. work in parentheses
   a) apply exponents
   b) do multiplications and divisions as they occur left to right
   c) do additions and subtractions as they occur left to right

2. apply exponents

3. multiply and divide *from left to right, in the order encountered*

4. add and subtract *from left to right, in the order encountered*

We have been working with examples that contain only two operations. We need to expand our thinking to include more complex problems while still keeping our conclusions in mind.

Sample Problems

1. Evaluate the following without a calculator. Then do each problem using a calculator. If your calculator answers do not agree with the solutions given below, check your key sequence.

   a) $3(6+2) \div 4$ b) $14 - 4^2 \div 2$ c) $25 - 10 \div 2 + 10 \bullet 5$ d) $\dfrac{15-3}{2+1}$

   Solutions:
   a) There are 3 operations shown: multiplication, addition, and division. We must do the addition first because it is within parentheses. The new problem will be $3(8) \div 4$. We do multiplication next, then division, because that is the order in which we encounter them from left to right.
   The final answer is 6.

   Note: When using your calculator for this problem, you *may* need to insert an "x" for multiplication before you key the first parenthesis.

   b) Of the three operations—subtraction, using an exponent, and division—we must apply the exponent first, then divide, then subtract.
   So $\quad 14 - 4^2 \div 2$
   $= 14 - 16 \div 2$
   $= 14 - 8$
   $= 6.$

c) The operations must be done in the following order: division, multiplication, subtraction, addition.

$$25 - 10 \div 2 + 10 \cdot 5$$
$$= 25 - 5 + 50$$
$$= 20 + 50$$
$$= 70.$$

d) This problem contains some hidden information. The fraction line tells us to divide the numerator (top value) by the denominator (bottom value), but because we have to *find* the two values first, the fraction line also acts like parentheses, telling us to do the addition and subtraction shown before we divide. Therefore, the intermediate answer is $12 \div 3$; the final answer is 4. When using your calculator, make sure to key in parentheses around the numerator and around the denominator.

2. Evaluate the following without a calculator, verifying results with a calculator:

   a) $4 \cdot (9 - 5 + 7)$

   b) $(3 + 2 \cdot 6)^2 - 20 \div 5$

   c) $[(6 + 4) \div (5 - 3)]^3$

Solutions:
a) Within the parentheses, subtract, then add; this will give us 11. Now multiply 4(11). The answer is 44.

b) Within the parentheses, multiply, then add; square the result; do the division, then the subtraction.

$$(3 + 2 \cdot 6)^2 - 20 \div 5$$
$$= (3 + 12)^2 - 20 \div 5$$
$$= (15)^2 - 20 \div 5$$
$$= 225 - 4$$
$$= 221.$$

c) Note that there are two sets of parentheses and one set of brackets. The brackets serve the same function as parentheses, but their use makes the problem visually easier to understand. Your calculator *may* show brackets on the same keys as the parentheses. Simplify within the inner sets first. Do the addition then the subtraction, followed by the division; finally raise the result to the third power using the y^x key.

$$[(6 + 4) \div (5 - 3)]^3$$
$$= [10 \div 2]^3$$
$$= 5^3$$
$$= 125.$$

3. In the expression $5(3 - 4 \div 2 + 6)^3$, which operation would be done first? Last?

   Solution: First: division within parentheses, that is, $4 \div 2$;
   Last: multiplication, that is 5 times the quantity which follows.

4. Write a calculator sequence for *your* calculator that gives the correct answer, .75, for the following expression: $\dfrac{14-5}{9+3}$

Solution: 1) (14 – 5) ÷ (9 + 3) =   or   14 – 5 = ÷ (9 + 3) =
Note that the "=" will cause the calculator to do the subtraction immediately.
or,   2) using the calculator's memory, 9 + 3 = STO   14 – 5 = ÷ RCL =

Challenge Problem

Insert parentheses to make the value of the following expression equal 10.
$$5 \cdot 2^2 + 10 \div 7$$
Solution: $5 \cdot (2^2 + 10) \div 7$

Translation and the Order of Operations

We have been finding the value of given expressions. Frequently in algebra, however, we must write a symbolic expression ourselves, translating into symbolic form information that is originally presented in words. For example, the phrase "the product of nine and two divided by the sum of six and three" could be translated symbolically into 9 • 2 ÷ (6 + 3).

Another wording for the same problem might be "the product of nine and two divided by the quantity six plus three." The phrase "the quantity" usually refers to the two values and one operation given immediately following. It is a useful phrase to avoid ambiguity or misunderstanding. For example, 2 • 5 + 1 is the correct translation of "two times five plus one", while 2 • (5 + 1) would be correct for "two times the quantity five plus one."

There is additional vocabulary with which we should be familiar. The following chart lists some of these important words and phrases along with the operation which each one specifies. You need to know these terms.

Addition: plus, more than, greater than, increased by, incremented by; sum

Subtraction: minus, decreased by, diminished by, difference between, less, less than; difference

Multiplication: times, of; product

Division: divided by, ratio of, per; quotient

Variables are also used in mathematical expressions. The expressions are sometimes given in symbolic form, sometimes in "word" form. We should be able to translate from one form to the other.

For example, 5n + 13 could be read as "five times some number *n* increased by 13.

The phrase "some number, *y*, decreased by twice itself" could be shown symbolically as "y – 2y."

Sample Problems

1. Translate the following from words to symbols. More than one answer may be correct.

   a) the difference between four and nine
   b) eight less than three
   c) three less eight
   d) the quotient of two and eight
   e) the sum of two and three multiplied by the difference between seven and one
   f) the difference between some number, m, and 17
   g) your age in five years if you are x years old now
   h) your age three years ago if you are x years old now
   i) the value in cents of n nickels, of d dimes, of q quarters

   Solutions: a) $4 - 9$ b) $3 - 8$ c) $3 - 8$
              d) $2 \div 8$ or $2/8$ e) $(2 + 3) \cdot (7 - 1)$ f) $m - 17$
              g) $x + 5$ h) $x - 3$ i) $5n, 10d, 25q$

2. Translate the following from symbols to words.

   a) $20 \cdot 5 + 4$ b) $20 \cdot (5 + 4)$ c) $27 - (-12)$ d) $2 \cdot 5 + 8 \div 4$
   e) $12n + 19(-2)$ f) $4(a - b)$ g) $(x + y) \div (m - n)$

   Possible solutions:
   a) twenty times five increased by four
   b) twenty times the sum of five and four
   c) the difference between 27 and negative twelve
   d) the product of two and five plus the quotient of eight and four
   e) the sum of twelve times some number, n, and nineteen times negative two
   f) four times the quantity a minus b
   g) the sum of x and y divided by the difference of m and n

Order of Operations and Informal Equation Solving

In section 1.1 we began informal equation solving by using the arithmetic operations on integers and trial and error. For example, in the problem $3 \cdot \square = -18$, we are looking for an integer that would make the statement true. We know $3 \cdot 6 = 18$, but because we get a negative product when we multiply a positive number by a negative number, we conclude that the missing number (the number in the window) is -6. We will now extend this reasoning to problems involving more than one operation.

In the current section we considered an expression similar to $3 + 3 \cdot 2$ and we found the final value to be $3 + 6$ or 9.

Let's view the same problem from a different perspective. Instead of finding the value of the *expression*, consider the related *equation*

$$3 + 3 \cdot \square = 9$$

The order of operations would call for the multiplication to be done first, but since one of the factors is unknown, we need a different strategy.

To find the value that would make the statement $3 + 3 \cdot \square = 9$ true (that is, to find the replacement for the window), let's develop more fully the concept of *equation* by examining the pan-balance below. A box, $\square$, contains an unknown number of marbles, and this represents the missing number in the equation. The circles are individual marbles and they represent the constant (or known-number) values.

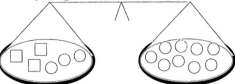

In other words, the three boxes on the left side, representing $3 \cdot \square$ in the equation, each contain the *same unknown number of marbles*, and whatever that number is, when added to 3, must equal a weight of 9.

It would seem reasonable that we could eliminate three marbles on each side without affecting the balance. We'll show this by crossing out what we remove:

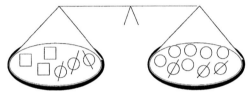

We can then write $3 \cdot \square = 6$ as the mathematical statement depicting what remains on the pan balance.

The last step of the problem is to find the missing factor that will generate a product of 6.

Symbolically, if 3 boxes on the left balance with six marbles on the right, then each box on the left must contain 2 marbles (we'll ignore the weight of the boxes) and we have

$$\square = 2$$

We have found the solution to an equation of 3 **terms**. **Terms of algebraic equations are separated by "plus", "minus", or "equal" signs.** Terms will be described in greater detail in section 1.3.

A more challenging question might be to find the value that makes the following statement true:

$$2 \cdot \square + 5 = 3 \cdot \square + 1$$

On the pan balance this equation translates to

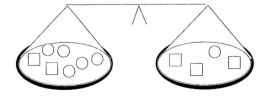

We'll again use "elimination of like quantities" on both sides. The steps are shown below.

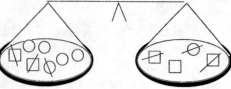

This gives us the statement

$$4 = \square$$

and the problem is solved directly for the unknown. That is, there are 4 "marbles" in each box and the number 4 makes the equation true.

Check: $2 \cdot 4 + 5 = 3 \cdot 4 + 1$

Sample Problem

1. Given the following pan balance:

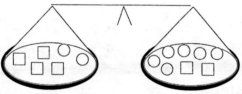

   a) Write the mathematical statement represented by this configuration.

   b) Find the value in each box.

Solution:

   a) The statement is $4 \cdot \square + 2 = 2 \cdot \square + 6$

   b) The process can be shown as

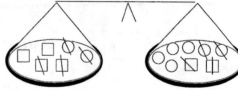

   implying $2 \cdot \square = 4$ and $\square = 2$.

There is another informal approach to solving equations. Let's look at the equation

$$3 + \square \cdot 2 = 17$$

This is essentially a two-step problem. Since the order of operations applies, the multiplication must be done first. We'll now use a circle to emphasize that fact, as shown below.

$$3 + \overparen{(\square \cdot 2)} = 17$$

This means that a quantity added to 3 gives a sum of 17.

Since $(\square \cdot 2)$ must equal 14, we'll write

$$3 + \overset{14}{(\square \cdot 2)} = 17$$

and conclude that

$$\square = 7 \text{ since } \boxed{7} \cdot 2 = 14 \text{ and } 3 + \boxed{7} \cdot 2 = 17 \text{ are true statements.}$$

If you "cover-up" the **term** containing the window, that is, $\square \cdot 2$, the order of operations called for in the equation may be more apparent. Let's try

$$\frac{2 \cdot \square + 3}{3} = 5$$

If we think of the numerator as a whole quantity, by using the circle method we have:

$$\frac{\overset{15}{(2 \cdot \square + 3)}}{3} = 5$$

and the whole numerator must = 15 since 15 ÷ 3 = 5.

This implies that another circle is needed containing just the term with the window, and this circle must have a value of 12 as shown (since 12 + 3 = 15.)

$$\frac{\overset{15}{(\overset{12}{(2 \cdot \square)} + 3)}}{3} = 5$$

Finally, since $2 \cdot \square = 12$, $\square$ must be 6 and we have found the solution.

Check:
$$\frac{2 \cdot \boxed{6} + 3}{3} = 5$$

Sample Problems

Find a solution for each equation.

a) $2 \cdot \square + 4 = 4$

b) $3 \cdot \square - 2 = -8$

c) $4 \cdot \square - 1 = 2 \cdot \square + 5$

Solutions: a) $(2\cdot\boxed{}^{\,0})+4=4$ b) $(3\cdot\boxed{}^{\,-6})-2=-8$

Therefore: $2\cdot\square = 0$ $3\cdot\square = -6$
  then, $\square = 0$ $\square = -2$

c) $4\cdot\square - 1 = 2\cdot\square + 5$

With the unknown on both sides of this equation, the "cover-up" (or "circle") method won't work here. Also, the concept of "negative" value extends beyond what we have done on the pan balance. Although we can use the model by employing different colored marbles, it probably is best to make a table and choose <u>trial</u> replacement values, *using the same value for both windows*. As we go along, we'll evaluate the left and right sides independently.

We'll begin by trying numbers such as 0, 1, 2, and if the left and right sides are getting closer in value, we'll keep using positive integers. If the values are getting further apart, we'll use a few negative numbers close to zero.

Replacement value	Left side	Right side	
0	−1	5	
1	3	7	Notice that the left side and right
2	7	9	side values are getting closer.
3	11	11	← The values match!!!

So the replacement value, or solution, is 3.

Check: $4\cdot\boxed{3} - 1 = 2\cdot\boxed{3} + 5$

Challenge Problems:

1. Solve for the unknown, x.

$$\frac{30-7x}{4\cdot 6-8}=1$$

Solution: The numerator and denominator must be equal in value in order for the quotient to be 1. The denominator equals $24 - 8$ or 16. Therefore, the numerator, $30 - 7x$, also must equal 16.

Using the circles, we have

$$\frac{\overset{16}{30}-\overset{14}{(7x)}}{4\cdot 6-8}=1$$

so $x = 2$

2. Find *two* replacement values for the windows in the following equation.
   $20-(\square^2+1)= \square\cdot 3 - 21$

Solution: Using a table and substituting values as listed for both boxes, we have the following:

	left	right
If ☐ = 0	19	−21
= 1	18	−18
= 2	15	−15
= 3	10	−12
= 4	3	−9
= 5	−6	−6

Therefore, when ☐ = 5, the left and right sides have the same value, and we know that 5 is one solution to the equation. Continuing to chose values for the left and right sides reveals that −8 is also a solution.

3. The sum of two consecutive integers is equal to three times the smaller integer. Letting *n* represent the smaller integer, translate the statement into symbols and find the two integers. A table might be useful.

Solution: If *n* is the first integer, *n*+1 is the next consecutive integer, and we have n + (n + 1) = 3•n

	left	right
If n = 0	1	0
= 2	5	6
= 1	3	3

So if n = 1, then n+1 = 1 + 1 or 2 and the integers are 1 and 2. To check, note that 1 + 2 = 3.

Alternate solution:
n + (n + 1) = (n + n) + 1 = 2n + 1   so our equation is simplified to

2n + 1 = 3n.

On the pan balance, using ☐ to represent *n*, this can be shown as

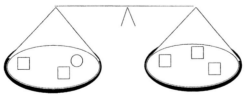

Eliminating like quantities on both sides

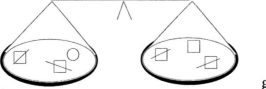   gives ☐ or n = 1

Thus, n + 1 = 2 so the integer solutions are 1 and 2.

Problem Set 1.2

1. Evaluate without using a calculator, then use a calculator to verify your answer.
   a) $14 \div 2 - 5 \cdot 3 + 1$
   b) $2 - 5^2 + 8 \div 4 \cdot 2$

   c) $3(5 - 7) \div 2(-4 + 3)$
   d) $(14 \div 7)^2 - 3(6 - 8)^3$

2. To evaluate each of the expressions shown, identify the operation which would be done first and the one which would be done last.

   a) $35 + 12 \div 2 - 5 \cdot 1$
   b) $100 - 2[5 + 4(17 - 8) + 1] + 15$

3. Explain why keying $7 + 1 \div 10 - 2 =$ will not give the correct value for:

   $$\frac{7 + 1}{10 - 2}$$

4. Write a calculator sequence which might be used to evaluate each expression.

   a) $(3 + 3^4) \div 7 \cdot 2 - 1$
   b) $\dfrac{4^2 + 6^2}{(4 + 6)^2}$ (Find more than one way to do this one.)

5. Insert parentheses in the expressions to give the value requested.

   a) value: 54; $4 + 5 \cdot 6$
   b) value: 70; $5 \cdot 3 \cdot 4 + 2$

   c) value: 32; $6 + 2 \cdot 8 \div 4 \cdot 2$
   d) value: 5; $13 + 10 + 7 \div 1 + 2 + 3$

   e) value: 35; $2 \cdot 5^2 + 10 \div 2$

6. Translate from words to symbols.

   a) five times four increased by three
   b) five times the sum of four and three
   c) the product of six and five decreased by two
   d) the product of six and the quantity five decreased by two
   e) the quantity three less than four times ten
   f) eleven divided by the sum of five and negative six
   g) the square of five times the cube of two
   h) three times five incremented by y
   i) the product of three and the quantity five plus m
   j) the difference between a temperature of twelve degrees below zero and a temperature of twenty degrees below zero

7. Translate from symbols to words.

   a) $12 + 15 \div 5$
   b) $(-2 - 5)^2$
   c) $2x - 7y$
   d) $6(x - y)^2$

   e) $\dfrac{42 + 38}{-12 - (-2)}$
   f) $(9 - 4)^2 + \left(\dfrac{1}{2}\right)^3$

8. Find a replacement value (solution) for each of the windows. (Remember that when more than one window appears in a problem, each window must contain the same value.) Use any method you wish.

   a) $235 + \square = 17$ b) $15 + (10 - \square) = 22$ c) $2(5 + \square) = 18$

   d) $\dfrac{2(\square - 4)}{5} = 6$ e) $\dfrac{20 + 3x}{3 \cdot 6 - 5} = 2$

9. For each diagram, write the mathematical equation and then solve the equation.

   a)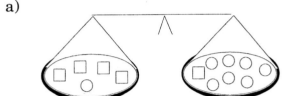

   b)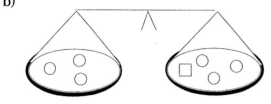

10. Draw a pan balance to represent the equation $5n = 3n + 6$ and find the number of marbles in each box.

11. Translate into equations and solve for the unknown values.

    a) Three times a number increased by two is seventeen.

    b) The product of two and the quantity x plus four equals twenty.

    c) The difference between a temperature of t degrees above zero and a temperature of ten degrees below zero is eighteen degrees.

Exploration Problems

1. Evaluate without using a calculator, showing steps using the order of operations.
   $16 + 2[8 - (-4 \cdot 3)(7 - 11)]^2$

2. Find all possible replacements for the window that will make the statement, $3 \cdot \square + 15 = \square^2 - 39$, true.

3. Rent-A-Frig offers college students a choice of rental plans for small refrigerators. Under Plan A, a student pays $50 down plus $2 per week. In plan B, a student pays $20 down and $3 per week.

   a) Which plan would you choose and why?

   b) Translate this problem into an equation, using x as the unknown number, and find when the costs under the plans are the same.

Solutions – Problem Set 1.2

1. a) –7   b) –19   c) 3   d) 28

2. a) First: 12 ÷ 2;   b) First: inner parentheses, 17 – 8,
   Last: subtract 5   Last: add 15

3. The first operation done will be 1 divided by 10 which results in a mixed number answer. The addition and subtraction should be done before the division. This can be accomplished if parentheses are used when keying both numerator and denominator, or STO, =, and RCL may be used. The result should be 1.

4. a) (3 + 3 y^x 4) ÷ 7 x 2 – 1 =   The result is 23.

   b) (4 + 6) x^2 STO 4 x^2 + 6 x^2 = ÷ RCL =
   The result is .52.
   Another possibility: (4 x^2 + 6 x^2) ÷ (4 + 6) x^2 =

5. a) (4 + 5) • 6
   b) 5 • (3 • 4 + 2)
   c) (6 + 2) • 8 ÷ 4 • 2
   d) (13 + 10 + 7) ÷ (1 + 2 + 3)
   e) 2 • (5^2 + 10) ÷ 2

6. a) 5 • 4 + 3   b) 5(4 + 3)   c) 6 • 5 – 2   d) 6(5 – 2)
   e) (4 – 3) • 10   f) 11/[5 + (–6)]   g) 5^2 • 2^3 h) 3 • 5 + y
   i) 3(5 + m)   j) –12 – (–20)

7. a) Twelve plus fifteen divided by five.
   b) The square of the quantity negative two minus five.
   c) Two times x decreased by seven times y.
   d) The product of six and the square of the quantity x minus y.
   e) The sum of forty-two and thirty-eight divided by the difference of negative twelve and negative two.
   f) The square of the quantity nine minus four increased by the cube of one-half.

8. a) ☐ = –218   b) ☐ = 3   c) ☐ = 4   d) ☐ = 19   e) x = 2

9. For x, the unknown,
   a) 4x + 1 = x + 7   b) 3 = x + 3
      x = 2   x = 0

10.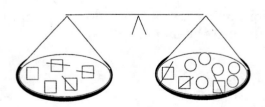

    so 2 • ☐ = 6 and there are 3 marbles per box.

11. a) $3n + 2 = 17$; $n = 5$

　　b) $2(x + 4) = 20$; $x = 6$

　　c) $t - (-10) = 18$; $t = 8$ degrees

Section 1.3 - Properties of Integers and Algebraic Expressions

There are several important properties of integer addition and multiplication that will be useful in our estimation and mental arithmetic work. Note that the variables a, b, and c will be used consistently in these properties to designate any integers.

1) **The commutative property:**

   for all integers a and b,

   $$a + b = b + a \quad \text{and} \qquad \textit{for addition}$$
   $$a \cdot b = b \cdot a \qquad \textit{for multiplication}, \text{ where the raised dot, } \cdot, \text{ is used to symbolize multiplication.}$$

   For example, $-2 + (-3) = -3 + (-2) = -5$
   and $(-3)(2) = (2)(-3) = -6$

2) **The associative property:**

   for all integers a and b,

   $$a + (b + c) = (a + b) + c \quad \text{and} \qquad \textit{for addition}$$
   $$a \cdot (b \cdot c) = (a \cdot b) \cdot c \qquad \textit{for multiplication}$$

   To illustrate: $2 + (-3 + 4) = (2 + -3) + 4$ (both sides = 3) and
   $2 \cdot (5 \cdot -3) = (2 \cdot 5) \cdot (-3)$ (both sides = -30)

3) **The identity property:**

   For all integers b, $\quad b + 0 = 0 + b = b$ and
   $\qquad\qquad\qquad\qquad\quad b \cdot 1 = 1 \cdot b = b$

   In each case, the "identity" of b remained unchanged. For addition, zero is the <u>additive</u> identity; for multiplication, one is the <u>multiplicative</u> identity.

4) **The inverse property:**

   For all integers a, $\quad a + (-a) = -a + a = 0.$

   $-a$ is referred to as the additive inverse of a (as well as the opposite of a) and a is the additive inverse of $-a$. By adding the inverses together, we get the identity for the operation for addition, namely 0.

   (Note: The inverse with respect to multiplication does not exist in the set of integers because there is no *integer*, $\square$, such that $a \cdot \square = 1$.)

5) **Zero property of multiplication:**

   For all integers a, $a \cdot 0 = 0 \cdot a = 0$. That is, zero times any number is zero.

6) **The distributive property of multiplication over addition:**

   $$a(b + c) = ab + ac$$

To illustrate this, if we want to find the area, A, of the largest rectangle below, where A is equal to length multiplied by width, we could do it in two ways.

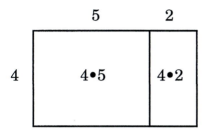

As shown above,
   a) $A = 4(5 + 2) = 4(7) = 28$ sq. units
or
   b) $A = 4 \cdot 5 + 4 \cdot 2 = 20 + 8 = 28$ sq. units

and, thus, $4(5 + 2) = 4 \cdot 5 + 4 \cdot 2$

In general, for a the length of one side, and b and c the components of the other side, where a, b, and c could represent any integers, we have

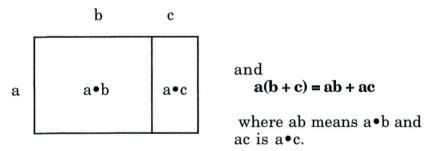

and
$a(b + c) = ab + ac$

where ab means $a \cdot b$ and ac is $a \cdot c$.

Also, since $b - c = b + (-c)$, we can write:

$a(b - c) = a(b + -c) = ab + (-ac) = ab - ac$ for all integers a, b, and c.

It is important to note that the **expression** $ab - ac$ consists of 2 **algebraic terms**, where a **term** is **a number, a variable, a product of variables, or the product of a number and variable(s)**. As previously noted, terms in an expression are separated by "plus" or "minus" signs.

We can use the distributive property to find the sum (or difference) of terms such as $3x$ and $2x$ in the following manner:

$3x + 2x = (3 + 2)x = 5x$ and
$3x - 2x = (3 - 2)x = 1x = x$ (since $1x$ means 1 <u>times</u> $x = x$)

When terms such as $3x$ and $2x$ can be combined in this way, we call them **like terms**. In **like** terms, the variable components are identical.

Sample Problems

1. Find the area of the largest square below.

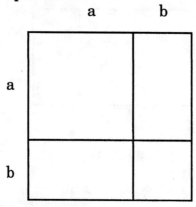

   Solution:
   The area of the largest square is the sum of the areas of the other rectangles and squares. Specifically,

   $$A_{total} = A_1 + A_2 + A_3 + A_4$$

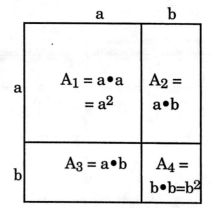

   Looking at the area model:
   $$(a + b)^2 = a \cdot a + ab + ab + b \cdot b$$
   $$= a^2 + (1 + 1)ab + b^2$$
   $$= a^2 + 2ab + b^2$$

2. Use the distributive property of multiplication over subtraction to calculate easily the total cost of 12 quarts of motor oil at $.97 per quart.

   Solution: $12(\$.97) = 12(1 - .03) = 12 - .36 = \$11.64$

3. Use the associative and commutative properties of addition to calculate easily the sum of $32 + 40 + 28$.

   Solution: $32 + (40 + 28) = 32 + (28 + 40)$ (using the commutative
   Then,   and associative properties)
   $(32 + 28) + 40 = 60 + 40 = 100.$ (the associative property)

4. Find the additive inverse of each of the following:
       a) -4 b) x c) $(a - b)$

Solutions: a) 4 b) –x c) –(a – b) or –a + b

5. a) Write 8ab as the sum of two terms.

   b) Write 3x + 3y as a product.

   Solutions: a) one possibility, 3ab + 5ab   b) 3(x + y)

Properties of Equality

In addition to the properties of integers, we need to consider the properties of equality. In a statement such as

$$3(2 + 4) = 3 \cdot 2 + 3 \cdot 4$$

illustrating the distributive property, both sides represent the same number. We can also say that the sides have the same "weight." If a number is added to both sides, the weights will now be different from the original weights but the two sides will remain in balance. For example, adding 4 to each side of the equality above gives

$$3(2 + 4) + 4 = 3 \cdot 2 + 3 \cdot 4 + 4 \quad \text{since}$$

$$3(6) + 4 = 6 + 12 + 4$$

and $\quad 22 = 22$.

Also, if we add a negative value to both sides, the balance is maintained. For example,

$$3(2 + 4) + (-5) = 3 \cdot 2 + 3 \cdot 4 + (-5)$$

$$18 + (-5) = 6 + 12 + (-5) \quad \text{and}$$

$$13 = 13.$$

We can generalize this property for all integers a, b and c:

$$\text{If} \quad a = b \quad \text{then} \quad a + c = b + c.$$

Now if we begin with a statement such as

$$6(2) = 3(4)$$

and multiply both sides by –2, for example, we have

$$6(2)(-2) = 3(4)(-2).$$

The product on both the left and right sides is –24, and it is reasonable that multiplying both sides of a statement of equality by an integer will change the value of each side but not affect the <u>equality</u> of the two sides.

Again, in general form, for all integers a, b, and c :

$$\text{If } \mathbf{a = b} \text{ then } \mathbf{ac = bc.}$$

Finally, let's consider the equality given by

$$3 \cdot 4 = 12.$$

If we divide both sides of the equation by 3, we have

$$\frac{3 \cdot 4}{3} = \frac{12}{3}.$$

Since any nonzero integer divided by itself is 1, and $1 \cdot 4 = 4$, the left and right sides are both equal to 4. In conclusion, for c a *nonzero* integer,

$$\text{if } \mathbf{a = b} \text{ then } \frac{\mathbf{a}}{\mathbf{c}} = \frac{\mathbf{b}}{\mathbf{c}}.$$

Introduction to Polynomials

In an example in this section, for any integers a and b, we found the area of a square with sides of length $(a + b)$ to be the expression

$$a^2 + 2ab + b^2.$$

We can also look at $(a + b)^2$ with respect to the distributive property. Follow the steps below:

$$\begin{aligned}(a+b)^2 &= (a+b)(a+b) \\ &= a(a+b) + b(a+b) \quad &&\text{(distributing both } a \text{ and } b \text{ to the 2nd quantity } (a+b)) \\ &= a^2 + ab + ba + b^2 \quad &&\text{(using the distributive property again)} \\ &= a^2 + 2ab + b^2 \quad &&\text{(reminder: } ab = ba \text{ so } ab + ba = 2ab)\end{aligned}$$

This is an example of a **polynomial** with 3 *terms*. Sometimes a polynomial consists of just one term with a number (called the **coefficient**) and variable(s), or it may be just a number, or constant. We have specific names for some of these polynomials based on the number of terms the polynomial contains. Consider these examples,

$3xy$ or 5 both have only one term and are called **monomials**;

$2x + 5y$ or $2x + 4$ both have two terms and are called **binomials**;

$3xy + 4x - y$ has three terms and is called a **trinomial**;

We can extend our work with the properties of integers to simplifying polynomials. For example, if we want to write the following sum in simplified form,

$$(3x + 2y - 4) \ + \ (4x - 5y + 3)$$

we eliminate the parentheses as appropriate, regroup the *like* terms and use the distributive property to combine them.

We then have
$$= 3x + 2y - 4 + 4x - 5y + 3$$
$$= (3x + 4x) + (2y - 5y) + (-4 + 3) = 7x + (-3y) + (-1)$$

We'll now use the property, $a + (-b) = a - b$, to rewrite the answer:

$$7x + (-3y) + (-1) = 7x - 3y - 1$$

To subtract two polynomials, say, $5x + 8y$ from $3x - 2y$, rewrite the problem as

$$(3x - 2y) - (5x + 8y) = (3x - 2y) + (-5x - 8y) = 3x - 2y - 5x - 8y,$$

where the last two terms represent the subtraction of the *quantity*, $(5x + 8y)$. (Recall from our work with the additive inverse that $-(5x + 8y) = -5x - 8y$.)

Now we reorder the terms, making sure the signs are appropriate, and combine like terms as shown below.

$$3x - 2y - 5x - 8y = 3x - 5x - 2y - 8y = -2x - 10y$$

(Again recall that $3x - 5x = 3x + (-5x)$, etc.)

Sample Problems

1. Subtract $(2x - 4)$ from $(5x + 4)$.

   Solution:
   $$\begin{array}{r} 5x + 4 \\ - 2x - 4 \\ \hline \end{array} \longrightarrow \begin{array}{r} 5x + 4 \\ + -2x + 4 \\ \hline 3x + 8 \end{array}$$

   or $(5x + 4) - (2x - 4)$
   $5x + 4 - 2x + 4$
   $3x + 8$

2. Perform the indicated operations and simplify:

   $$(4x - 2y + 5) + (2x + 3y) - (5x - 5y - 8)$$

   Solution:
   Remove parentheses and use additive inverses as appropriate:

   $$4x - 2y + 5 + 2x + 3y - 5x + 5y + 8$$

   Then reorder and combine like terms:

   $$(4x + 2x + (-5x)) + (-2y + 3y + 5y) + (5 + 8) = x + 6y + 13$$

In an expression such as
$$3(x + 4)$$
we used the distributive property to remove the parentheses. We can now think of this problem as the multiplication of a monomial, 3, by a binomial, $x + 4$, and we generate the same result
$$3(x + 4) = 3x + 3 \bullet 4 = 3x + 12.$$

We can also extend the distributive property to simplify the following:

$$2(3x + 5y - 2) = 6x + 10y - 4$$

Finally, to illustrate the product of two binomials, $(x + 2)$ and $(x + 1)$, for example, we will use algebra tiles and a mat with positive and negative regions as shown below.

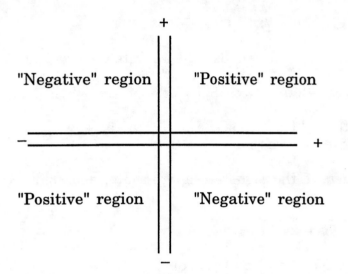

Note that the value of a region is determined by multiplying the values of the axes forming the region.

To use the algebra tiles, we need the following representations:

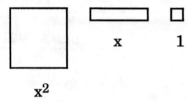

The model of $(x + 2)(x + 1)$ is on the mat below.

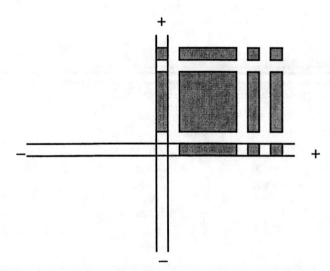

Note that $(x + 2)$ is positioned horizontally, $(x + 1)$ is positioned vertically, and since all are positive values, the entire resultant rectangle (with spaces to help identify the individual components only), is located in the upper right "+" region.

The result is: $(x + 2)(x + 1) = x^2 + 3x + 2$

We can also verify this by again using the distributive property:

$$(x + 2)(x + 1) = x(x + 1) + 2(x + 1) = x^2 + x + 2x + 2 = x^2 + 3x + 2$$

After working more with exponents, we will expand on the multiplication of polynomials in chapter 6.

Finally, to divide the polynomial, $3x - 15$, by 3, we again employ the distributive property and the fact that a quantity divided by itself is 1. Follow the steps below.

$$3x - 15 = 3(x - 5) \quad \text{(This is now in \textbf{factored} form because it is written as a \textbf{product}.)}$$

and

$$\frac{\cancel{3}(x-5)}{\cancel{3}} = x - 5$$

It is important to remember that in an expression like

$$\frac{3-6}{3}$$

we cannot merely "cross-out" the 3's because we would then have

$$\frac{\cancel{3}-6}{\cancel{3}} = \frac{1-6}{1} = -5$$

This does not generate the same value as the correct procedure,

$$\frac{3-6}{3} = \frac{-3}{3} = -1$$

However, if we isolate the factor of 3 in the top of our original expression, we have

$$\frac{3(1-2)}{3} = \frac{\cancel{3}(-1)}{\cancel{3}} = -1$$

Similarly, in our first example,

$$\frac{3x - 15}{3}$$

the 3's cannot be merely crossed-out, although it is more difficult to check the error because we do not know the value of the variable.

As we have illustrated, *it is very important that in the division of polynomials, common factors are isolated before any dividing (or cancelling) is done.*

Again, as we work with rational numbers, we'll expand our discussion of the division process.

Sample Problems

1. Use algebra tiles to model and find the product, $(x + 3)(x - 1)$.

   Solution:

   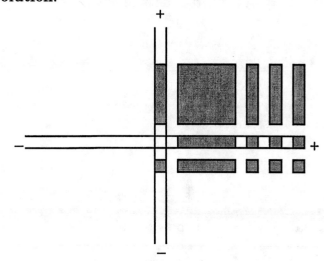

   Keeping the positive and negative regions in mind,
   $$(x + 3)(x - 1) = x^2 + 3x - x - 3 = x^2 + 2x - 3$$

2. Perform the indicated operations and simplify.

   a) $(5x - 2)3$ b) $\dfrac{2x - 18}{2}$ c) $6(x - 2) - 5(x + 1)$

   Solutions:

   a) $(5x - 2)3 = 15x - 2 \cdot 3 = 15x - 6$

   b) $\dfrac{2x - 18}{2} = \dfrac{\overset{1}{\cancel{2}}(x - 9)}{\underset{1}{\cancel{2}}} = x - 9$

   c) Think of $6(x - 2) - 5(x + 1)$ as $6(x - 2) + (-5)(x + 1)$ and then

   $$6(x - 2) + (-5)(x + 1) = 6x - 12 + (-5x) + (-5)$$
   $$= (6x + (-5x)) + (-12 + (-5)) = (6x - 5x) + (-12 - 5)$$
   $$= x + (-17) = x - 17$$

Challenge Problem

1. Write $3(2x + y) + 2(2x + y)$ as a product.

    Solution: $(3 + 2)(2x + y) = 5(2x + y)$

Problem Set 1.3

1. Use the properties of integers to compute each of the following as easily as possible.

   a) $-9 + 10 + 9 + 22$ b) $45 + 34$ c) $13(8)$

   d) $65 + 20 + 35$ e) $\dfrac{(-3)(5)(0)(-1)}{(-5)(3)}$

2. Find the additive inverse of each of the following.

   a) -4 b) $(x + y)$ c) $|-6|$ d) $-a(-b)$

3. Find the perimeter, or distance around, the largest rectangle. Then find the area of the largest rectangle in two ways.

   a) 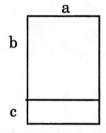   b)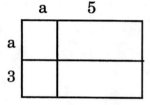

4. Use the distributive property of multiplication over addition to compute easily the cost of 4 pounds of apples at $1.04 a pound.

5. Write the properties and the operations illustrated in each of the following statements. Some may illustrate more than one property.

   a) $3(4 + 2) = (4 + 2)3$ b) $-5 \cdot 1 = 1 \cdot -5 = -5$

   c) $6 + (5 + 4) = (6 + 5) + 4$ d) $-2 + (3 + -6) = -2 + (-6 + 3)$

6. When you were a child in elementary school learning to add columns of numbers, you started at the top and added your way down. The teacher probably told you to check your answer by adding the numbers in the opposite order, from the bottom up to the top. What property(ies) of addition were you using in the check?

7. Beginning with $6 + 3(4) = 18$ as the base statement of equality, write another statement of equality by

   a) adding a negative integer to both sides

   b) multiplying both sides by a positive integer.

   Find the value of each side in both cases.

8. Given the following statements, find the value(s) that make each true.

   a) $6 \cdot \Box = -6 \cdot 4$

   b) $-3 + \Box = 0$

   c) $3(12) = 3(7 + \Box)$

   d) $2(\Box)(-5) = 3(0)(-1)$

9. Use the distributive property to combine quantities, if possible.

   a) $-5y + 8y$
   b) $-6x - 4x$
   c) $8a^2 + 12a^2$
   d) $4x - 6x^2$

10. Write $9xy$ as a sum.

11. Use the distributive property to write as a product, i.e., put the expression in factored form.

    a) $3a + 4ab$
    b) $6x - 2xy$
    c) $6x + 9y - 3$

12. Use the distributive property to remove parentheses.

    a) $3(x + 5)$
    b) $2(2y - 1)$
    c) $5x(a + b)$
    d) $(a + 3)(a - 3)$

13. Perform the indicated operations and simplify, if possible.

    a) $(5x + 3) + (2x - 1)$

    b) $(-6x - 5) - (3x + 2)$

    c) $(2x + 5xy - 6y) + 2(3xy - 3y)$

    d) $(5a - 2ab + b) - (-3a + 5ab + b)$

    e) $3(5x - 8) - 2(x + 4)$

    f) $\dfrac{15ab + 20a - 30ax}{5a}$

14. Sketch a model of $(x + 2)(x - 3)$ on an algebra tile mat. Then use the model to find the product.

15. Write the product of the binomials illustrated below. Find the solution using the mat and then verify your results by using the distributive property.

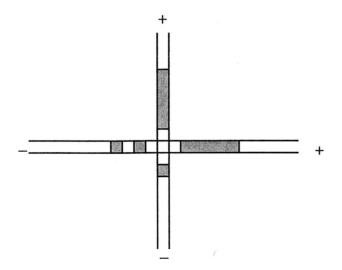

16. Write $7(a-2b) - 5(a-2b)$ as a product.

Exploration Problems

1. Use the distributive property and combine like terms to simplify.

    a) $4(x^2 - 3x - 1) - 3(2x^2 - 3x + 2)$

    b) $(3a + 2)(2a - 5)$

2. Draw an area model to show $(3a + 2)(6a + 1)$ and find the product.

3. For this triangle

    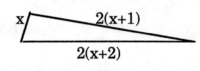

    a) Give at least 3 sets of numbers that could represent the lengths of the sides.

    b) Design another triangle (to scale, if possible), with x as the length of the shorter side and the other sides given in terms of x.

Solutions - Problem Set 1.3

1. a) $0 + 32 = 32$ b) $40 + 30 + 9 = 79$ c) $8(10 + 3) = 80 + 24 = 104$

   d) $100 + 20 = 120$ e) $0/(-15) = 0$

2. a) 4 b) $-(x + y) = -x - y$ c) -6 d) $-(ab)$ or $a(-b)$ or $-ab$

3. a) $P = a + b + c + a + b + c = 2a + 2b + 2c$ or $2(a + b + c)$
      $A = a(b + c) = ab + ac$

   b) $P = a + 3 + a + 5 + a + 3 + a + 5 = 4a + 16$
      $A = a \cdot a + 5a + 3a + 15 = a^2 + 8a + 15$

4. $4(1 + .04) = \$4 + .16 = \$4.16$

5. a) Commutative, multiplication b) multiplicative identity

   c) associative, addition d) commutative, addition

6. The commutative and associative properties

7. a) $6 + 3(4) = 18$; $6 + 3(4) + (-3) = 18 + (-3)$ and $15 = 15$ ⟵ examples

   b) $6 + 3(4) = 18$; $4[6 + 3(4)] = 4(18)$ and $72 = 72$ ⟵

8. a) $\square = -4$ b) $\square = 3$ c) $\square = 5$ d) $\square = 0$

9. a) $(-5 + 8)y = 3y$ b) $(-6 - 4)x = -10x$ c) $(8 + 12)a^2 = 20a^2$

   d) can't be done because the exponents on the variable x are not the same and <u>one</u> quantity cannot be isolated.

10. $(xy + 8xy)$, for example.

11. a) $a(3 + 4b)$ b) $2x(3 - y)$ c) $3(2x + 3y - 1)$

12. a) $3x + 15$ b) $4y - 2$ c) $5ax + 5bx$ d) $a^2 - 9$

13. a) $7x + 2$ b) $-9x - 7$

    c) $2x + 5xy - 6y + 6xy - 6y = 2x + 11xy - 12y$

    d) $5a - 2ab + b + 3a - 5ab - b = 8a - 7ab$

    e) $15x - 24 - 2x - 8 = 13x - 32$ f) $3b + 4 - 6x$

14.

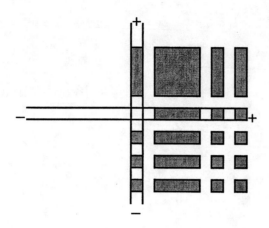

$(x + 2)(x - 3) = x^2 - x - 6$

15. The binomials are $(x - 2)$ and $(x - 1)$ and the product, $(x - 2)(x - 1) = x^2 - 3x + 2$, as shown below.

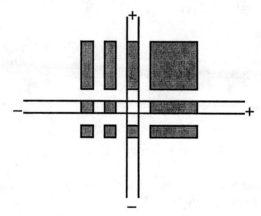

Using the distributive property:

$(x - 2)(x - 1) = x(x - 1) - 2(x - 1) = x^2 - x - 2x + 2 = x^2 - 3x + 2$.

16. $2(a - 2b)$

Section 1.4 – Integers in Problem Solving

Finding Midpoints

In Section 1.1, we first found the length of a line segment by counting the number of intervals separating the two endpoints of the segment. We then found the length by subtracting the two endpoints, specifically by taking the smaller integer from the larger. If the endpoints of an interval are 5 and –4, we can also take the **absolute value** of the <u>difference</u> of 5 and –4, and we do not have to be concerned about which integer comes first. Using this method, the distance between these two integers is written

$$|5 - (-4)| = |9| = 9 \text{ or } |-4 - 5| = |-9| = 9$$

This result can be verified on the number line below.

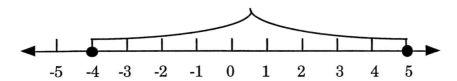

It is important to note that the subtraction can be done in any order since taking the absolute value will assure the distance will be positive.

We can also use subtraction and/or addition of integers to find the midpoint of a line segment. (We will restrict our discussion to horizontal and vertical line segments at this time.) Let the endpoints of a given line segment be –5 and +3. To find the point midway between the endpoints, we can find the distance between the two points and divide this value by two.

In this case, $|(-5) - 3| = 8$ and $8/2 = 4$. However, it is apparent from the number line below that 4 is not the midpoint of this line segment. We need to add this amount (4) to the left endpoint (or subtract it from the right endpoint) to find the actual <u>location</u> of the midpoint. Thus, the midpoint of the line segment generated by –5 and 4 is

$$-5 + 4 = -1 \text{ (or } 3 - 4 = -1)$$ which can be verified below.

An easier way to find the midpoint directly is to "**average**" the two numbers. This is done by adding the two integer coordinates and dividing the sum by two. Thus, the point midway between –5 and +3 can be found by calculating as follows:

$$\frac{-5 + 3}{2} = \frac{-2}{2} = -1$$

and this agrees with our earlier result. In general, to find the **midpoint** of the line segment between points a and b, find the average of a and b, that is:

$$\frac{a + b}{2}$$

Sample Problems

1. Find the midpoint of the line segment connecting –5 and –10 on the number line.

   Solution: Using the "average" method:
   $$\frac{-5 + (-10)}{2} = \frac{-15}{2} = -7.5$$

   Note that here the midpoint is not an integer, but this value is still the location midway between the two integers.

2. Find the midpoint (as a set of coordinates) of the line segment connecting the points (1,4) and (1,–2) on the coordinate system.

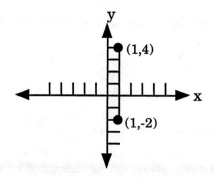

   Solution: Since the x-coordinates are the same, these points generate a vertical line. We only need to find the distance between the y–coordinates, 4 and –2. Thus, the <u>y-coordinate</u> of the midpoint of this line segment is:

   $$y = \frac{4 + (-2)}{2} = \frac{2}{2} = 1$$

   To be on the same line the x-coordinate must be 1, so the midpoint of this line segment is (1,1).

Number Sequences

Consider the numbers: 2, 4, 6, 8, 10, . . . Note that the three dots (called ellipsis dots) indicate the numbers continue on in the same manner. This is an example of a number sequence. There are many different kinds of number sequences. We could describe the numbers in the sequence above as the set of even positive integers, each one differing from the next one by two.

When a sequence such as this has what we call a "common difference", that is, when *any* two consecutive terms of the sequence always generate the same *difference*, we call it an **arithmetic sequence**.

Now let's examine a different kind of sequence: 1, 2, 4, 8, 16, . . . Note here that there is no common difference (4 – 2 = 2 but 16 – 8 = 8), but the numbers appear to be related. In fact, if you multiply any one of the numbers by 2, you can generate the next number in the sequence. This type of sequence which has a "common ratio," i.e., a common multiplier, is called a **geometric sequence**.

Another interesting sequence is the sequence of triangular numbers.

This sequence can be depicted by arrays of dots:

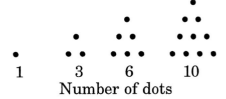

1   3   6   10
Number of dots

Can you draw the next array of dots? How many dots will the array contain? Let's determine if a pattern exists here. Note first that there is not a common difference nor is there a common ratio. Making a table will help to find the pattern if one exists.

Term no.	Triangular Number (Number of dots)	
1	1	1
2	3	$= 1 + 2$
3	6	$= 1 + 2 + 3$
4	10	$= 1 + 2 + 3 + 4$
5	?	

If we examine the sequence 1, 3, 6, 10, . . . and note that the difference between the first two terms is 2, the difference between the second and third terms is 3, the difference between the third and fourth terms is 4, we can picture the pattern as

1,   3,   6,   10,   ...

   2   3   4   ?   etc.

Assuming the pattern holds, the next difference is 5 and the fifth term should then be $10 + 5 = 15$.

How would we find the 20th triangular number? One way, of course, is to continue the table down to the 20th term and use the pattern we generated to find the 20th triangular number. Another way would be to generalize and develop a "formula" that we could then apply to find any triangular number.

Consider the relationship between the numbers across from each other in the table above and look at the arrangement of dots in each case. Note that the first term has 1 dot, the second term has $3 = 1 + 2$ dots, the third term has $6 = 1 + 2 + 3$ dots, and the fourth term has $10 = 1 + 2 + 3 + 4$ dots. It would seem then that the 20th term would have

$$1 + 2 + 3 + 4 + \ldots + 19 + 20 \text{ dots.}$$

We could certainly use a calculator to compute this sum. But let's try another method. Writing the sums as

$$1 + 2 + 3 + 4 + 5 \ldots + 18 + 19 + 20 \text{ and}$$
$$20 + 19 + 18 + 17 + 16 \ldots + 3 + 2 + 1$$

53

results in vertical pairs of numbers, and each pair, in this case, has a sum of 21. Since we have 20 groups (or sums) of 21, we should have a total of (20 x 21) dots. However, these rows are exactly alike, and to get the sum of the numbers in *one* row we must divide by 2.

Therefore, the 20th triangular number must be:

$$\frac{20 \times 21}{2} = 210$$

In other words, the 20th arrangement of dots will contain 210 dots.

Using this same method, let's find the 100th triangular number. Extending the sum of dots arrangement we now have:

```
  1 +  2 +  3 +  4 + ...    + 98 + 99 + 100
100 + 99 + 98 + 97 + ...    +  3 +  2 +   1
------------------------------------------------
```

Here we have 100 vertical sums of 101 each so the 100th triangular number is

$$\frac{100 \times 101}{2} = 5050$$

> Karl Friedrich Gauss is considered to be the greatest German mathematician of the 19th century. At age 3 he corrected an error in his father's calculation of a worker's pay. When Gauss was 10, his teacher asked the class to find the sum of the first 100 natural numbers. He thought this would take the students a long time, but Gauss solved the problem before the teacher could sit down. Gauss had used the method above to derive the solution. After this episode the teacher bought Karl the best arithmetic text available at that time and encouraged him to get a private tutor. Gauss went on to make major contributions in number theory, differential geometry, and statistics.

To generalize these examples, let's use **n** to represent any term number. From our reasoning above, the **nth** triangular number can be found by the formula

$$\frac{n(n+1)}{2}$$

Note that *n* is really the number of sums and *(n + 1)* represents the amount of each sum.

Using this formula to reaffirm our answer for the 5th triangular number, we'll substitute 5 for *n*. This gives us

$$\frac{5(5+1)}{2} = \frac{5(6)}{2} = 15$$

Similarly, to check our answer for the 100th triangular number, substitute 100 for *n*.

$$\frac{100(100+1)}{2} = \frac{100(101)}{2} = 5050$$

Again, the 100th triangular number implies that there are 5050 dots in the 100th array.

Sample Problems

1. Given the arithmetic sequence: 33, 29, 25, 21, ... Find the common difference and find the 10th term.

   Solution: Checking the difference between any two consecutive terms, 29 – 33, for example, the common difference is –4. So we can generate the sequence by adding –4 (or subtracting 4) each time and we have: 33, 29, 25, 21, 17, 13, 9, 5, 1, –3, ..., and the 10th term is –3.

2. Given the geometric sequence: 2, –2, 2, –2, ... Find the common ratio and the 8th term.

   Solution: The common multiplier or ratio here is –1. (Divide –2 by 2, 2 by –2, etc.) More terms of the sequence are:
   2, –2, 2, –2, 2, –2, 2, –2, 2, –2 ... and the 8th term is –2.
   Also note that –2 is the value of each of the even-numbered terms.

3. Draw a picture of the next figure in this sequence:

    , , , _____

   Solution:

Problem Solving Strategies

When we **looked for a pattern** in the sequence of triangular numbers, we really used one of the strategies for problem solving. Let's examine a problem where a different strategy might be used.

A log 10 feet long and 18 inches in diameter is to be cut into 1-foot sections. If each cut takes 1 minute, how long will it take to cut this log into 1-foot sections?

Our first reaction may be that this will take 10 minutes. Does the diameter of the log affect the time here? Why or why not?

By **drawing a picture** of the log and the cuts needed,

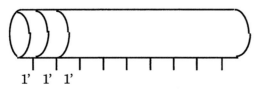

we realize that only 9 cuts are required, and so it will take 9 minutes to cut this log into 1-foot sections. [Note that the time for making a cut is 1 minute, and that the diameter of the log has already been taken into account. If the diameter of the log was 24 inches, for example, the given cutting time per cut would probably have been greater.]

Another strategy for problem solving is **guess-and-check**. Let's try another problem. Can you arrange the digits 1 through 9, inclusive, in the circles below using each digit exactly once and using all the digits so that the sum along each side of the triangle is 17?

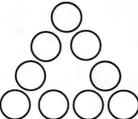

The easiest way to approach this is to make slips of paper, each containing one of the digits that can be used, put the pieces in the same arrangement as the triangle above and manipulate the numbered pieces. After a little guessing and checking, it might be helpful to think about the vertices first. Try putting the larger numbers, specifically, 7, 8, and 9, into the vertex circles. Does this work? Why or why not? Now try putting the smaller numbers, 1, 2, and 3, into the vertex circles. This appears to give more options for a possible solution. Now do more experimenting with the remaining numbers. Finally, record your answer and test to see if another arrangement of the numbers will generate a solution to the problem. (One possible answer is included in the solutions for problem set 1.4.)

Sample Problems

1. This square arrangement has three rows and three columns. Can you put the digits 1 – 9, inclusive, using each digit only once, in the squares so that all the rows, columns and diagonals have a sum of 15? Check to see if there is more than one solution to this problem.

One solution:

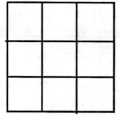

2.[1]   As part of an environmental project, a landscaper plans to place sprinklers at A, B, C, D, and E in a park. M is the point of connection to the water main.

---

[1] *Mathematical Investigations*, Book Two, Souviney, Britt, Gargiulo, Hughes, page 10

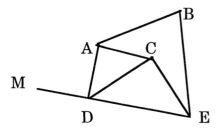

AB = 50m; AC = 41m;
AD = 22m; BE = 80m;
CD = 36m; CE = 40m;
MD = 27m; DE = 48m.

Jan, an accountant for the project, realizes that a number of pipes are unnecessary and just add to the cost.

a) She removes pipe BE from the plan. Explain her reasoning.

b) The longest pipe left is AB, but she cannot remove it. Why not?

c) The next longest pipe is DE. Can she remove it?

d) Jan continues to remove pipes until she reaches the connected network shown below.

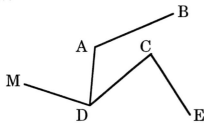

e) Jerry, Jan's assistant, suggests an alternative plan. Will Jerry's connected network of pipes get water to all sprinklers?

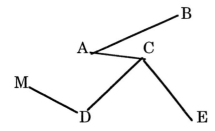

f) Which plan has the shorter connecting lengths and is, therefore, cheaper? Be sure to justify your answer.
Solution:

a) BE is the longest pipe and its removal still leaves the sprinklers connected.

b) It will disconnect B from the network.

c) Yes, E will still be connected.

e) Yes.

> f) The total length of Jan's design is 175 units while Jerry's is 194 units, and thus, Jan's design will cost less.

We will continue problem solving throughout the text and encourage you to use these different strategies as appropriate. Later we'll extend the use of tables and charts to the development of algebraic equations, another tool for solving problems.

Problem Set 1.4

1. Use the subtraction of integers method to find the midpoint of the line segment defined by:

   a) −10 and −4 on the number line.

   b) (6,−1) and (6,5) on the coordinate plane. Write the coordinates of the midpoint.

   c) Check your answers to parts a) and b) using the "averaging" method.

2. Find the perimeter and area of the triangle on the coordinate system below.

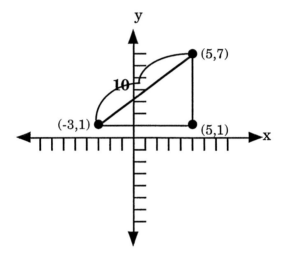

3. The high temperature in New York City last February 5 was 2°C. The low temperature for New York City that day was −8°C. Find the average temperature in New York City last February 5.

4. One endpoint of a vertical line segment is (−2,5) and the midpoint is (−2,1). Give the coordinates of the other endpoint of this vertical line segment.

5. Write the first five terms of any arithmetic sequence that has −3 as a common difference.

6. Write the first five terms of the geometric sequence that has a common ratio of 3 and a first term of 2.

7. Find the missing numbers.

   a) $\dfrac{\square + 4}{2} = 3$

   b) $|\square - 6| = 5$

   c) $\dfrac{|\square + 3|}{2} = 5$

   d) $\dfrac{2 \cdot \square - 3}{3} = 3$

   e) $7 - 2 \cdot \square = -1$

8. Noncollinear points are points in a plane that are not all in a straight line. If pairs of points are connected with line segments, use the diagrams below to help you decide the number of line segments that can be drawn in each case and complete the following table. (Note: If all points were collinear, only one straight line could be drawn. We make the points noncollinear to take care of this trivial case.)

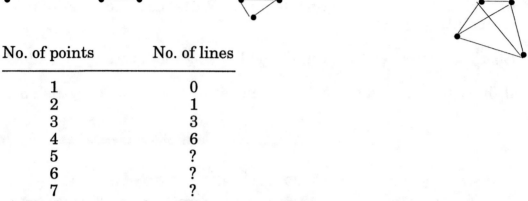

No. of points	No. of lines
1	0
2	1
3	3
4	6
5	?
6	?
7	?

a) Find a pattern that will help you to determine the number of line segments that can be drawn if there are 15 noncollinear points on a plane.

b) Find the rule of this sequence to describe the nth term.

9. If the nth term of a sequence is $n - 3$, generate the first five terms of the sequence.

10. Find the next 3 terms for each sequence. Label whether the sequence is arithmetic, geometric, or neither.

a) 6, 9, 12, 15, __, __, __
b) 2, 6, 18, __, __, __
c) 1, 5, 12, 22, 35, __, __, __
d) 5, 7, 10, 14, __, __, __
e) 10, 8, 6, 4, __, __, __

11. Given the letters H, A, P and Y in the configuration below, in how many ways can the word "HAPPY" be spelled if the given paths must be followed and only paths that are angled downward are permitted?

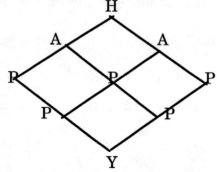

Hint: Traversing the paths and keeping track of the number is reasonable but another approach is to label the number of ways to get to each intersection and see whether a pattern exists.

12. The map below shows the <u>distances</u> between towns. Find the best route from A to F by cutting down the distance traveled. What is the longest route? Note: The distances are not necessarily measured along straight lines. That is, the route from B to C is a winding road.[1]

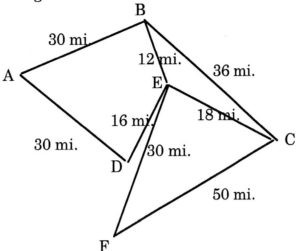

13. Use the digits 0, 1, 2, 3, 6, 7, and 9 in this cryptogram to make the following sum a true statement. Each letter represents only one digit.[2]

    ```
        S U N
    +   F U N
    ─────────
      S W I M
    ```

14. A frog is at the bottom of an 8-foot hole. During each day he can ease up 2 feet but each night he drops back 1 foot. How many days will it take him to get out of the hole? (Note that if the frog can get to the rim or be high enough to jump right out, he'll be out of the hole.)

15. Make a table to find how many ways you can make change for a $20 bill by using $1, $5, and $10 bills only.

16. In Montana, fence posts are to be placed in a row 4 meters apart. How many posts are needed for 200 meters of fence? Hint: A diagram may help.

17. Find the 36th triangular number. Use any method you wish.

18. A student needs an "80" for a grade of B in a college biology course. The grade is based on three test scores. His first test scores were 84 and 72. What score does he need on the third exam to guarantee him a B in the course?

---

[1] *Mathematical Investigations, Book Two*, Souviney, Britt, Gargiulo, Hughes, page 11
[2] *Mathematics for Elementary Teachers, A Contemporary Approach*, Musser and Burger, page 9

Exploration Problems

1. Using the sequence 1, 4, 9, 16, 25 . . . , find the 10th term and then use a chart to help find a formula for determining the nth term.

2. Find the **total** number of rectangles in the following diagram.

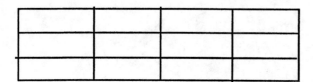

3. Find the midpoint of the line connecting points (5,7) and (–3,1) in the triangle pictured on the coordinate system in problem 2.

4. If a very wealthy person gave you a choice of accepting $1,000,000 in one lump sum or taking 1 cent the first day, 2 cents the second day, 4 cents the third day, 8 cents the fourth day, etc. for 31 days, which offer would you choose? Why? How much more would the better offer pay you? Explain how you derived your answer.

Solutions - Problem Set 1.4

1. a) |6|/2 = 3, –10 + 3 = –7   b) Coordinates of midpoint: (6,2).
   c) [–10 + (–4)] ÷ 2 = –7;   [5 + (–1)]÷ 2 = 2 so midpoint is (6,2)

2. P = 24 graph units;   A = 24 square graph units

3. The average temperature was –3°C.

4. (–2,–3)

5. One example: 5, 2, –1, –4, –7, ...

6. 2, 6, 18, 54, 162, ...

7. a) □ = 2   b) □ = 1, 11   c) = 7, -13   d) = 6□   e) = 4

8. | No. of points | No. of lines |
    |---|---|
    | 1 | 0 |
    | 2 | 1 |
    | 3 | 3 |
    | 4 | 6 |
    | 5 | 10 |
    | 6 | 15 etc. |
    | 7 | 21 |
    | 8 | 28 |
    | 9 | 36 |
    | 10 | 45 |
    | 11 | 55 |
    | 12 | 66 |
    | 13 | 78 |
    | 14 | 91 |
    | 15 | 105 |

   b) Rule:

   nth term = $\dfrac{n(n-1)}{2}$

9. –2, –1, 0, 1, 2

10. a) 18, 21, 24 (arithmetic)   b) 54, 162, 486 (geometric)   c) 51, 70, 92 (neither)
    d) 19, 25, 32 (neither)   e) 2, 0, -2 (arithmetic)

11. 6 paths

12. The best route from A to F is 72 miles (A to B to E to F).
    The longest route (only hitting each town one time) is 144 miles
                    (A to D to E to B to C to F).

13.   1 3 6
        + 9 3 6   so N = 6, U = 3, I = 7, S = 1, W = 0 , F = 9 and M = 2
         1 0 7 2

14. 7 days

15.

No. of 1's	No. of 5's	No. of 10's	Value
20	0	0	20
15	1	0	20
10	2	0	20
10	0	1	20
5	1	1	20
5	3	0	20
0	2	1	20
0	0	2	20
0	4	0	20

It appears there are 9 ways.

16. 51 fence posts are needed.

17. The 36th triangular number is 666.

18. He needs at least an 84 to get a *B* in the course.

One possible solution to the triangle of circles on page 56 is

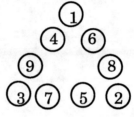

Section 1.5 - Connecting Integers and Geometry

Transformations in the plane

In section 1.1 we did some work on the Cartesian coordinate system. We'll now explore geometric figures on this system and observe some transformations of those figures.

Graphing the points A = (1,1), B = (4,1), C = (4,6) and D = (1,6) on the coordinate system below,

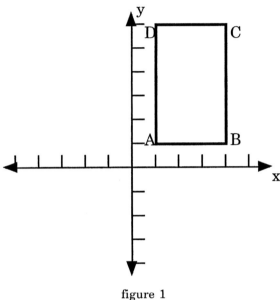

figure 1

we have a rectangle. If we add 2 to the x-coordinate of each vertex (the rectangle "corners"), can you predict what might happen to the rectangle on the coordinate system? The new vertices are: A' = (3,1), B' = (6,1), C' = (6,6) and D' = (3,6), where A' is the image of point A, etc. Plotting these points we have

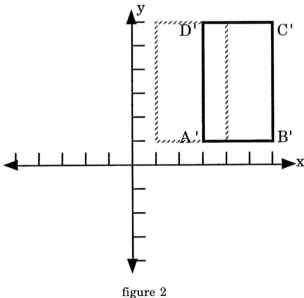

figure 2

Comparing this rectangle to the original in figure 1, it appears that the rectangle has the same shape and size and same orientation on the graph but that it has "slid over" along the x-axis. This motion, or **transformation,** in the plane is called a **translation** or **slide.** Think about what will happen if we add 2 to the y-coordinate of each vertex of the original rectangle. If the same reasoning applies, it would seem that the rectangle should "slide" up two units along the y-axis but should stay the same size and the same distance from the y-axis. Confirm this conclusion by drawing the new rectangle and comparing it to the original in size, shape and orientation.

A triangle is drawn on the coordinate system below with vertices at the points E = (2,1), F = (−3,1) and G = (2,3)

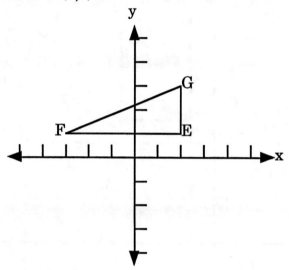

If we multiply the x-coordinate of each vertex by −1, what do you think will happen? Let's examine this on the graph below. The new triangle will have vertices: E' = (−2,1), F' = (3,1) and G' = (−2,3). [Again note the use of point E', read as "E prime", to correspond to the original point E, etc.] Both triangles are plotted below. (The new one is in darkened lines.)

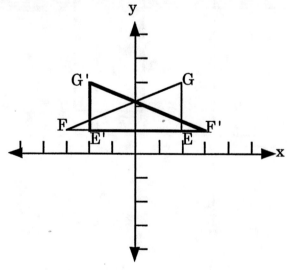

It is important to realize that a slide did not occur this time. Again the triangle has the same shape and size as the original, but its orientation on the graph is different.

If we think of the y-axis as a "fold-line", vertex G' would map onto (or coincide with) vertex G, vertex F' would map onto vertex F and vertex E' would map onto vertex E. This transformation is called a **reflection or flip**. In this case it is a reflection in the y-axis. Where would the triangle be on the coordinate system if instead we multiplied each of the y-coordinates of the vertices by –1? Try this to justify your answer. In the following exercises you will explore other transformations, including size changes.

Sample Problems

1. On a coordinate system, draw the rectangle defined by the vertices D = (3,–2), E = (3,1), F = (0,1), and G = (0,–2) and label the points. Add 1 to each of the x-coordinates and 2 to each of the y-coordinates. Label the image points and then explain how the image is related to the original rectangle in terms of shape, size, and position on the coordinate system.

   Solution: The image points will be: D' = (4,0), E' = (4,3), F' = (1,3) and G' = (1,0) as shown on the coordinate system below.

   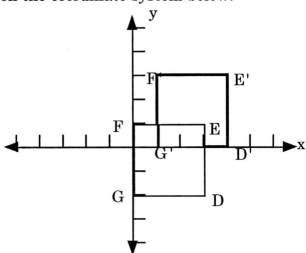

   The image rectangle has the same size and shape as the original and has the same orientation on the graph, but its position has changed. The image rectangle is the same as the original but the image has moved 1 unit to the right and 2 units up.

2. Multiply each of the coordinates of all the vertices of the original rectangle DEFG by 2. Explain how the image relates to the original rectangle in terms of shape, size, and position on the graph.

   Solution: The coordinates will now be: D' = (6,–4), E' = (6,2), F' = (0,2), and G' = (0,–4).

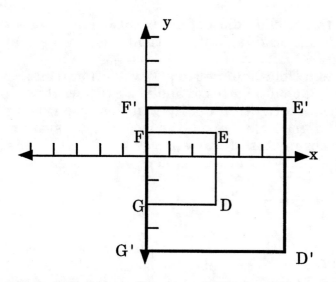

The rectangle image has the same orientation as the original with respect to the x and y axes and it hasn't moved, but it has changed size. In fact, the rectangle has quadrupled in size—that is, the image is 4 times the size of the original!

Exploring the Relationship Between Perimeter and Area

On the Coordinate System

Given the rectangle with vertices (1,1), (4,1), (1,6) and (4,6), and the absolute value method we've used previously to find the length of any line segment, we can easily find the perimeter, or distance around, this rectangle..

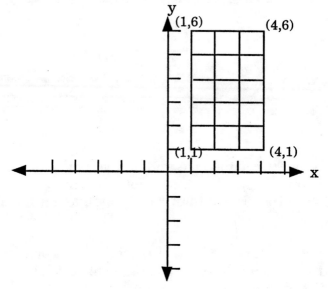

Note that we do not need to find the distance between (4,1) and (1,6) nor do we need to find the distance between (4,6) and (1,1) since these represent lines called **diagonals** whose lengths do not affect the perimeter of the rectangle. To find the perimeter we need the following distances:

between (1,6) and (4,6) which is $|1-4| = |-3| = 3$;
between (4,6) and (4,1) which is $|6-1| = 5$;
between (4,1) and (1,1) which is $|4-1| = 3$ and
between (1,1) and (6,1) which is $|1-6| = |-5| = 5$.

The perimeter of this rectangle is the sum of these four distances, so letting P be the perimeter,

$$\begin{aligned} P &= 3 + 5 + 3 + 5 \\ &= 3 + 3 + 5 + 5 \\ &= 6 + 10 \\ &= 16 \text{ graph units.} \end{aligned}$$

Rearranging the numbers in this step illustrates that we're really adding two equal widths and two equal lengths. This is true for the perimeter of *any* rectangle as we'll discuss shortly.

To find the area of this rectangle, count the number of unit squares within the rectangle as shown above. In this case we get 15 square units. By the way, it is not coincidental that the product of 3, the length of one side, and 5, the length of an adjacent side, is 15.

Sample Problem

1. Find the area and perimeter of a rectangle defined by the points (–4,3), (–4,–1), (2,–1), and (2,3).

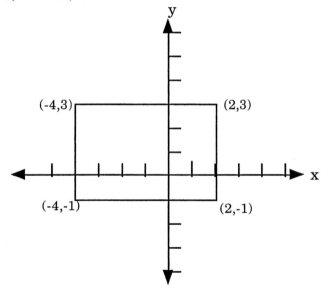

Solution: Each vertical side of the rectangle has length $|3-(-1)| = |3+1| = 4$. Each horizontal side has length $|2-(-4)| = |2+4| = 6$. To find the perimeter of this rectangle,

$$P = 4 + 4 + 6 + 6 = 8 + 12 = 20 \text{ units.}$$

The area occupied by this rectangle is (4)(6) or 24 square units.

Using Tiles

Another way to develop the understanding of perimeter and area, the difference between the two as well as their relationship, is to use unit square tiles, independent of a coordinate system. If one unit square looks like □

where each side has a length of one unit, we could determine the number of <u>square</u> units it would take to fill up a given rectangle, allowing no gaps and no overlaps. For example, the 7 x 3 rectangle shown below

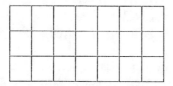

occupies an area of 21 square units because 21 unit squares would be needed to fill the rectangle. This is consistent with our previous work on the coordinate system, and, in general, the **area of any rectangle** can be found by multiplying the length, L, and the width, W, not necessarily in that order:

$$A = l \cdot w \text{ or } lw$$

The perimeter is 7 + 7 + 3 + 3 or 2(7) + 2(3) since there are two equal lengths and two equal widths. Therefore:

$$P = 2(7) + 2(3) = 14 + 6 = 20 \text{ units.}$$

In general terms, the **perimeter of any rectangle** can be written as

$$P = (2 \cdot l) + (2 \cdot w) = 2l + 2w$$

where again L is the length and W is the width of the rectangle.

It is crucial to understand that the distance around the rectangle is measured in units of length while the space occupied by the rectangle is measured in square units.

Sample Problems

For each of the figures, find the outside perimeter and the area.

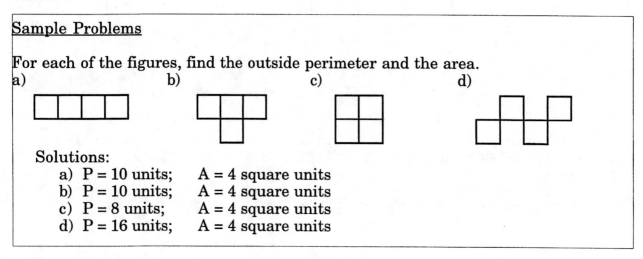

Solutions:
 a) P = 10 units;   A = 4 square units
 b) P = 10 units;   A = 4 square units
 c) P = 8 units;   A = 4 square units
 d) P = 16 units;   A = 4 square units

On the Geoboard

The geoboard is a square array of pegs on a piece of plastic or board. The geoboard looks like

and typically has 5 rows with 5 pegs in each row. Rubber bands are used to make figures on the geoboard.

The perimeter of any figure containing horizontal and vertical lines (for now) can be found by counting the number of intervals each segment contains. (The distance between two adjacent horizontal or vertical pegs, shown as •—•, is 1 unit of length.)

One square unit, with P = 4 units, on the geoboard is shown here:

The area of any figure can be found by counting the number of these square units which the figure contains.

Sample Problems

Make the following figures on a geoboard, or on dot paper symbolizing a geoboard, and find the area and perimeter of each figure. What do these figures have in common? What is different about these figures?

a)   b)   c)   d)

Solutions: a) P = 10 units, A = 6 sq. units   b) P = 10 units, A = 5 sq. units

c) P = 10 units, A = 4 sq. units   d) P = 10 units, A = 4 sq. units

The perimeters of these figures are all 10 units but the areas are not all the same.

Given the following rectangle on the geoboard model, if we draw the diagonal connecting opposite vertices, we separate the rectangle into two right triangles of equal area. (Note that right triangles contain a "square" corner.)

If the area of the rectangle is 8 square units, then the area of each of these triangles is 4 square units.

Using this model to find the area of a right triangle is called the "rectangle method." We'll extend this model to find the area of a triangle which does *not* contain a right angle in one of the exploration problems at the end of this section.

Sample Problems

1. Find the area of each of the triangles below using the rectangle method.

a)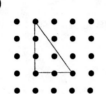
b)
c)
d)

Solutions:

a) $\frac{1}{2}(3 \cdot 2) = \frac{1}{2}(6) = 3$ sq. units   b) $\frac{1}{2}(4 \cdot 4) = \frac{1}{2}(16) = 8$ sq. units

c) $\frac{1}{2}(3 \cdot 4) = \frac{1}{2}(12) = 6$ sq. units   d) $\frac{1}{2}(4 \cdot 2) = \frac{1}{2}(8) = 4$ sq. units

2. Find the perimeter and area of each of the following figures. Non-vertical and non-horizontal lengths are shown.

a)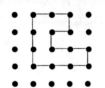
b)
c)
d)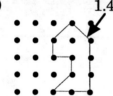

Solutions:
   a) P = 3 + 3 + 1 + 2 + 1 + 1 + 1 + 2 = 14 units;   A = 6 sq. units
   b) P = 3 + 4 + 5 = 12 units;   A = (3 • 4) ÷ 2 = 6 sq. units
   c) P = 4 + 1 + 1 + 3 + 2 + 3 + 1 + 1 = 16 units;   A = 10 sq. units
   d) P = 3(1.4) + 2 + 3 + 1 + 1 + 1 = 12.2 units;   A = 5.5 sq. units

3. On the geoboard make all possible rectangles, each with an area of 4 square units. Record your answers.

Solutions:

a)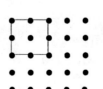
b)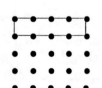
c)

4. On the geoboard make at least 3 triangles, each with an area of 2 square units.

Possible Solutions:

a) b) c)

Applications

Let's explore a rectangle with a perimeter of 20 units. There are many such rectangles, but in particular we'll examine one with a length of 8 and a width of 2 units, drawn below.

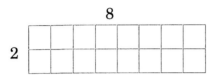   P = 2(8) + 2(2) = 20 units

Note that the area of this rectangle is 16 square units.

If we made a different shaped rectangle, again keeping the perimeter constant at 20 units, do you think the area would change?

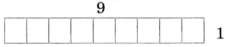

For a 9 by 1 rectangle,
    P = 2(9) + 2(1) = 20 units and A = 9 • 1 = 9 square units.

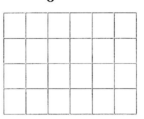

For a 6 by 4 rectangle,
    P = 2(6) + 2(4) = 20 units and A = 6 • 4 = 24 square units.

Sketching a 5 by 5 rectangle and calculating perimeter and area as before:

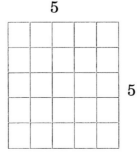

P = 2(5) + 2(5) = 20 units and A = 5 • 5 = 25 square units.

In all cases above the perimeter was held constant at 20 units, but note what happened to the areas. Try other rectangles with *integer* dimensions and with perimeter equal to 20 units. How do the areas compare?

There are ways to maximize the area of a rectangle while keeping the perimeter constant. The best way to visualize this is by drawing a graph which relates the area of a rectangle to its width (or length).

Consider a gardener who has 100 feet of fencing and wants to enclose a rectangular garden and obtain the largest planting area possible. We begin our analysis by making a table where we can mathematically change the shape of the garden. We'll select various widths, find the corresponding lengths to make sure the perimeter of that particular garden is still 100 feet, and then calculate the corresponding areas.

We first select a width of 0 feet, which means that we have put all our fencing into the two lengths and we have a straight line where $A = 0 \cdot 50 = 0$. This is considered to be a trivial case because we will not have any garden enclosed with fencing. If we select a width of 5, we have used 10 feet of fencing for the two widths and have 90 feet left to divide evenly between the two lengths. So, in this case, each length must be 45 feet. Note, in fact, that the width plus the length always equals half of the perimeter. This measurement is sometimes referred to as the semi-perimeter. The table below has more length, perimeter, and area values calculated in the same fashion for selected values of the width.

Width (in ft.)	Length (in ft.)	Perimeter (in ft.)	Area (in sq. ft.)	Approximate Shape
0	50	100	0	
5	45	100	225	
10	40	100	400	
15	35	100	525	
20	30	100	600	
25	25	100	625	
30	20	100	600	
35	15	100	525	

As we continue with this table of values, it should be apparent that, for a constant perimeter, as we *increase* the width, the length of the rectangle *decreases*, and we start to get values for area that occurred earlier in the table. This happens because there are a finite number of rectangles with integer values for length and width, and the rectangles with the widths larger than lengths are the very same rectangles in size and shape as those with lengths larger than widths. [The diagrams above confirm this statement. In fact, it is only by convention that we call the larger side of a rectangle the length and the shorter side the width.] Also note that there is a pattern in the change in areas as we chose different values for length and width, even though the perimeter was kept constant at 100 feet. The areas increased to a point and then decreased again. From the table, there does appear to be a particular rectangular garden where the area is the largest possible.

Graphing the width of the rectangle on the horizontal axis and the area on the vertical axis we have a better picture of the pattern.

Note that the graph is restricted to the first quadrant. (Why?) Since the numbers we're graphing are large, the intervals are larger than 1 so that the values of width and area fit onto the coordinate system. Also note that the interval values for area are larger than the interval values for width.

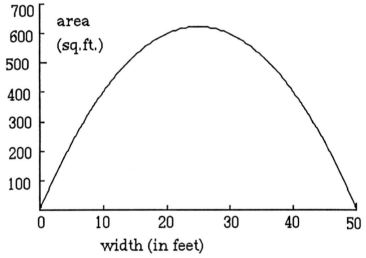

From the graph, and from the table, the rectangle with dimensions 25 feet by 25 feet generated the largest area. This, in fact, is *a square - a rectangle with equal length and width.*

We can also use the graph to approximate the dimensions of a rectangle whose perimeter is 100 feet and whose area is 300 square feet. As shown at the top of the next page, we first find 300 sq. ft. on the vertical axis, sketch a horizontal line through 300 and note where it intersects the graph. Then by sketching vertical lines from the points of intersection down to the horizontal axis, we should find the width to be approximately 7 feet or 42.5 feet. Note that if width is 7 feet, length is 42.5 feet and vice versa.

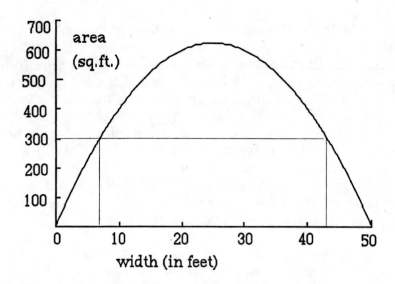

Sample Problems

1. Find the dimensions (integer values only) of all possible rectangles with an area of 30 square units and find the perimeter of each of these rectangles. For which rectangle is the perimeter the smallest?
   Solution:

   (Note these are factors of 30)
   - 1 by 30,   P = 2(30) + 2(1) = 62 units
   - 2 by 15,   P = 2(15) + 2(2) = 34 units
   - 3 by 10,   P = 2(10) + 2(3) = 26 units
   - 5 by 6,   P = 2(6) + 2(5) = 22 units

   If we were to include 30 x 1, 15 x 2, etc., we would have the same rectangles we have already listed. So we have found all of the <u>distinct</u> rectangles. The one with the smallest perimeter is the 5 x 6 rectangle. In this case we kept the area constant and compared the perimeters. Does it surprise you that, of the rectangles above, the one closest to a square shape has the smallest perimeter?

2. Give integer dimensions of at least three rectangles having a perimeter of 80 units.

   Possible solutions:

   - 20 by 20, P = 2(20) + 2(20) = 80 units; A = 20 • 20 = 400 square units
   - 10 by 30, P = 2(30) + 2(10) = 80 units; A = 30 • 10 = 300 square units
   - 5 by 35,   P = 2(35) + 2(5) = 80 units;   A = 35 • 5 = 175 square units
   - 15 by 25, P = 2(25) + 2(15) = 80 units; A = 25 • 15 = 375 square units

   Note that the rectangle with a maximum area of 400 sq. units is a 20 by 20 square.

Problem Set 1.5

1. a) Find the coordinates of the rectangle after a horizontal slide of +4 graph units.

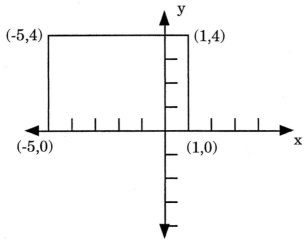

   b) Find and compare the perimeter and the area both of the original rectangle and of its image.

2. Given the triangle as shown. Describe the transformation on the original triangle.

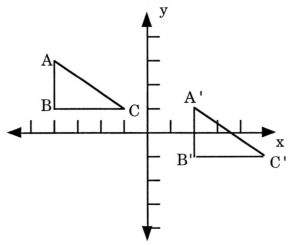

3. The image triangle after a slide of 3 graph units right and 1 unit down has vertices at the points (−1,2), (−4,−2), and (5,−3). Find the coordinates of the vertices of the original triangle.

4. A four-sided figure, called a quadrilateral, is pictured on the next page. Multiply each of the y-coordinates of the vertices by −1 and plot the image quadrilateral on the coordinate system. How does the new quadrilateral compare to the old with respect to size, shape, and orientation on the graph?

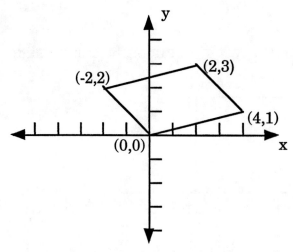

5. Given the triangle defined by the vertices (0,0), (3,1), and (3,–3), multiply both coordinates of each vertex by –2 and draw the new triangle on the same system. How does the new triangle compare to the old with respect to size, shape, and orientation on the graph?

6. On a coordinate system draw a rectangle with integer vertices. Multiply the x-coordinates of each vertex by –1 and the y-coordinates by +2. Predict the shape, size, and orientation of the resulting figure and then draw it on the same axes. Describe the similarities and differences between the two figures.

7. Using 9 square unit tiles, arrange the tiles in several different patterns, each of which has a different perimeter from the others. Find the arrangement with the smallest possible perimeter and find one with the largest perimeter possible. Tiles should touch each other at least at a corner.

8. Find the perimeter and area of each figure below. The rectangle method may help here.

   a)   b)   c)

                2.2   4.5
                            4.1

9. Use the formulas to find the perimeter and area of each figure below. Assume apparent square corners are right angles. Measurements are in inches.

   a)   b)
                                  2.2 2 2.2
      3   1 2   1
         5
                                (find the outside perimeter here)

10. Find the area of each shaded region below. Assume corners are right angles. Measurements are in meters.

78

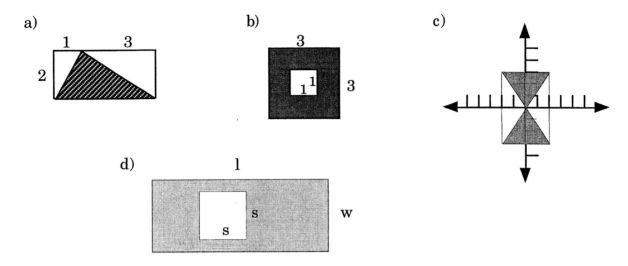

11. Suppose a pig breeder has 200 feet of fencing to enclose a rectangular "play" area for the pigs.
    a) Using the values given as a guide, complete the table for additional widths through 100 feet.

Possible width	0	10	20	--	--	--	--	--	--
Corresponding length	100	90	80	--	--	--	--	--	--
Resulting Perimeter	200	200	200	--	--	--	--	--	--
Resulting Area	0	900	1600	--	--	--	--	--	--

    b) Graph the data above to show how the area depends on the width selected. (Put the width on the horizontal axis and the area on the vertical axis.)

    c) From the graph determine the maximum play area that can be generated for the pigs using 200 feet of fencing.

    d) From the graph find the width and corresponding length of the rectangle that has an area of 2475 square feet.

12. Find the integer dimensions for all possible rectangles having an area of 24 square feet.

13. Use the "rectangle method" to find the area of each of the triangles below.

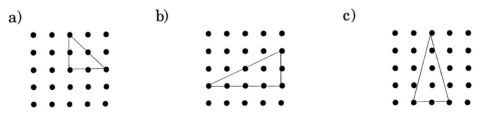

14. On dot paper, draw at least two different rectangles with an area of 4 square units.

79

15. On dot paper, draw at least two different figures with a perimeter of 10 units.

Exploration Problems

1. Using a geoboard, make each of the following figures and record your answers on dot paper.

   a) a rectangle with an area of 2 square units;
   b) a square with an area of 4 square units;
   c) a square with an area of 2 square units;
   d) a triangle with an area of 4 square units;
   e) an isosceles triangle with an area of 1 square unit;
   f) at least 3 figures, each with a perimeter of 16 units.

2. Find the area of the following triangle.

3. You have 100 feet of fencing with which to enclose a garden planted next to a barn. Fencing is needed for only the three sides not bordered by the barn. Draw a graph which shows how the area of the garden varies depending upon the width. *On your graph* show how to determine the dimensions of the garden with the greatest area.

Solutions – Problem Set 1.5

1. a) (–1,4), (5,4), (–1,0), (5,0)
   b) Perimeter = 20 units and area = 24 square units for both figures.

2. A slide 6 units to the right and 2 units down.

3. (–4,3), (–7,–1), (2,–2)

4. The new vertices are: (–2,–2), (2,–3), (4,–1), and (0,0). The figure is a quadrilateral with the same size and shape. However, it is in a different position. In fact, it is the image of the original reflected in the x-axis.

5. The new vertices are: (0,0), (–6,–2), (–6,6). This triangle has the same shape as the original but it is larger (four times larger) and is in a different position on the coordinate system, represented by a rotation about the point (0,0).

6. Drawings will vary. The new rectangle will be reflected about the y-axis. Perimeter will increase because height is doubled; area will double.

7. Possible solutions:

   P = 20 units
   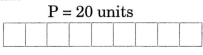

   P = 18 units
   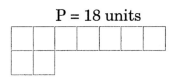

   P = 16 units
   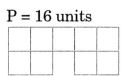

   P = 12 units (smallest perimeter)

   P = 36 units (largest perimeter)

8. a) P = 16 units   b) P = 6.4 units   c) P = 11.6 units
   A = 7 sq. units   A = 2 sq. units   A = 6 sq. units

9. a) P = 16 inches, A = 15 sq. inches   b) P = 10.4 inches, A = 6 sq. inches

10. a) A = 4 sq. meters   b)   A = 8 sq. meters
    c) A = 12 sq. meters   d)   A = (lw − s^2) sq. meters

11.
W	0	10	20	30	40	50	60	70	80	90	100
L	100	90	80	70	60	50	40	30	20	10	0
P	200	200	200	200	200	200	200	200	200	200	200
A	0	900	1600	2100	2400	2500	2400	2100	1600	900	0

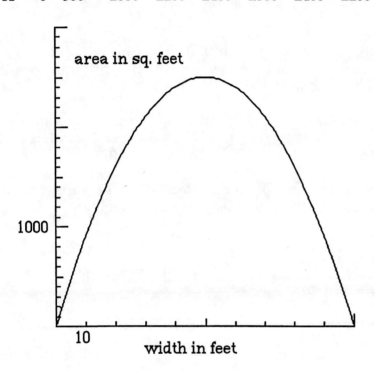

   c) 50' by 50'   d) 55' by 45'

12. 1' by 24'; 2' by 12'; 3' by 8'; 4' by 6'.

13. a) 2 sq. units   b) 4 sq. units   c) 4 sq. units

14. Examples:

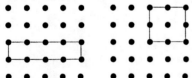

15. Examples:

Chapter One Review Problems

1. Write out the meaning of $-8 - (-4)$ in terms of positive and negative charges and give the answer.

2. Give the coordinates of the point on a vertical line through the point (4,2) and 5 units down from the x-axis.

3. Give the coordinates of a point on a horizontal line through the point (2,–1) and 3 units to the right of the y-axis.

4. Write a word problem in terms of temperature that uses the numbers 5 and –8 and the operation of addition. Give the mathematical sentence and the answer to your problem.

5. The highest point around the periphery of Lake Superior is Eagle Mountain, MN, which is approximately 2301 feet above lake level. The deepest point in Lake Superior is 1333 feet below lake level. What is the difference in altitude between these two points?

6. Evaluate each of the following. Check your answers with your calculator.

    a) $\dfrac{(-3)(-2)(4)(5)}{(-6)(2)}$

    b) $4 + 8 \div 4 - (-3) - 3$

    c) $\dfrac{8 - 10}{1 - (-1)}$

    d) $3(6 + 2)^2 - 4 \cdot 2$

7. Show two ways to find the area of the largest rectangle below using the distributive property.

    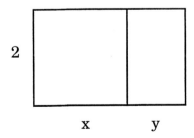

8. Give at least two meanings of $-(a)$.

9. Describe what is meant by $-10 < -5$ with respect to a checking account.

10. Use some of the properties of integers to help you easily compute each of the following. Show your work and tell which properties you used.

   a) 54 + (–20) – (30)   b) 6 • 12   c) 62 + 38

11. There are chickens and pigs only in a barnyard. There is a total of 13 heads and 40 legs. Complete this chart to help you find the number of chickens and the number of pigs in this barnyard.

No. of chickens	No. of pigs	No. of heads	No. of legs
0	13	13	13(4) = 52
1	12	13	1(2) + 12(4) = 50

12. Give the coordinates (ordered pairs) of the midpoint of each side of this rectangle.

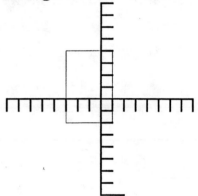

13. Complete the next 3 terms of each of the following sequences.

   a) 11, 17, 23, 29, ___, ___, ___

   b) –10, –6, 1, 12, 28, ___, ___, ___

   c) 1, 3, 9, 27, ___, ___, ___

14. Find the 75th triangular number.

15. Write the mathematical equation depicted by the pan balance below, using n as the variable. Then solve the equation using any method you wish.

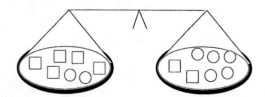

16. Draw the pan balance representation of the equation, $3x + 1 = 2x + 6$, and find the solution to the equation.

17. Draw diagrams that show all reasonable routes between A and E below. What is the shortest route between A and E.[1]

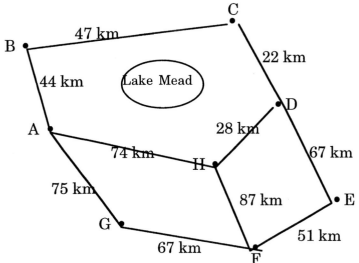

18. Find a replacement(s) for each window.

   a) $\square^2 = 81$ (two solutions here)

   b) $2 \cdot \square - 1 = 9$

   c) $\dfrac{6 \cdot \square - 2}{4} = 7$

19. Insert parentheses in the following so that the value of the expression is −96.

   $24 - (-18) - 3 + 3 \cdot (-5)$

20. Translate from words to symbols. Use n to represent an unknown if needed.
   a) Six increased by two times four.

   b) The difference of eight and two squared.

   c) The product of a number and that number increased by one.

21. Explain why the area of the large rectangle below is <u>not</u> (4a + 2). Keep the order of operations in mind.

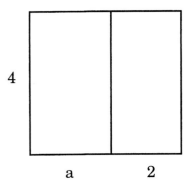

---

[1] *Mathematical Investigations, Book Two*, Souviney, Britt, Gargiulo, Hughes, page 10

22. Find the image of the following rectangle under a slide 3 units to the right and 2 units down. Compare the perimeters and areas of the rectangles.

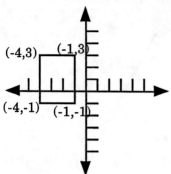

23. Explain <u>how</u> to form the image of the following triangle (same size and shape) so that the image is a reflection of the original triangle in the y-axis.

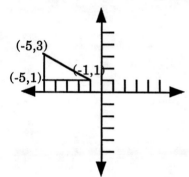

24. Find the perimeter and area of each of the following figures.
    a) b) c)

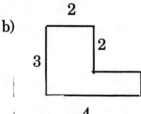

25. Use the formulas to find the perimeter and area of the following figures. Assume measurements are in centimeters and that angles are 90°.
    a) b)

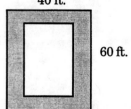

26. Given this rectangular plot containing a rectangular pool that is 30 ft. by 40 ft., find the area occupied by the sidewalk (shaded in the diagram) around the pool.

40 ft.

60 ft.

86

27. Below is a graph that shows the relationship between <u>area</u> and <u>length</u> of a rectangular garden with a perimeter of 240 feet. Answer the questions with respect to this graph.

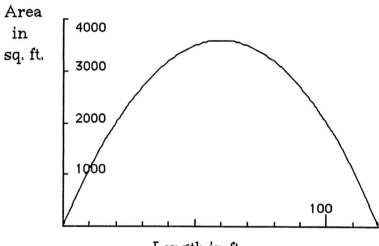

Length in ft.

a) What is the largest area possible for this rectangular garden with P = 240 feet? What is the length of the garden in this case?

b) From the graph, what is the area for a rectangular garden with the same perimeter but a length of 85 feet?

28. On dot paper, draw

a) a rectangle with a perimeter of 12 units and an area of 9 square graph units.

b) a triangle with an area of 3 square graph units.

29. Perform the indicated operations and simplify each of the following, if possible.

a) $5(c + 2) - 4c$

b) $3a + 5a^4$

c) $(6b^3 + 3b^2) - 3(b^3 - b^2)$

d) $(2x - 3xy + 5y) + (8x + 5xy - 2y)$

e) $\dfrac{30s - 20t}{10}$

f) $3(x^2 - 2x + 1) + 4(x^2 - x - 1)$

g) $6(x^2 + y^2) - 3(x^2 - y^2)$

30. Find the coordinates of the midpoint of the line segment connecting the points (2,4) and (−6,4).

31. Use 6 tiles, touching by edge only, and make an arrangement so that the perimeter is a) a maximum; b) a minimum.

Exploration Problems

1. Construct on a geoboard and record on dot paper:

   a) a triangle with an area of 6 square units.

   b) two nonrectangular figures each with a perimeter of 18 units.

2. Draw a four-sided figure on a coordinate system. Then

   a) determine which operations on which coordinates would double the size of the figure <u>and</u> reflect it in the *x*-axis;

   and

   b) give the coordinates of the vertices of the image figure.

3. Find the areas of regions I and II and show that the sum equals $b^2 - a^2$.

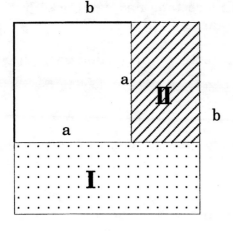

Solutions - Chapter One Review Problems

1. 4 negative charges subtracted from 8 negative charges is 4 negative charges.

2. (4,–5)

3. (3,–1)

4. Example: At 6 A.M. in January at Hibbing, MN, the temperature was 5°. The temperature then dropped 8 degrees in two hours. What was the temperature in Hibbing that day at 8 A.M.?
   $5 + (-8) = -3$ so the temperature at 8 A.M. in Hibbing was 3° below zero.

5. $2301 - (-1333) = 3634$ feet.

6. a) –10   b) 6   c) –1   d) 184

7. $2(x + y) = 2x + 2y$

8. The opposite of a; the additive inverse of a.

9. –10 < –5 means that you have less money when you are overdrawn by $10 than when you are overdrawn by $5.

10. a) $54 + (-20 - 30) = 54 + (-50) = 4$

    b) $6(10 + 2) = 60 + 12 = 72$

    c) $62 + 38 = 60 + 30 + 2 + 8 = 90 + 10 = 100$

11. 6 chickens and 7 pigs.

12. Starting at the top and going clockwise, midpoints are: (–1,4), (1,1), (–1,–2), and (–3,1).

13. a) 35, 41, 47   b) 50, 79, 116   c) 81, 243, 729

14. 2850

15. $5n + 2 = 2n + 5$, $n = 1$

16.
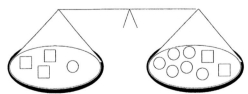

17. The shortest route is 169 km from A to H to D to E.

18. a) 9, –9   b) 5   c) 5

19. $24 - [-18 - (3 + 3)] (-5)$

20. a) $6 + 2(4)$ b) $8 - 2^2$ c) $n(n + 1)$

21. Area $= 4(a + 2) = 4a + 4(2)$. The area of the second rectangle is $4 \cdot 2$, not just 2.

22.

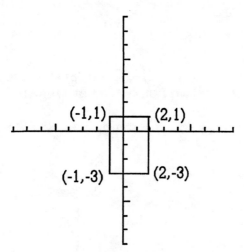

The perimeters and areas of the original and image rectangles are the same.

23. Multiply the x-coordinates by -1.

24. a) $P = 11.32$ units $A = 8$ sq. units

    b) $P = 12$ units $A = 6$ sq. units

    c) $P = 14.2$ units $A = 8.5$ sq. units

25. a) $P = 20$ units $A = 25$ sq. units

    b) $P = 14$ units $A = 8$ sq. units

26. The area of the sidewalk is $(40 \cdot 60) - (30 \cdot 40) = 1200$ sq. ft.

27. a) 3600 sq. ft., $L = 60$ feet   b) Area is approx. 3000 sq. ft.

28. a)   b)

29. a) $c + 10$ b) not possible   c) $3b^3 + 6b^2$ d) $10x + 2xy + 3y$ e) $3s - 2t$
    f) $3x^2 - 6x + 3 + 4x^2 - 4x - 4 = 7x^2 - 10x - 1$ g) $6x^2 + 6y^2 - 3x^2 + 3y^2 = 3x^2 + 9y^2$

30. $(-2, 4)$

31. Examples: a) 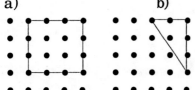 $P = 14$ units   b) $P = 10$ units

Chapter 1 Worksheet 1

1. Draw a number line and locate the points –3, –5, and 2. Then find the distance between the points –3 and –5.

2. Write a temperature problem that would require the addition of –18 and 5 as the solution.

3. Put "<", ">", or "=" signs between each of the following pairs.

   a) –4   –7
   b) |–2| 2
   c) –|3| 3
   d) –5   0

4. Find a value (or values) that make(s) each of the following statements true.

   a) ☐ + 5 = –3
   b) $\dfrac{\square}{-3} = -4$
   c) 3 • ☐ = –18
   d) | 3 – ☐ | = 5
   e) 8 + ☐ = 4
   f) –4 – ☐ = –15

5. Evaluate each of the following without your calculator. Then check using your calculator.

   a) –13 + (–31) – 45 + (3)(–2)(–4)

   b) –8 • 8 – (–4)(4)

   c) –3[5 – (–3)] ÷ (–2 • 12)

6. Label the points on the coordinate system below.

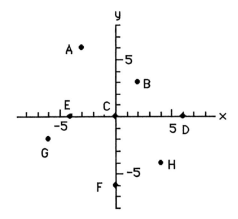

7. Mr. Smith had a checking account balance of $156.78 as of 1/14/00. Over the weekend he wrote two checks, each for $15, and one check for $23.54. On Monday, he deposited a paycheck for part-time work in the amount of $235.98. Find the amount in his checking account on Monday evening.

8. What is the temperature difference between –8° and 16°?

9. The sum of two integers is 72 and their difference (the smaller from the larger) is 28. Find the integers.

10. The Dead Sea is 395 meters below sea level and Mt. McKinley is 6,194 meters above sea level. Find the difference in altitude between the Dead Sea and Mt. McKinley.

11. I have 16 coins in my purse – dimes, pennies, and nickels only. I have 3 more dimes than pennies. If I have n pennies, find the number of dimes and the number of nickels with respect to n. Then list some possible combinations of dimes, pennies, and nickels that satisfy these requirements.

12. The charged particle model below illustrates the problem $-6-4$. Write a related addition problem illustrated by the same diagram.

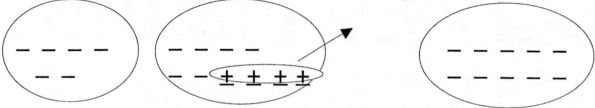

13. Given the following figure. Find its perimeter and put in simplified form.

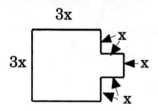

Chapter 1 Worksheet 2

1. Evaluate without using a calculator, showing intermediate steps.

$$3[5 + 6(7 - 3) + 7]$$

2. Write a calculator sequence which you could use with your calculator to evaluate the following expression. Then find the value.

$$\frac{5^3 - 1}{3^2 - 5}$$

3. Given the following algebraic expression:
$$2 - [(5 + 3 \div 4 \bullet 6^2) \div 18 \bullet 6]^3$$

   a) Which OPERATION would be done first?
   b) Which OPERATION would be done second?
   c) Which OPERATION would be done last?

4. What result does your calculator give with the following key sequences?

   a) $2 \bullet (27 - 36) =$

   b) $2 (27 - 36) =$

5. It's important to avoid ambiguous statements in mathematics as well as in other areas of life. Two translations of the following statement, *the sum of three and five squared*, might be considered correct. What are they?

6. Complete the table if the values refer to a rectangle.

width	length	area	perimeter
25 in.	25 in.		
6 in.		30 sq. in.	
	8 in.		24 in.

7. Given X, B, and A are single digit positive integers.
   Solve the cryptogram: XXX + B = BAAA

8. The second of two numbers is 20 minus the square of the first number. Complete the table:

first number	second number	sum	product
2	$20 - (2)^2 = 16$	18	32
−2			
3			
−3			
5			
−5			

9. Draw a pan-balance to illustrate each of the following equations. Then find the solution to each equation.

   a) $3x + 4 = x + 8$

   b) $x + 5 = 5x + 1$

10. Use the "cover-up" or "circle" method to solve each of the following.

   a) $4 \cdot \square - 5 = 15$

   b) $\dfrac{3x + 2}{4} = 8$

   c) $\dfrac{2 \cdot \square + 1}{3} = 5$

   d) $3(\square - 4) = 18$

11. Identify the properties of integers used in each of the following.

   a) $4(5 + 2) = (5 + 2)4$

   b) $(a + 2b) + b = a + (2b + b)$

   c) $4(x + y) = 4x + 4y$

12. Simplify each of the following.

    a) $(2x + 3) + (6x - 2)$

    b) $6(x - 1) - (3x + 1)$

    c) $(6a + 3b) - (2a - 3b)$

    d) $\dfrac{14x + 49}{7}$

13. Use algebra tiles to model the following products.

    a) $(x + 1)(x + 1)$

    b) $(x + 1)(x - 1)$

Chapter 1 Worksheet 3

1. For the given figure:

   a) Divide this shape into 3 equal parts of the same size and shape.

   b) Divide this shape into 4 equal parts of the same size and shape.

   c) Can this shape be divided into 5 equal parts? Explain.

   d) Divide this shape into 6 equal parts of the same size and shape.

   e) Are there other numbers of equal parts into which the original shape can be divided?

2. Give the first five terms of an arithmetic sequence that has a common difference of 4.

3. Write the next three terms of this geometric sequence:

    $$-3, 6, -12, 24, -48, ...$$

4. While I was reading a book, I discovered that the product of the number of the page I was reading and the next page number was 15,252. Find the number of the page I was reading.

5. Find the midpoint of the horizontal and vertical lines on the coordinate system below.

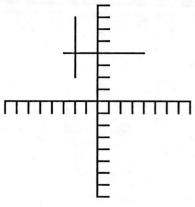

6. Solve each of the following.

   a) $\dfrac{6 \cdot \square - 4}{2} = 10$

   b) $8 - 3 \cdot \square = 11$

7. Find the perimeter and area of each figure below.
   a)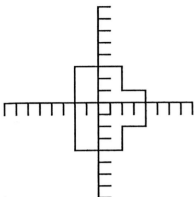
   b)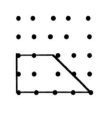

8. Describe the transformation on the original trapezoid ABCD.

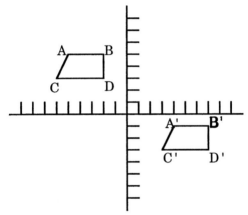

9. Below is a graph of a projectile that is thrown straight up into the air with a velocity of 128 feet per second. Answer the questions from the graph.

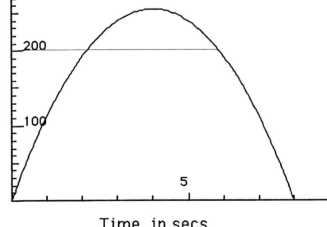

Time, in secs.

   a) How many seconds does it take the projectile to reach its maximum height?

   b) The object is 200 feet up after how many seconds? Explain your answer(s).

10. To put decorative fencing along a sidewalk, John decided to use the 32' of picket fencing that he had. He put up one post at each end. If the posts are to be 4' apart, how many posts will he need between the two outer ones?

CHAPTER TWO – The Role of Rational Numbers in Algebra

Section 2.1 – Rational Numbers: Concepts, Models, Addition and Subtraction, Comparison

Definition of a rational number

A rational number is the comparison, or **ratio**, of two integers. Examples are $3/4$, $-1/2$, or $6/5$. (Recall that $3/4$ means $3 \div 4$.)

Mathematically, we write a **rational number** as

$$a/b \text{ for } b \neq 0, a \text{ and } b \text{ any integers}$$

where a is the <u>numerator</u> and b is the <u>denominator.</u>

[Note that if a and b are both positive, the rational number is often called a <u>fraction</u>. When there is no misrepresentation, these terms will be used interchangeably.]

To understand why b must not be zero, consider the relationship between the operations of multiplication and division of integers as illustrated in the last chapter.

$$6/3 \text{ (or } 6 \div 3 \text{)} = 2 \quad \text{because } 2 \cdot 3 = 6.$$

In fact, you were probably taught to check your long division by multiplying the quotient by the divisor, and then adding the remainder, if it existed, to generate the dividend. Similarly, $16/2 = 8$ since $2 \cdot 8 = 16$. What happens when we try to divide by 0?

a) If $a/0 = 0$ for any nonzero integer a, then checking by multiplication, $0 \cdot 0 = a$. However, this statement is never true for $a \neq 0$.

b) If $a/0 = a$ for any nonzero integer a, then $a \cdot 0 = a$. But by the zero property of multiplication, this again is never a true statement when $a \neq 0$. Thus,

division of a nonzero integer by zero is undefined.

c) Finally, if $a = 0$, then $0/0$ could $= 1$ since $0 \cdot 1 = 0$, but $0/0$ could also equal -2 since $0 \cdot -2 = 0$. There is no <u>unique</u> answer to $0/0$ since every number that could be selected as a quotient satisfies the multiplication check and we say that

division of zero by zero is indeterminate.

<u>Sample Problems</u>

1. Give an example of a number that is rational but not an integer.

    Possible answers: $1/2$, $-2/3$, $8/5$

2. Can you find an integer that is not a rational number? Why or why not?

Since all integers, a, can be put in the form $a/1$, every integer <u>is</u> a rational number. Thus, no such integer exists.

3. If n is a rational number and $n \leq -2$, find all points on the number line that make $n \leq -2$ a true statement.
Solution:

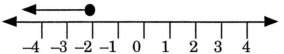

Note here that a directed line segment is used because there are rational numbers "in between" the integers that satisfy the inequality.

Concepts

Now that we have a formal definition of a rational number, it is important to understand exactly what a rational number really represents.

Consider a unit interval defined by this length on the number line:

To represent $1/3$, we first visualize each unit interval divided into <u>three</u> equal parts.

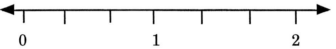

The numerator of the rational number tells us how many of these parts we are from 0. Labeling the smaller intervals we have:

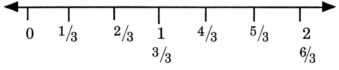

We can also divide each of the segments of length $1/3$ in half and we would have

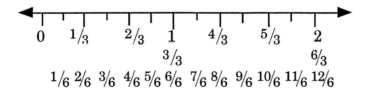

since each whole unit is now divided into <u>sixths</u>. Several other observations should be noted here:
  1) $1/3$ and $2/6$, for example, name the same point on the number line. All points, in fact, can be identified by many rational number names;
  2) At 1, the numerator and denominator are equal; and
  3) for numbers greater than 1, the numerator of the rational number is larger than the denominator.

There are three common models for representing rational numbers: the **number line** as we have illustrated above, the **set** model and the **area** model.

If we have 12 cupcakes and divide the **set** into four equal parts, each set of three is ¼ of the total:

        o o o | o o o | o o o | o o o
          ¼   ¼   ¼   ¼

Eating ¼ of the cupcakes corresponds to eating 3 cupcakes; eating 2/4 of the cupcakes corresponds to eating two groups of ¼, or 6 cupcakes, which is also ½ of the total number of cupcakes.

To represent 2/5 using an **area** model, we sketch a unit rectangle (any rectangle we define as one square unit of area), divide it into 5 equal parts and then shade 2 of those 5 parts as indicated below.

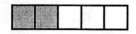

Using a unit circle, 2/5 looks like

From a different perspective, if we were to define this line segment, ─────── , as representing ⅓, what would the unit (interval of length 1) look like? The given line segment represents <u>one</u> of <u>three</u> equal parts. If we put three of these line segments together, we should have three out of three equal parts, 3/3, or one whole interval. Therefore, the unit, based on this segment is:

        |──── ⅓ ────|──── ⅓ ────|──── ⅓ ────|

                      3/3 = 1

As another example, consider this hexagon as representing 3/2, or three halves.

Again, this given area contains three equal parts, each of which represents one-half. Dividing the area accordingly and labeling, we have

(½, ½, ½ labeled inside hexagon)

Since 2 halves make 1 whole, the unit here must consist of 2 of these parts and will look like

Comparing the unit rectangle, unit square and unit circle, it is important to realize that ¼, for example, of one unit may not be the same size or amount as ¼ of another unit. This is because the amount represented by the fraction is given with respect to the *unit to which it refers*. As another example, consider the two unit intervals below and ½ of each interval.

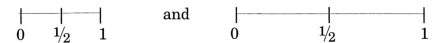

Even though ½ of the unit interval on the left is a shorter distance from zero when compared to the ½ unit on the right, each represents ½ <u>of its respective whole</u>. It is not unlike the comparison of half an orange to half a watermelon. The sizes are different, and therefore, ½ of an orange is much smaller than ½ of a watermelon. However, in concept and value, *½* stands for *one of two equal parts* of each.

Sample Problems

1. Shade ⅚ of this region.

   One possible solution: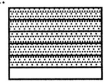

2. Mark the position on this number line that represents 3/2.

   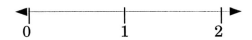

   Solution:
       Divide each unit interval in half and count 3 halves.

   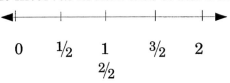

3. Give a fraction that represents the area shaded in this rectangle.

   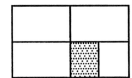

   Solution: ⅛

94

Estimation with Fractions

Compare the shaded regions of the following unit rectangles.

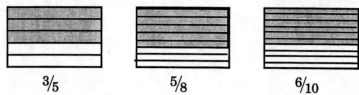

3/5   5/8   6/10

As you can see, each shaded area represents close to ½ of its respective rectangle. Examine the fractions themselves. Can we discern a pattern here?

> **When the denominator is close to twice the value of the numerator, the value of the fraction is close to ½.**

Similarly, in these unit rectangles, examine the shaded regions and fractions approximating the areas shaded.

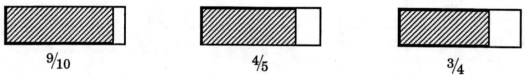

9/10   4/5   3/4

Here a very significant portion of each rectangle is shaded.

> **When the numerator and denominator of a fraction are close in value, the value of the fraction is close to 1.**

Finally, compare the position of each fraction on the number line below.

1/9   1/12   2/15

These fractions represent a length that is small compared to the unit length.

> **When the denominator of a fraction is very large compared to the numerator, the value of the fraction is close to 0.**

We can round each fraction to the nearest half (or whole which represents 2 halves) to estimate the sums and differences of fractions. Work through the following examples:

⅓ + 3/5   is approximately 1 since both ⅓ and 3/5 are close to ½ and two halves make 1 whole.

¾ + 1/16   is approximately 1 since ¾ is close to 1 and 1/16 is close to 0.

7/8 + 8/9   is approximately 2 since 7/8 and 8/9 are each close to 1 in value.

2 − 3/5   is close to 1½ since 3/5 is close to ½ and backtracking ½ unit from an interval length of 2 units leaves an interval length of 1½ units.

Sample Problems

1. Give an example of a fraction close to:
   a) $\frac{1}{2}$ b) 0   c) 1

   Solutions: A few possibilities for each:

   a) $\frac{3}{5}, \frac{4}{7}, \frac{5}{9}$ b) $\frac{1}{1000}, \frac{3}{20}, \frac{1}{10}$ c) $\frac{8}{9}, \frac{15}{16}, \frac{7}{8}$

2. Round each fraction to the nearest half (or whole, as appropriate) and estimate the answer for each of the following.
   a) $\frac{1}{3} + \frac{3}{8}$ b) $3 - \frac{5}{6}$ c) $2\frac{3}{4} + \frac{4}{5}$

   Solutions:
   a) 1 since $\frac{1}{3}$ and $\frac{3}{8}$ are both close to $\frac{1}{2}$.
   b) 2 since $\frac{5}{6}$ is close to 1 and $3 - 1 = 2$
   c) 4 because $2\frac{3}{4}$ is close to 3 and $\frac{4}{5}$ is almost 1; then $3 + 1 = 4$.
   Note that 3/4 is exactly halfway between 1/2 and 1. We typically "round up" to the larger value.

3. On March 2, Hope put about $2\frac{1}{8}$ pounds of birdseed in her outdoor feeder. The next day she added approximately $3\frac{1}{4}$ pounds of birdseed to the feeder. Estimate the amount of birdseed she put it the bird feeder in those two days.

   Solution: Round $2\frac{1}{8}$ to 2 and $3\frac{1}{4}$ to $3\frac{1}{2}$ so she dispensed about $5\frac{1}{2}$ pounds of birdseed in those two days.

Equivalent Fractions

On the number line, we have shown that there are several fractional names, such as $\frac{1}{3}$ and $\frac{2}{6}$, for the same point. Returning to the area model, let's divide each of the two unit fraction bars in half and label as below.

1	$\frac{1}{2}$	$\frac{1}{2}$
1	$\frac{1}{2}$	$\frac{1}{2}$

If we divide the bottom fraction bar into fourths by cutting each $\frac{1}{2}$ in half, we have:

1	$\frac{1}{2}$		$\frac{1}{2}$	
1	$\frac{1}{4}$	$\frac{1}{4}$	$\frac{1}{4}$	$\frac{1}{4}$

Note that the same area is occupied by 1, $\frac{2}{2}$ (two halves) and $\frac{4}{4}$ (four fourths) and that $\frac{2}{4}$ represents the same area as $\frac{1}{2}$. This again implies that *rational numbers can have the same value and designate the same amount of area, even though they have different names.*

Mathematically, this result can be achieved through the

> **Fundamental Law of Fractions:**
>
> For a/b a rational number and k any nonzero integer,
>
> $$\frac{a}{b} = \frac{ak}{bk}$$

Thus, we can make a rational number **equivalent** (having the same value) to another rational number by <u>multiplying</u> both the numerator and denominator by the same value. As examples:

$$\frac{1}{2} = \frac{1 \cdot 2}{2 \cdot 2} = \frac{2}{4} \quad \text{and} \quad \frac{3}{4} = \frac{3 \cdot (-2)}{4 \cdot (-2)} = \frac{-6}{-8} \quad \text{and} \quad \frac{3}{2} = \frac{3 \cdot 5}{2 \cdot 5} = \frac{15}{10}$$

(Note: In the model, when we cut 1/2 in half, we doubled the number of sections, as shown in the denominator above, but it then took two of them to equal 1/2, as illustrated in the numerator.)

The **Fundamental Law of Fractions** also gives us a way to *simplify* rational numbers; we *divide numerator and denominator by k, $k \neq 0$*.

A fraction such as $1/2$ is said to be in **simplest form** because 1 is the only divisor of both 1 and 2, but $12/48$ is not in simplest form because both 12 and 48 can be divided by 2, 3, 4, 6, or 12. Selecting a divisor of 4, for example, gives:

$$\frac{12 \div 4}{48 \div 4} = \frac{3}{12}.$$

However, note that 3 and 12 still have a divisor of 3 in common. Thus, $12/48$ in simplest form is:

$$\frac{12}{48} = \frac{3}{12} = \frac{3 \div 3}{12 \div 3} = \frac{1}{4}$$

It obviously would have been easier to select a divisor of 12 (called the <u>greatest common divisor</u>) initially so that we would have generated the simplest form of $1/4$ in one step.

<u>Sample Problems</u>

1. Divide a unit bar into 3 equal parts and label. Shade 2 of these parts. Then divide each of the 3 original parts into four equal parts and label. Give another name for $2/3$ based on the new divisions.

   Solution:

   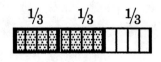

   Since the smallest regions are $1/12$ of the original,

   $$2/3 = 8/12$$

97

2. Give three rational numbers equivalent to $2/5$.

   Examples: $4/10$ (multiplying numerator and denominator by 2);
   $200/500$ (multiplying numerator and denominator by 100);
   $-2/-5$ (multiplying numerator and denominator by –1).

3. Put each rational number into simplest form.

   a) $14/16$ b) $24/36$ c) $-25/75$

   Solutions: a) $7/8$ (divide numerator and denominator by 2);
   b) $2/3$ (divide numerator and denominator by 12);
   c) $-1/3$ (divide numerator and denominator by 25)

Adding and Subtracting Rational Numbers

We must first note that, for a and $b > 0$, a rational number has a negative value when it is in one of the following forms:

$$\frac{-a}{b}, \frac{a}{-b}, -\frac{a}{b}$$

Note: $\frac{-a}{-b}$ is positive since the division of two negative integers is a positive integer.

To add rational numbers with denominators that are identical, or **like**, let's use the number line and directed arrows.

For $1/5 + 2/5$, draw one arrow the distance $1/5$ and *from there* draw another arrow of length $2/5$. We end up at $3/5$ on the number line

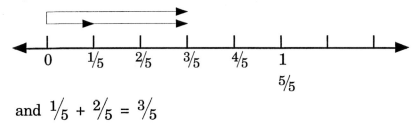

and $1/5 + 2/5 = 3/5$

The size of each piece, as indicated by the value in the denominator (here it's fifths), *does not change in the addition process. What does happen is that the number of those sized pieces, as indicated by the values of the numerators, does accumulate.*
What about $1/3 + 1/2$? If a very hungry person ate $1/3$ of one pie and $1/2$ of another pie of the same size as shown below,

it is difficult to ascertain the total amount of pie that person has eaten because the pieces are not the same size. By our estimation, however, between $1/2$ and 1 whole pie was eaten.

How can we cut these pies so that each would have the same number of pieces, all of equal size? Since 2 and 3 both divide 6, we could cut both pies into 6 equal pieces and we would solve our problem. Thus, we must cut each $1/3$-size piece in the first pie in half, and each $1/2$-size piece in the second pie into 3 equal pieces as shown below.

The shaded region on the left shows that $1/3 = 2/6$ (2 pieces of the "sixths" size) and $1/2 = 3/6$ (3 pieces of the "sixths" size).

Now we can add to get:
$$1/3 + 1/2 = 2/6 + 3/6 = 5/6$$
and this agrees with our estimation that the result should be between $1/2$ and 1.

What we did here to add two rational numbers with **unlike** denominators was to **rename** each rational number (that is, form rational numbers equivalent to the original ones) so that the denominators are the same (analogous to adding pieces that are the same size). This is called "finding a common denominator" and is the common method used for adding and subtracting rational numbers.

As another example, consider

$$6 - 5/6 \text{ which we can rewrite as } 6/1 - 5/6$$

[Don't forget that any integer can be written as a rational number with a denominator of 1.]

We can estimate that the result should be a little more than 5 (since we're subtracting a value smaller than 1). Using a common denominator of 6 and equivalent fractions,
$$6 - 5/6 = \frac{6 \cdot 6}{1 \cdot 6} - 5/6 = 36/6 - 5/6 = \frac{36-5}{6} = 31/6$$

(which is a little more than 5!)

As we discussed earlier, a fraction such as $31/6$, in which the numerator is larger than the denominator, does represent a value larger than 1. It may be more meaningful if we convert this **improper fraction** to a different form, commonly called a **mixed number**.

To convert, merely do the division called for in the notation $31/6$ or

```
       5
   6 ⟌ 31      is 5 and 1 part out of six as a remainder.
       30      We can write this as 5 1/6
        1
```

Summarizing the operations of addition and subtraction of rational numbers, for any integers a, b, c, and d, where b and d are nonzero, we have:

$$\frac{a}{b} \pm \frac{c}{d} = \frac{ad}{bd} \pm \frac{bc}{bd} = \frac{ad \pm bc}{bd}$$

Here **bd** represents the product of the denominators. In fact, a common denominator can always be generated by multiplying the denominators together. Finding a smaller number, however, called the **lowest common denominator**, that is divisible by all denominators in the problem, may save some work at the end. In the example of adding $\frac{1}{3}$ and $\frac{1}{2}$, the "LCD" and "bd" are both equal to six. But if we were adding $\frac{1}{4}$ and $\frac{3}{8}$, instead of using 32 as a common denominator, the number "8" would do and would alleviate simplifying the sum at the end as shown.

$$\frac{1}{4} + \frac{3}{8} = \frac{2}{8} + \frac{3}{8} = \frac{5}{8}$$

instead of
$$\frac{1}{4} + \frac{3}{8} = \frac{8}{32} + \frac{12}{32} = \frac{20}{32} = \frac{5}{8}$$

Finally, consider the problem above as shown below.

$$\frac{1}{4} + \frac{3}{8} = \frac{1 \bullet 8 + 4 \bullet 3}{32} = \frac{20}{32}$$

Note that we generate the same numerator, $(8 + 12) = ad + bc$ and the denominator, bd, by using this shortcut.

Finally, when performing addition in problems such as
$$\frac{x}{2} + \frac{x}{4}$$

the same process applies. The common denominator is *4* and the equivalencies and sum are shown below.

$$\frac{x}{2} + \frac{x}{4} = \frac{2x}{4} + \frac{x}{4} = \frac{2x + x}{4} = \frac{3x}{4}$$

Finally, the operation of addition provides a way to convert a mixed number back into a fraction. To put $3\frac{1}{2}$ in fraction form, add 3 and $\frac{1}{2}$ and we get

$$3 + \frac{1}{2} = \frac{6}{2} + \frac{1}{2} = \frac{7}{2}$$

Starting at the bottom arrow and working clockwise, this is shown in short-cut form as

$$3\frac{1}{2} = \frac{6+1}{2} = \frac{7}{2}$$

then add / first multiply

Sample Problems

1. a) Estimate the sum of $\frac{1}{4}$ and $\frac{1}{8}$.
   b) Use an area model to illustrate the mathematical steps needed to add $\frac{1}{4}$ and $\frac{1}{8}$.

Solutions:
a) The sum will be a little less than $1/2$.

b)
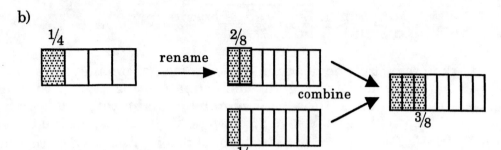
The result is $2/8 + 1/8 = 3/8$.

2. Use a number line to illustrate the mathematical steps needed to subtract $1/2$ from $2/3$.

Solution:
First, we must divide the unit interval into sixths (since both $1/2$ and $2/3$ can be expressed in sixths). Since
$$1/2 = 3/6 \text{ and } 2/3 = 4/6,$$
we'll use directed arrows to indicate the subtraction

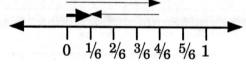

The boldest resultant arrow shows that $2/3 - 1/2 = 1/6$.

3. Perform the following operations. Make sure the answers are in simplest form. If you have a scientific or graphing calculator with fraction capability, check your solutions to parts a) and b) with your calculator.

a) $4\frac{2}{3} + 2\frac{1}{5}$

Solution: $4\frac{2}{3} = \frac{14}{3}$ and $2\frac{1}{5} = \frac{11}{5}$ so

$$\frac{14}{3} + \frac{11}{5} = \frac{70}{15} + \frac{33}{15} = \frac{103}{15} \text{ or } 6\frac{13}{15}$$

b) $4\frac{1}{5} - 2\frac{2}{5}$

Solution: Rewrite the problem as:
$$\left(4\frac{1}{5} + \frac{3}{5}\right) - \left(2\frac{2}{5} + \frac{3}{5}\right) = 4\frac{4}{5} - 3 = 1\frac{4}{5}$$
or subtract
$$\frac{21}{5} - \frac{12}{5} = \frac{9}{5} = 1\frac{4}{5}$$

c) $y/3 - 2y/4$

Solution:
$$y/3 - 2y/4 = 4y/12 - 6y/12 = -2y/12 = -y/6$$

4. Sam cut a lemon meringue pie into 8 equal pieces. His sister ate 1 piece and his brother ate 2 pieces. What fraction of the pie was left? Write out your solution using fractions.

   Solution: 8 – 1 – 2 = 5 out of 8 pieces or ⁵⁄₈ of the total pie is left.

Ordering Rational Numbers

When comparing two rational numbers such as ⁴⁄₅ and ²⁄₅, we can use the number line to show which is the farther from zero. Since both numbers are positive and involve parts of the unit interval that are the same size (fifths), from the diagram below it is apparent that ⁴⁄₅ > ²⁄₅.

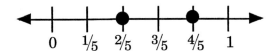

[Note that we can also write ²⁄₅ < ⁴⁄₅]

However, when comparing two rational numbers such as ⁴⁄₆ and ⁴⁄₁₂, there are several methods that can be used.

  a) Reduce each rational number to its simplest form and then make the comparison.

   In this case, ⁴⁄₆ = ²⁄₃ and ⁴⁄₁₂ = ¹⁄₃ and since ²⁄₃ > ¹⁄₃, ⁴⁄₆ > ⁴⁄₁₂.

  b) Use common denominators.

   Since ⁴⁄₆ = ⁸⁄₁₂, and ⁸⁄₁₂ > ⁴⁄₁₂, then ⁴⁄₆ > ⁴⁄₁₂.

  c) Use the cross-products.

   When making rational numbers equivalent to $\frac{a}{b}$ and $\frac{c}{d}$ with common denominators, respectively, we will compare

   $$\frac{a}{b} = \frac{ad}{bd} \quad \text{and} \quad \frac{c}{d} = \frac{bc}{bd}$$

With common denominators, we only need to compare the numerators to determine which rational number is larger. Thus, we are really comparing the product **ad** to the product **bc**. Note that in the original rational numbers

$$\frac{a}{b} \quad \text{and} \quad \frac{c}{d}$$

ad and *bc* represent what we call **cross-products**, often shown as

$$\frac{a}{b} \bowtie \frac{c}{d}$$

What this gives us is a shortcut method for checking equality, as well as the inequality, of rational numbers.

To illustrate this method, let's compare
$$3/6 \text{ and } 4/8$$
The cross products are 3•8 and 4•6. Since these products are equal, the rational numbers 3/6 and 4/8 are equal.

In the case of
$$2/3 \text{ and } 3/4$$
$$2\cdot4 < 3\cdot3 \text{ so } 2/3 < 3/4.$$

[To be consistent, do the <u>ad</u> product and record *on the left* ; do the <u>bc</u> product and record *on the right*. Then put in the appropriate symbol of equality or inequality.]

<u>Sample Problems</u>

1. Order the following rational numbers by using common denominators.
$$3/4, \ 5/8, \ 7/12, \ 5/6$$

   Solution: Using a common denominator of 24, we have
   $$3/4 \ = \ 18/24, \quad 5/8 \ = \ 15/24, \quad 7/12 \ = \ 14/24, \quad 5/6 \ = \ 20/24$$

   and, thus, $\qquad 7/12 < 5/8 < 3/4 < 5/6$

2. Use cross-products to compare 4/9 and 3/7 .

   Solution: Since (4 • 7) > (9 • 3),   4/9 > 3/7

Problem Set 2.1

1. What fraction of an hour is:

   a) 15 minutes?

   b) 20 minutes?

   c) 80 minutes?

2. An inch is what fraction of

   a) a foot?

   b) a yard?

   c) a mile?

3. Shade 5/6 of this hexagon.

4. A summer course meets 90 minutes, 4 days per week for 8 weeks.

   a) How many hours are spent in this class?

   b) If you miss 2 classes, what fraction of the total class time would you miss?

5. In geometry, starting at a point A as below and rotating around a full circle back to point A is a rotation of 360°. If in the rotation, we stop at point B, what portion of the full circle

   a) has been transversed?

   b) is left to be transversed?

   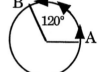

6. The following area represents 5/4. Draw a picture of the corresponding unit area.

   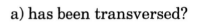

7. The point on the number line below represents what fraction of the unit interval?

   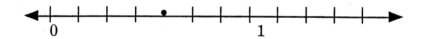

8. Rename the fraction -5/7 with a denominator of 42. Give two other names for -5/7.

9. Grandma Marinelli's recipe for old-fashioned molasses cookies calls for 2 eggs, 3½ cups of flour and ¾ cup of molasses, among other ingredients. If you decide to make twice as many cookies, how much flour and how much molasses would you need?

10. Explain how you would use an area model to add ½ and ⅕.

11. Maria typically rides her bike from her apartment to campus which is a distance of $1\frac{1}{4}$ miles. One day her bike broke down after she had ridden $\frac{5}{6}$ mile. If she walked her bike the rest of the way, how far did she walk that day to campus?

12. Name at least 3 fractions that are *approximately* equal in value to $\frac{3}{4}$.

13. Round each fraction to the nearest half and estimate the answers to each of the following.

    a) $\frac{1}{10} + 3\frac{7}{8}$ b) $\frac{7}{10} + \frac{4}{9}$ c) $2\frac{5}{6} - \frac{4}{9}$ d) $7 - \frac{9}{10}$

14. As shown in one of the sample problems in this section, rewrite the problem below so that we can subtract a whole number instead of a mixed number.

    $$4\frac{2}{7} - \frac{5}{7}$$

15. Perform the following algebraic computations.

    a) $\frac{a}{4} + \frac{2a}{4}$ b) $\frac{2x}{3} - \frac{3x}{2}$ c) $\frac{x}{3} - \frac{x}{4} + \frac{x}{2}$

16. Find the perimeter of the following figure. Assume the measurements are in cm.

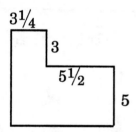

Exploration Problems

1. A newspaper headline writer assigns a value to each character for the purpose of spacing headlines.[1] These values are:

    punctuation, I, i, the letter l and the number 1 count $\frac{1}{2}$;

    m and w count $1\frac{1}{2}$;

    all other digits and lower case letters count 1;
    spaces and most capital letters count 2;
    capitals M and W count $2\frac{1}{2}$.

    Find the count value for each of the following headlines:

    a) MATHEMATICS IS IMPORTANT

    b) 1993, Another Year of Budget Cuts

---
[1] NCTM Sourcebook of Applications, 1980, pg. 27.

2. Use the Internet and find a recipe for chocolate chip cookies. Write out the recipe, then triple the recipe and record the amounts of the ingredients. How many cookies will this recipe make? Finally, halve the originally recipe and record the amounts of the ingredients. How many cookies will a half-recipe make?

Solutions to Problem Set 2.1

1. a) $15/60$ or $1/4$ b) $20/60 = 1/3$ c) $80/60 = 4/3 = 1\,1/3$

2. a) $1/12$ b) $1/36$ c) $1/63360$

3.

4. a) 48 hrs.   b) $3/48 = 1/16$

5. a) $120/360 = 1/3$ b) $240/360 = 2/3$

6. represents $5/4$ so $4/4 = 1$ is ▯▯▯▯

7. $4/7$

8. $-30/42$, $-5/7 = -10/14 = -50/70$

9. 7 cups of flour, $1\,1/2$ cups of molasses

10. Draw 2 unit rectangles, each cut into 10 pieces. Since $1/2$ is the same as $5/10$ and $1/5$ is the same as $2/10$, adding these pieces together gives a total of $7/10$ of a unit rectangle.

11. $5/12$ mile

12. Examples are $2/3$, $7/10$, $5/7$

13. a) 4   b) 1   c) $2\,1/2$ d) 6

14. $4\,2/7 - 5/7 = \left(4\,2/7 + 2/7\right) - \left(5/7 + 2/7\right) = 4\,4/7 - 1 = 3\,4/7$

15. a) $3a/4$ b) $2x/3 - 3x/2 = 4x/6 - 9x/6 = -5x/6$
    c) $x/3 - x/4 + x/2 = 4x/12 - 3x/12 + 6x/12 = 7x/12$

16. $33\,1/2$ cm

Section 2.2 - Rational Numbers: Multiplication, Division, and Applications

Multiplying Rational Numbers

Examine the unit rectangle below where the width is divided into 2 equal parts and the length is divided into 3 equal parts.

Now highlighting the rectangle that represents dimensions of ½ by ⅓

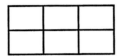

we can see that the area occupied by the ½ by ⅓ rectangle is really ⅙ of the unit rectangle. In other words,

$$\frac{1}{2} \cdot \frac{1}{3} = \frac{1}{6}$$

Let's consider the product of ⅔ and ¾. If we divide the length of the rectangle below into fourths and the width into thirds, we can outline and shade the area occupied by the rectangle that is ⅔ by ¾.

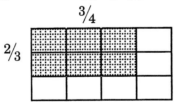

The area of this outlined rectangle occupies 6 out of a total of 12 pieces (of equal size) in the unit rectangle. We have found that

$$\frac{2}{3} \cdot \frac{3}{4} = \frac{6}{12} = \frac{1}{2}$$

Looking at the numbers and the products in these examples, the rule for multiplying rational numbers can be justified:

For all integers, a, b, c, and d, where b and d are nonzero,

$$\boxed{\frac{a}{b} \cdot \frac{c}{d} = \frac{ac}{bd}}$$

In simple terms,
the product of two rational numbers is the product of the numerators over (or divided by) the product of the denominators.

Sample Problems

1. Draw a diagram to illustrate the product of $\frac{3}{2}$ and $\frac{1}{5}$ and then give the product.

   Solution:

   $\frac{1}{2}$
   $\frac{1}{2}$
   $\frac{1}{2}$

   The unit is outlined and contains 10 parts.
   3/2 • 1/5 = 3/10 since the three squares that are shaded must be compared to the number of parts in the unit, which is 10. Note that another 1/2 had to be added to the unit since $3/2 = 1\frac{1}{2}$.

2. Perform the following computation:
   $$2\tfrac{2}{5} \cdot 1\tfrac{3}{4}$$
   We need fractional forms, so $2\tfrac{2}{5} = \tfrac{12}{5}$ and $1\tfrac{3}{4} = \tfrac{7}{4}$; then
   $$\tfrac{12}{5} \cdot \tfrac{7}{4} = \tfrac{84}{20} = \tfrac{21}{5} = 4\tfrac{1}{5}$$

   or $\tfrac{12}{5} \cdot \tfrac{7}{4}$ can be simplified first and we can rewrite the problem as $\tfrac{12}{4} \cdot \tfrac{7}{5} = \tfrac{3}{1} \cdot \tfrac{7}{5} = \tfrac{21}{5}$ (since 5 • 4 = 4 • 5)

   This process is sometimes shown as:
   $$\frac{\overset{3}{\cancel{12}}}{5} \cdot \frac{7}{\underset{1}{\cancel{4}}} = \frac{21}{5}$$

   where a common factor in a numerator and denominator is divided out. Remember that this process really is no different from reducing the *product* to simplest terms.

3. A t-shirt originally priced at $20.59 is marked $\tfrac{1}{3}$ off. **Estimate** the amount of discount on this t-shirt.
   Solution: $\tfrac{1}{3}$ of $21 is $7. So the amount of discount is *about* $7.

4. Find the product of
   $$\frac{y}{2} \text{ and } 4$$

   Solution:
   $$\frac{y}{2} \cdot 4 = \frac{y}{2} \cdot \frac{4}{1} = \frac{\overset{2}{\cancel{4}}y}{\underset{1}{\cancel{2}}} = 2y$$

Rational Number Division

Let's return to the rectangle model and consider $2 \div 1/3$. The division sign means we're really trying to determine the number of one-thirds in 2. First, we divide the unit rectangles into thirds so we know what $1/3$ looks like and we have

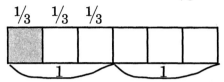

Since each unit rectangle contains 3 one-thirds, there are 6 one-thirds in two. Therefore,

$$2 \div \frac{1}{3} = 6$$

Let's try $1/2 \div 1/8$ using this model. Remember, we're looking for the number of eighths in $1/2$. We divide the unit rectangle into eighths and shade $1/2$ of it as shown below

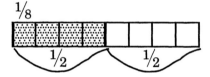

We can see that $1/2$ contains 4 pieces, each of which is $1/8$ and so

$$\frac{1}{2} \div \frac{1}{8} = 4$$

Mathematically extending the operation of division on the set of integers to the division of rational numbers let's rewrite

$$1/2 \div 1/8$$

as the complex fraction $\dfrac{\frac{1}{2}}{\frac{1}{8}}$

To simplify this fraction, we'll multiply both numerator and denominator by the same value (using the Fundamental Law of Fractions), a value that will generate a denominator of 1. This eliminates some of the "complexity" of the fraction.

$$\frac{\frac{1}{2} \cdot 8}{\frac{1}{8} \cdot 8} = \frac{4}{1}$$

Note that the denominator of this complex fraction is made equal to 1 by using 8 as the multiplier. We now have rewritten

$$1/2 \div 1/8 \quad \text{as} \quad 1/2 \cdot 8$$

and what has really happened here is that the *division* of $1/2$ and $1/8$ has been converted to the *multiplication* of $1/2$ by 8. The $1/2$ (dividend) did not change but instead of <u>dividing by</u> $1/8$, we <u>multipled by 8</u>. We call 8 and $1/8$ **reciprocals** of each other because $8 \cdot 1/8 = 1$. [Note that the reciprocal of a fraction is generated by the numerator and denominator changing places.]

As another example, $3/4 \div 2/3$ can be written as

$$\frac{\frac{3}{4}}{\frac{2}{3}} = \frac{\frac{3}{4} \cdot \frac{3}{2}}{\frac{2}{3} \cdot \frac{3}{2}} = \frac{\frac{9}{8}}{1} = \frac{9}{8}$$

To achieve a denominator of 1 here, the reciprocal of $2/3$, that is, $3/2$, was used as the multiplier.

Note again that the division problem $3/4 \div 2/3$ became $3/4$ *times* $3/2$!

In general, the division of rational numbers can be accomplished through the multiplication process, but with the use of the reciprocal. For all integers a, b, c, and d, where b, d, and c are nonzero,

$$\boxed{\frac{a}{b} \div \frac{c}{d} = \frac{a}{b} \cdot \frac{d}{c} = \frac{ad}{bc}}$$

Once the operation is changed and the reciprocal is used, we use the rule for multiplication of rational numbers.

Note that multiplication and division of rational numbers do not require the use of a common denominator which does make things a little easier. However, common denominators can be used. Study the examples below.

$$\frac{1}{2} \cdot \frac{1}{3} = \frac{3}{6} \cdot \frac{2}{6} = \frac{6}{36} = \frac{1}{6}$$

$$\frac{1}{2} \div \frac{1}{8} = \frac{4}{8} \div \frac{1}{8} = \frac{4}{8} \cdot \frac{8}{1} = \frac{32}{8} = 4$$

Both agree with our previous results.

[Note that in the third step above, specifically, the step, $\frac{4}{8} \cdot \frac{8}{1}$, the first denominator, 8, and second numerator, 8, can be divided to form a factor of 1. (With common denominators, this will always happen.) We can actually eliminate this step and just divide across, $4 \div 1 = 4$, in the second step. Try this method with other examples.]

Also, when multiplying a fraction by a whole number, especially in cases such as $3/4 \cdot 20$, the following shortcut is often used:

$$\frac{3}{4} \cdot \overset{5}{\cancel{20}} = 15$$
$$\underset{1}{}$$

Sample Problems

1. a) Mentally calculate the number of $1/4$'s in three quarters.
   b) Draw a diagram to illustrate $3/4 \div 1/4$ and give the quotient.

Solution:
There are 3 one-quarters in ¾, so the quotient is 3, as shown in the diagram below.

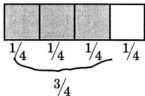

2. Find the quotient of
$$3\tfrac{1}{2} \text{ and } 2\tfrac{2}{3}$$

   Solution:

   Fractions are needed and $3\tfrac{1}{2} = \tfrac{7}{2}$ and $2\tfrac{2}{3} = \tfrac{8}{3}$

   so the quotient is $\dfrac{7}{2} \div \dfrac{8}{3} = \dfrac{7}{2} \cdot \dfrac{3}{8} = \dfrac{21}{16} = 1\dfrac{5}{16}$.

3. Simplify each of these complex fractions.

   a) $\dfrac{\tfrac{3}{5}}{\tfrac{3}{4}}$ b) $\dfrac{\tfrac{1}{3} + \tfrac{1}{2}}{\tfrac{5}{6} - \tfrac{2}{3}}$

   Solutions:
   a) First method (multiplying the numerator and denominator of the complex fraction by the reciprocal of the denominator):

   $$\dfrac{\tfrac{3}{5}}{\tfrac{3}{4}} = \dfrac{\tfrac{3}{5} \cdot \tfrac{4}{3}}{\tfrac{3}{4} \cdot \tfrac{4}{3}} = \dfrac{3}{5} \cdot \dfrac{4}{3} = \dfrac{4}{5}$$

   Second method: (multiplying numerator and denominator of the complex fraction by 20 (a **multiple** of both 4 and 5)),

   $$\dfrac{\tfrac{3}{5} \cdot 20}{\tfrac{3}{4} \cdot 20} = \dfrac{12}{15} = \dfrac{4}{5}$$

   b) (adding the fractions in the numerator) $\dfrac{1}{3} + \dfrac{1}{2} = \dfrac{2}{6} + \dfrac{3}{6} = \dfrac{5}{6}$

   and

   (subtracting the fractions in the denominator) $\dfrac{5}{6} - \dfrac{2}{3} = \dfrac{5}{6} - \dfrac{4}{6} = \dfrac{1}{6}$

   Then
   $$\dfrac{5}{6} \div \dfrac{1}{6} = \dfrac{5}{6} \cdot \dfrac{6}{1} = 5$$

4. Simplify by performing the operation required.

$$\frac{\frac{x}{2}}{\frac{y}{4}} \quad \text{(assume } y \neq 0\text{)}$$

Solution:

Using the reciprocal of the denominator as a multiplier,

$$\frac{\left(\frac{x}{2}\right)\left(\frac{4}{y}\right)}{\left(\frac{y}{4}\right)\left(\frac{4}{y}\right)} = \frac{\frac{4x}{2y}}{1} = \frac{\frac{2x}{y}}{1} = \frac{2x}{y}$$

5. Show that $\frac{1}{2}x = \frac{x}{2}$

Solution:

$$\frac{1}{2}x \text{ means } \frac{1}{2} \cdot \frac{x}{1} = \frac{x}{2}$$

Challenge Problems

1. Perform the following computations. Assume the variables are nonzero.

   a) $\frac{4}{x} + \frac{3}{y}$ b) $\frac{x}{2} \cdot \frac{y}{3}$ c) $\frac{x}{2a} \div \frac{3x}{a}$

   Solutions:

   a) $\frac{4}{x} + \frac{3}{y} = \frac{4y}{xy} + \frac{3x}{xy} = \frac{4y + 3x}{xy}$

   b) $\frac{x}{2} \cdot \frac{y}{3} = \frac{xy}{6}$

   c) $\frac{x}{2a} \div \frac{3x}{a} = \frac{\cancel{x}^1}{\cancel{2a}_2} \cdot \frac{\cancel{a}^1}{\cancel{3x}_3} = \frac{1}{6}$

Variables and Rational Numbers in Problem Solving

An array of dots is covered as shown below.

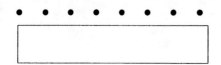

If we represent the number of dots in the original configuration as x and we know that $2/3$ of the dots are covered, we can represent the amount covered as

$$\frac{2}{3}x.$$

We also know that the 8 dots in the top row represent $1/3$ (since $1/3 = 1 - 2/3$) of the total number of dots. Thus, we can write the equation

$$\frac{1}{3}x = 8$$

and using our informal method to solve for x,

$$1/3 \text{ of } 24 = 8 \quad \text{so}$$

$$x = 24.$$

In other words, 8 dots are shown, and $24 - 8 = 16$ dots are covered by the box.

As another example, if the capacity of a compact car gas tank is unknown, we can use variables to estimate the number of gallons of gas the tank will hold. Suppose the gas gauge indicates the tank is $3/4$ full (or down $1/4$) and to fill the tank you need 3 gallons of gas. Letting n represent the tank's capacity, in gallons, and since $1/4$ of the tank represents 3 gallons of gas, we have

$$\frac{1}{4}n = 3$$

Again, using our informal equation solving method, $n = 12$ and the gas tank holds approximately 12 gallons of gasoline.

Sample Problems

1. A cake had been cut into n equal sized pieces. One-third of the cake, or 5 pieces, remained. Into how many pieces had the cake been cut originally?

   Solution: $\frac{1}{3}n = 5$ so $n = 15$. The cake had been cut into 15 pieces.

2. Estimate:

   a) $2/3$ of 49   b) $8 \div 6/7$

   Solutions:

   a) compute $2/3$ of $48 = 32$ since 48 is "compatible" with 3.

   b) use $8 \div 1$ since $6/7$ is close to 1 and the result is 8.

Problem Set 2.2

1. Round each fraction to the nearest half and estimate.
   a) $\frac{4}{7} \cdot 18$
   b) $\frac{8}{9} \cdot 23$
   c) $14 \div \frac{3}{7}$
   d) $\frac{4}{5} \div \frac{5}{9}$

2. Using unit rectangles, model each of the following and use the model to find the answer.
   a) $\frac{1}{3}$ of $\frac{3}{5}$
   b) $\frac{2}{3} \div \frac{1}{6}$

3. Estimate the amount of discount on a jacket if the original selling price is $49.95 and the jacket is marked $\frac{1}{4}$-off.

4. Find the number of desks in a classroom if $\frac{3}{4}$ of the desks are occupied by students and there are 4 desks unoccupied.

5. In a recent financial crunch a community college reduced its staff by $\frac{1}{9}$. If there were 261 staff members before, how many were left after the reduction?

6. A store had a winter clearance sale where everything was marked $\frac{1}{3}$-off. After several weeks, the store marked clearance items down again another $\frac{1}{2}$-off. What fraction of the original price did customers pay for items on the clearance rack?

7. By midnight of June 16 of one year Marquette, MI had received about $13\frac{1}{2}$ inches of precipitation. The yearly average for Marquette, including rain, melted down snow, sleet, etc., is about $37\frac{1}{10}$ inches! What fraction of the annual precipitation had Marquette received that year by June 16? Be sure to put the answer in simple fraction form.

8. A 30 foot by 60 foot area in a hotel will be turned into a lounge. The dance floor will occupy a 18 foot by 42 foot area. What fraction of the area of the lounge will be occupied by the dance floor?

9. When driving from Chicago, IL to Minneapolis, MN, a distance of approximately 416 miles, a compact car used approximately 11 gallons of gasoline. How many miles per gallon did this compact car get on this trip?

10. A board for shelving is $15\frac{3}{4}$ feet long. If 3 shelves of equal length are to be made from this board, first estimate the length of each shelf, and then calculate each length, first in feet and then in feet and inches.

11. Explain how you would reason out $4 \div \frac{1}{4}$ without doing any computation.

12. Use common denominators to multiply $\frac{2}{3}$ and $\frac{1}{5}$.

13. Use common denominators (and the "divide across" method) to divide each of the following.

    a) $9/5$ by $1/3$ b) $21/6$ by $1/2$

    c) Examine problem b) and the solution and write your observations.

14. Draw the rectangle formed by the points (2,1), (2,3), (−2,1) and (−2,3) on a coordinate system. Then take $1/2$ of each x-coordinate and draw the new rectangle. Compare the original and the new rectangle in size. What do you find?

15. Find the area of the following figure.

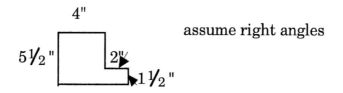

assume right angles

16. Perform the following computations:

    a) $\dfrac{a}{4} \cdot \dfrac{b}{4}$ b) $\dfrac{2x}{3} \cdot \dfrac{3}{2}$ c) $\dfrac{3}{a} \div \dfrac{5}{a}$

17. During one week, Raul ran the following distances (in miles):
$$2\tfrac{1}{4},\ 3\tfrac{3}{10},\ 3\tfrac{2}{5},\ 4\tfrac{7}{10}$$
    <u>Estimate</u> the total number of miles his brother ran if he ran $1/2$ the total distance Raul did. Explain your estimation process.

18. Simplify these complex fractions using any method you wish.

    a) $\dfrac{\frac{3}{4}+\frac{1}{2}}{\frac{1}{3}}$ b) $\dfrac{\frac{3x}{4}}{x}$ c) $\dfrac{\frac{x}{2}+\frac{2x}{3}}{\frac{2}{3}}$ d) $\dfrac{\frac{2}{3}}{\frac{x}{2}+\frac{2x}{3}}$

Exploration Problems

1. A proponent of natural foods in 1980 claimed that each day there are 25 tons of aspirin consumed in the U.S. What does this mean in terms of aspirin tablets per person per day? Assume that 1 aspirin tablet contains 5 grains of aspirin, 1 lb. is 7000 grains (avoirdupois units) and U.S. population was 210 million at that time.[1]

2. Use a yardstick and measure the dimensions of your math classroom (to the nearest $1/4$-inch). Then find the exact area of the floor of your classroom.

---

[1] *NCTM Sourcebook of Applications*, 1980, page 34

Solutions to Problem Set 2.2

1. a) 9   b) 23   c) 28   d) 2

2. a) The shaded area represents 3 parts out of 15, $3/15$ or $1/5$ of the total area.

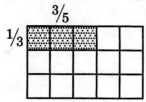

b) There are 4 (one-sixths) in $2/3$ as shown below.

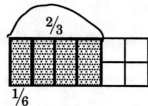

3. Using "compatible" numbers, $1/4$ of $48 is $12, so the discount will actually be a little more than $12.

4. $1/4$ of the desks represents 4 desks so 16 desks are in the classroom.

5. $261 - 1/9 (261) = 232$ left

6. Customers <u>paid</u> $1/2$ of $2/3$ which is $2/6$ or $1/3$ of the original price.

7. $\frac{135}{371}$

8. $\frac{756}{1800} = \frac{21}{50}$

9. $37\,9/11$ miles per gallon

10. Estimate: 5 plus feet. Each shelf actually will be $5\,1/4$ feet or 5 feet, 3 inches long. (Or: $15 \div 3$ and $3/4 \div 3 = 5$ and $1/4$!)

11. There are 4 one-quarters in every whole so there are $4 \cdot 4 = 16$ one-quarters in 4.

12. $2/3 \cdot 1/5 = 10/15 \cdot 3/15 = 30/225 = 2/15$

13. a) $6/5 \div 1/3 = 18/15 \div 5/15 = 18/5$

    b) $21/6 \div 7/2 = 21/6 \div 21/6 = 21/21 = 1$

117

c) Note here that dividing both the numerators and denominators (across) we have $3/3 = 1$.

14. The image is $1/2$ the size of the original.

15. A = 25 sq. inches

16. a) $ab/16$ b) $6x/6 = x$ c) $3/5$

17. $1/2$ of $13 1/2$ is not quite 7 miles.

18. a) $\dfrac{15}{4}$ b) $\dfrac{3}{4}$ c) $\dfrac{7x}{4}$ d) $\dfrac{4}{7x}$

Section 2.3 - Development and Use of the Properties of Rational Numbers

The properties of integers under the operations of addition and multiplication that we discussed in section 1.3 can be extended to the rational numbers.

1) **The commutative property:**

   For all rational numbers, a/b and c/d where $b, d \neq 0$,

   $$a/b + c/d = c/d + a/b \quad \text{for addition and}$$
   $$a/b \cdot c/d = c/d \cdot a/b \quad \text{for multiplication}$$

   For example, $1/2 + 1/3 = 3/6 + 2/6$
   $$= 2/6 + 3/6 = 1/3 + 1/2$$

   and $\quad 1/2 \cdot 1/3 = 1/3 \cdot 1/2 = 1/6$

2) **The associative property:**

   For all rational numbers, a/b, c/d, e/f where $b, d,$ and $f \neq 0$,

   $$(a/b + c/d) + e/f = a/b + (c/d + e/f) \quad \text{for addition and}$$
   $$(a/b \cdot c/d) \cdot e/f = a/b \cdot (c/d \cdot e/f) \quad \text{for multiplication}$$

   To illustrate: $\left(1/2 + 1/3\right) + 3/4 = 5/6 + 3/4$
   $$= 20/24 + 18/24 = 38/24 = 19/12$$
   $$= 1/2 + \left(1/3 + 3/4\right) = 1/2 + 13/12$$
   $$= 6/12 + 13/12 = 19/12$$

   and $\left(1/2 \cdot 1/3\right) \cdot 3/4 = 1/6 \cdot 3/4 = 3/24 = 1/8$
   $$1/2 \cdot \left(1/3 \cdot 3/4\right) = 1/2 \cdot 3/12 = 3/24 = 1/8$$

3) **The identity property:**

   For all rational numbers a/b, $b \neq 0$,

   $$a/b + 0/1 = a/b + 0 = 0 + a/b = a/b$$
   and $\quad a/b \cdot 1 = 1 \cdot a/b = a/b$

   In each case, the <u>identity</u> of the rational number is unchanged.

4) **The additive inverse:**

   For all rational numbers a/b, $b \neq 0$,

$$\tfrac{a}{b} + (-\tfrac{a}{b}) = 0$$

In other words, the additive inverse for a rational number is the same rational number but with opposite sign.

5) **The multiplicative inverse** (also called **reciprocal**):

For all rational numbers except 0, $\quad \tfrac{a}{b} \cdot \tfrac{b}{a} = 1$

As mentioned before, the reciprocal of any rational number can be found by changing the positions of the numerator and denominator. The <u>sign</u> of the rational number remains <u>unchanged</u>. Since division by zero is undefined, it should be clear that $\tfrac{0}{1}$ does not have a reciprocal.

6) **The zero property**:

For all rational numbers, $\tfrac{a}{b}$, $b \neq 0$,

$$\tfrac{a}{b} \cdot 0 = 0 \cdot \tfrac{a}{b} = 0$$

7) **The distributive property of multiplication over addition**

For all rational numbers $\tfrac{a}{b}$, $\tfrac{c}{d}$ and $\tfrac{e}{f}$ where b, d and $f \neq 0$,

$$\tfrac{a}{b}(\tfrac{c}{d} + \tfrac{e}{f}) = \tfrac{a}{b} \cdot \tfrac{c}{d} + \tfrac{a}{b} \cdot \tfrac{e}{f} = (\tfrac{c}{d} + \tfrac{e}{f})\tfrac{a}{b} = \tfrac{c}{d} \cdot \tfrac{a}{b} + \tfrac{e}{f} \cdot \tfrac{a}{b}$$

Also, consistent with our work with integers,

$$\tfrac{a}{b}(\tfrac{c}{d} - \tfrac{e}{f}) = \tfrac{a}{b} \cdot \tfrac{c}{d} - \tfrac{a}{b} \cdot \tfrac{e}{f} = (\tfrac{c}{d} - \tfrac{e}{f})\tfrac{a}{b} = \tfrac{c}{d} \cdot \tfrac{a}{b} - \tfrac{e}{f} \cdot \tfrac{a}{b}.$$

To illustrate:
$\tfrac{1}{2}(34)$ can be thought of as $\tfrac{1}{2}(30 + 4) = 15 + 2 = 17$

8) **The closure property**:

For all rational numbers $\tfrac{a}{b}$, $\tfrac{c}{d}$, $\tfrac{e}{f}$ and $\tfrac{g}{h}$, where $b, d, f, h \neq 0$,

$\tfrac{a}{b} + \tfrac{c}{d} = \tfrac{e}{f}$ where $\tfrac{e}{f}$ is rational and unique
and $\tfrac{a}{b} \cdot \tfrac{c}{d} = \tfrac{g}{h}$ where $\tfrac{g}{h}$ is rational and unique.

This last property was not mentioned in section 1.3 because we discussed the properties in the context of mental arithmetic and estimation, but it does apply to the set of integers under the operations of addition and multiplication.

<u>Sample Problems</u>

1. Use the properties of rational numbers to compute:

    a) $\dfrac{1}{3} + \left(\dfrac{1}{2} + \dfrac{2}{3}\right)$ b) $\dfrac{1}{2}\left(16\dfrac{3}{4}\right)$

Solutions:

a) $\frac{1}{3} + \left(\frac{1}{2} + \frac{2}{3}\right) = \frac{1}{3} + \left(\frac{2}{3} + \frac{1}{2}\right)$

(using the commutative property)

and $= \left(\frac{1}{3} + \frac{2}{3}\right) + \frac{1}{2}$

(using the associative property)

$= 1 + \frac{1}{2} = 1\frac{1}{2}$

b) $\frac{1}{2}\left(16 + \frac{3}{4}\right) = 8 + \frac{3}{8} = 8\frac{3}{8}$

2. The distributive property can help to combine any **like** terms such as $3x$ and $\frac{1}{2}x$ in the following way:

$$3x + \frac{1}{2}x = \left(3 + \frac{1}{2}\right)x = \left(3\frac{1}{2}\right)x = \frac{7}{2}x$$

Use this procedure to combine

a) $4x^2 - 5x^2$ and b) $3x - \frac{1}{4}x$

Solutions:

a) $(4 - 5)x^2 = -1x^2$ b) $\left(3 - \frac{1}{4}\right)x = \left(2\frac{3}{4}\right)x$ or $\frac{11}{4}x$

We can extend our initial work with polynomials to include simplifying expressions like

$$\frac{1}{2}\left(2x^2 - 4x + 1\right) - \frac{1}{3}\left(6x^2 + 1\right)$$

Using the distributive property and the rules for multiplying rational numbers we have

$$\frac{1}{2}\left(2x^2 - 4x + 1\right) - \frac{1}{3}\left(6x^2 + 1\right) = \frac{2x^2}{2} - \frac{4x}{2} + \frac{1}{2} - \frac{6x^2}{3} - \frac{1}{3}$$

$$= x^2 - 2x + \frac{1}{2} - 2x^2 - \frac{1}{3} = (1-2)x^2 - 2x + \frac{3}{6} - \frac{2}{6} = -x^2 - 2x + \frac{1}{6}$$

$$= x^2 - 2x + \frac{1}{2} - 2x^2 - \frac{1}{3} = (1-2)x^2 - 2x + \frac{3}{6} - \frac{2}{6} = -x^2 - 2x + \frac{1}{6}$$

In division, we can use the distributive property to isolate common factors and simplify rational expressions.

For example, in

$$\frac{4ab + 8a - 16b}{8}$$

we write the numerator as a product (in other words, we use the distributive property from right to left!) and then simplify by canceling out factors that are common to both the numerator and denominator as shown.

$$\frac{\overset{1}{\cancel{4}}(ab+2a-4b)}{\underset{2}{\cancel{8}}} = \frac{ab+2a-4b}{2}$$

Finally, the distributive property can provide a convenient method for simpliying complex fractions.

Follow the steps below where both the fundamental law of fractions and the distributive property are used.

$$\frac{\frac{3}{4}+\frac{1}{2}}{\frac{1}{3}} = \frac{\left(\frac{3}{4}+\frac{1}{2}\right)12}{\left(\frac{1}{3}\right)12} = \frac{9+6}{4} = \frac{15}{4}$$

For an algebraic rational expression, we have

$$\frac{\frac{x}{4}+\frac{2x}{5}}{\frac{x}{2}} = \frac{\left(\frac{x}{4}+\frac{2x}{5}\right)20}{\left(\frac{x}{2}\right)20} = \frac{5x+8x}{10x} = \frac{13x}{10x} = \frac{13}{10}$$

Sample Problems

1. Using the properties discussed in this section, simplify each of the following.

   a) $\dfrac{3/5}{3/4}$ b) $\dfrac{1/2 + 1/3}{5/6 - 2/3}$ c) $6\left(\frac{1}{2}+\frac{1}{3}\right)$ d) $\dfrac{6x^2+4x+8}{2x}$

   Solutions:

   a) $\dfrac{3/5}{3/4} = \dfrac{(3/5)\,20}{(3/4)\,20} = \dfrac{12}{15} = \dfrac{4}{5}$

   b) $\dfrac{1/2+1/3}{5/6-2/3} = \dfrac{(1/2+1/3)\,12}{(5/6-2/3)\,12} = \dfrac{6+4}{10-8} = \dfrac{10}{2} = 5$

   c) $6\left(\frac{1}{2}+\frac{1}{3}\right) = 3 + 2 = 5$

   d) $\dfrac{6x^2+4x+8}{2x} = \dfrac{\overset{1}{\cancel{2}}(3x^2+2x+4)}{\underset{1}{\cancel{2}}x} = \dfrac{3x^2+2x+4}{x}$

Problem Set 2.3

1. Use the properties of rational numbers to compute each of the following *mentally*. Remember that an exact answer is needed here.

    a) $\frac{3}{4}(424)$

    b) $\frac{1}{2}\left(8\frac{2}{3}\right)$

    c) $\frac{3}{4} + \left(\frac{4}{5} + \frac{1}{4}\right)$

    d) $\frac{1}{2} - 6$

2. Simplify:

    a) $\frac{1}{3}x - \frac{1}{2}x$

    b) $\frac{3}{2}x^2 - \frac{1}{2}x^2$

    c) $5x^3 - 7x^3$

    d) $\frac{1}{4}y + \left(\frac{1}{2}y + \frac{3}{4}y\right)$

3. Tell which property on which operation is indicated in each of the following.

    a) $\frac{1}{2} + \frac{1}{3} + \frac{1}{4} = \frac{1}{2} + \frac{1}{4} + \frac{1}{3}$

    b) $2\left(\frac{2}{3} + 4\right) = \left(\frac{2}{3} + 4\right)2$

    c) $\frac{3}{4} \cdot \frac{4}{3} = 1$

4. Use the distributive property to make each computation easier.

    a) $\frac{1}{4} \cdot 8\frac{4}{5}$

    b) $\frac{1}{2} \cdot 6\frac{1}{4}$

5. The area of the rectangle shown below is $\frac{2}{3}$ the area of a larger rectangle. Find the area of the larger rectangle.

    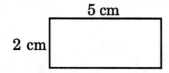

6. A jacket was originally priced at $99.95. If the store is offering a discount of $\frac{1}{5}$ on all items, estimate the sale price of the jacket.

7. In one week, a certain stock gained $\frac{1}{2}$ point, gained another $\frac{5}{8}$ point, and then lost $\frac{1}{8}$ of a point. Use the properties and the most efficient way possible to find the net change in the stock that week.

8. If denim fabric is selling for $4 a yard, find the cost of $6\frac{1}{2}$ yards of this fabric. (Use the properties and mental arithmetic.)

9. At Wildcat College, the fall enrollment one year showed a drop of $1/10$ from the previous semester's enrollment. This represented the loss of 224 students. Find the total number of students enrolled at Wildcat College in the fall.

10. Perform the indicated operations and simplify, if possible.

   a) $\dfrac{3x}{5} - 2x$

   b) $\dfrac{1}{4}(8x + 3y)$

   c) $\dfrac{2}{3}(9a + 3b - 4)$

   d) $-\dfrac{3}{4}(8x + 3y) - \dfrac{1}{2}(x + y)$

   e) $\dfrac{25x + 10y - 30}{10}$

   f) $\dfrac{9x^2y + 6y^2x}{3x}$

11. Use the distributive property and the fundamental law of fractions to simplify each of the following complex rational expressions.

   a) $\dfrac{\dfrac{x}{4} + \dfrac{2x}{5}}{\dfrac{x}{2}}$

   b) $\dfrac{\dfrac{y}{2} - \dfrac{y}{3}}{\dfrac{y}{2}}$

   c) $\dfrac{5 + \dfrac{y}{8}}{\dfrac{3y}{2}}$

   d) $\dfrac{\dfrac{y}{2} + \dfrac{y}{3}}{\dfrac{y}{3} - \dfrac{y}{2}}$

Exploration Problems

1. Find two rational numbers between $3/4$ and $4/5$. Discuss whether or not you could find more than two numbers. Explain your reasoning and illustrate with examples.

2. Ashley has $1/4$ of her college credits in social studies and $1/3$ of her credits in English. She has 30 credits outside of social studies and English. What is her total number of credits at this point in her college career?

Solutions to Problem Set 2.3

1. a) $\frac{3}{4}(400 + 24) = 318$ b) $\frac{1}{2}\left(8 + \frac{2}{3}\right) = 4\frac{1}{3}$ c) $1\frac{4}{5}$ d) $-5\frac{1}{2}$

2. a) $-\frac{1}{6}x$ b) $\frac{2}{2}x^2 = x^2$ c) $-2x^3$ d) $1\frac{1}{2}y$ or $\frac{3}{2}y$

3. a) Commutative property of addition
   b) Commutative property of multiplication
   c) Multiplicative inverse (reciprocal)

4. a) $\frac{1}{4}\left(8 + \frac{4}{5}\right) = 2 + \frac{1}{5} = 2\frac{1}{5}$ b) $\frac{1}{2}\left(6 + \frac{1}{4}\right) = 3 + \frac{1}{8} = 3\frac{1}{8}$

5. 15 sq. cm.

6. The sale price is approximately $80.

7. A 1-point gain.

8. $\$4\left(6 + \frac{1}{2}\right) = \$26$

9. $2240 - 224 = 2016$ students

10. a) $\frac{-7x}{5}$ b) $2x + \frac{3y}{4}$ c) $6a + 2b - \frac{8}{3}$

    d) $\frac{-13x}{2} - \frac{11y}{4}$ or $-6\frac{1}{2}x - 2\frac{3}{4}y$ or $\frac{-26x - 11y}{4}$

    e) $\frac{5x + 2y - 6}{2}$ or $\frac{5}{2}x + y - 3$

    f) $3xy + 2y^2$ or $y(3x + 2y)$

11. a) Use 20 as a multiplier and the result is $\frac{13x}{10x} = \frac{13}{10}$.

    b) Use 6 as the multiplier and the result is $\frac{1}{3}$.

    c) Use 8 as the multiplier; $\frac{40 + y}{12y}$

    d) Use 6 as the multiplier and the result is -5.

Section 2.4 – Informal Equation Solving

In the last section, we learned how to make fractions equivalent to a given fraction. If we are given two equivalent rational numbers in a statement of equality, such as

$$\frac{3}{5} = \frac{6}{10}$$

we have set up a **proportion**. This equivalency can be verified by <u>multiplying</u> both the numerator and denominator of the first fraction by 2, by <u>dividing</u> the numerator and denominator of the second fraction by 2, or by checking the cross-products.

If we have a similar statement set up in proportion form,

$$\frac{-4}{3} = \frac{\square}{9}$$

can we find the missing numerator? Examining the relationship between the denominators, the multiplier that would make the fraction on the left equivalent to the fraction on the right is the number 3. The diagram below will help to illustrate this.

$$\frac{-4 \times 3}{3 \times 3} = \frac{\boxed{-12}}{9}$$

Thus, the missing number is –12.

Using cross-products, since
$$9(-4) = -36, \text{ then } 3 \cdot \square = -36$$
$$\text{and } \square = -12.$$

Similarly, if we set up a proportion such as

$$\frac{7}{\square + 2} = \frac{14}{10}$$

what number can be put into the window to make the equation true? Analyzing the numerators given and the placement of the missing denominator, a divisor of 2 (going from right to left) would be appropriate. Dividing the denominator of 10 by 2 to maintain the equivalency implies that the denominator on the left should be 5. But that denominator is given as a sum. If the total must be 5, and 2 is already an addend, then 3 must be the missing addend. A circle is shown below to indicate the steps considered.

$$\frac{7}{\underbrace{\square + 2}_{5}} = \frac{14 \div 2}{10 \div 2}$$

Now place 3 in the window and make sure this solution works.

Sample Problems

1. Find the missing numbers in each of the following.

a) $\dfrac{\square}{4} = \dfrac{21}{28}$ b) $\dfrac{8}{6} = \dfrac{12}{\square}$ c) $\dfrac{3 \cdot \square}{18} = \dfrac{-5}{2}$

Solutions:

a) Examining the denominators from right to left, the divisor needed is 7. Thus, the missing number is $21 \div 7 = 3$.

b) Reduce $8/6$ down to $4/3$. We then have
$$\dfrac{4}{3} = \dfrac{12}{\square}$$
Comparing numerators from left to right, we need a multiplier of 3. So $3 \cdot 3 = 9$ and $\square = 9$.

c) The number 18 is generated from the right denominator of 2 by multiplying by 9.
Here are the steps with the circles included.

$$\dfrac{\overbrace{3 \cdot \square}^{-45}}{18} = \dfrac{-5 \cdot 9}{2 \cdot 9}$$

so $\square = -15$

As an extension of some of the work in section chapter 1, let's try a few more equations. For example,
$$3 \cdot \square - 1 = 10$$
does not generate an integer answer. We can estimate the *nearest integer* solution to be 4, using circles to show the order of operations:

$\overbrace{3 \cdot \square}^{11} - 1 = 10$ and since there is no integer times 3 that equals 11, we must use 4 as the *nearest integer* solution.

We can extend our work with rational numbers and reciprocals, however, and determine the actual solution to this equation.

Using the circle, we know that $3 \cdot \square = 11$, and understanding how multiplication and cancellation of factors work, it is not difficult to see that

$$3 \cdot \boxed{\dfrac{11}{3}} = 11 \text{ and } \square = \dfrac{11}{3}$$

Instead of the window, let's use a variable, x, in the equation

$$\dfrac{5x + 3}{4} = 3$$

(Don't forget, $5x$ means 5 times x.)

Here the whole numerator must be 12 as shown in the circle below, and this means that 5x must be 9. We can approximate the solution as x = 2.

$$\frac{\overset{912}{\overbrace{(5x)+3}}}{4} = 3$$

Again, we will find the solution to the equation if we can find the solution to

$$5x = 9.$$

But $\quad 5\left(\dfrac{9}{5}\right) = 9 \text{ and } x = \dfrac{9}{5}$

Sample Problems

1. Find the solution to the nearest integer. If x occurs more than one time in an equation, be sure to use the same replacement number for each x in that equation. Then find the actual rational number solution.

    a) $\dfrac{x+11}{x} = 5$ \qquad\qquad b) $2 = \dfrac{4x-1}{3}$

    Solutions:

    a) There are several ways to get a quotient of five: $10 \div 2$, $15 \div 3$, etc. Trial and error is called for here, but we can focus on the fact that the numerator and denominator differ in value by 11.
    If we let x = 2, then we have
    $$\frac{2+11}{2} = 6\frac{1}{2}$$
    If we let x = 3,
    $$\frac{3+11}{3} = 4\frac{2}{3}$$

    If x = 4, then we have
    $$\frac{4+11}{4} = 3\frac{3}{4}$$
    which is further away from the desired result of 5. Apparently the nearest integer answer is x = 3 since the result of $4\frac{2}{3}$ is the value closest to 5.

    To find the actual solution, use cross products (and $\dfrac{5}{1}$) to generate
    $$5x = x + 11$$

    From our work on the pan balance, this generates the equation:
    $$4x = 11$$

    Now we can see that $4\left(\dfrac{11}{4}\right) = 11$ and so $x = \dfrac{11}{4}$.

b) In this case, the entire numerator must be 6. Using circles:

$$2 = \frac{\overset{7\nearrow\quad\overset{6}{\frown}}{(4x) - 1}}{3}$$

If 4x is 7, 2 is the nearest integer solution.

Finally, 4x = 7 implies that $4\left(\dfrac{7}{4}\right) = 7$ and $x = \dfrac{7}{4}$.

2. Translate into an equation and solve:

Four more than three times a number is five. Find the number.

Solution:
Letting n be the number,

$$3n + 4 = 5 \text{ is the equation.}$$

Using a circle,

$\overset{1}{\frown}$
$(3n) + 4 = 5$ and since 3n = 1, the nearest integer answer is 0,

but 3n = 1, and n must be the reciprocal of 3, or ⅓. Note that ⅓ does make the statement true.

Problem Set 2.4

1. In each of the problems below, use the properties of proportions to find a replacement value for the window or variable.

   a) $\dfrac{10}{4} = \dfrac{25}{\square}$

   b) $\dfrac{12}{7} = \dfrac{\square}{28}$

   c) $\dfrac{16}{6} = \dfrac{\square}{9}$

   d) $\dfrac{-35}{14} = \dfrac{5}{\square}$

   e) $\dfrac{\square + 2}{3} = 6$

   f) $\dfrac{\square - 3}{-1} = \dfrac{2}{\square}$

   g) $\dfrac{2 \cdot \square + 5}{9} = 3$

   h) $\dfrac{8}{\square} = 24$

   i) $\dfrac{4}{\square} = \dfrac{8}{1}$

2. Determine whether $\frac{1}{4}$, $\frac{1}{3}$ or $\frac{1}{2}$ is the solution.

   a) $2 \cdot \square - 1 = 0$

   b) $3(\square + 1) = 4$

   c) $20x - 6 = -1$

   d) $\dfrac{1}{3}x = 5$

   e) $\dfrac{x}{2-x} = \dfrac{1}{3}$

3. Find the *nearest integer* answer **and** the exact solution to each of the following equations.

   a) $6x + 3x = 35$

   b) $4x + 15 = 0$

   c) $8 + y = 8y$

   d) $2d = 36 + 7d$

   e) $\dfrac{6a + 1}{4} = 9$

4. Translate each of the following into a mathematical equation. Use n as the variable, then solve for n.

   a) The product of 6 and a number is −36. Find the number.

   b) Twice a number, decreased by 5, equals 25. Find the number.

   c) If 3 times a number is decreased by 10, the result is the same as when twice the number is increased by 5.

   d) Four less than 3 times a number is 20. Find the number.

   e) One quarter of a number is 4. Find the number.

5. Solve each of the following word problems and write the answer in a complete sentence.

   a) After Sue lost $10\frac{1}{2}$ pounds, she weighed 123 pounds. Find Sue's original weight.

b) One-half of an integer is −12. Find the integer.

c) The width of a rectangle is $\frac{1}{4}$ of its length. If the width of the rectangle is 10 feet, what is the length of the rectangle?

d) One number is 5 times another number. If their difference is 24, find the two numbers. (Hint: There are two possible pairs of numbers here.)

e) If the driver of a car travels at an average rate of 52 mph, find the distance traveled by this driver (distance = rate • time)

   1) in 2 hours

   2) in $3\frac{1}{3}$ hours

   3) in x hours

f) Given the rectangular garden shown below. Find the length of this garden.

$$A = lw = 85 \text{ sq. ft.} \qquad 8\frac{1}{2} \text{ ft.}$$

Exploration Problems

1. Using a table or chart, determine whether a nearest integer solution exists for each of the following equations. Justify your conclusions.

   a) $\dfrac{1}{x} + \dfrac{x}{1} = 10$

   b) $\dfrac{x}{x+1} = 1$

2. Write an equation containing at least three terms, using n as the variable, that has a **solution** of $\frac{2}{3}$.

Solutions – Problem Set 2.4

1. a) 10   b) 48   c) 24   d) −2   e) 16
   f) 1,2   g) 11   h) $1/3$ i) $1/2$

2. a) $1/2$
   b) $1/3$
   c) $1/4$
   d) None of these is a solution.
   e) $1/2$

3. a) $x = 4$; $35/9$
   b) $x = -4$; $-15/4$
   c) $y = 1$; $8/7$
   d) $d = -7$; $-36/5$
   e) $a = 6$; $35/6$

4. a) $6n = -36$
      $n = -6$

   b) $2n - 5 = 25$
      $n = 15$

   c) $3n - 10 = 2n + 5$
      $n = 15$

   d) $3n - 4 = 20$;
      $n = 8$

   e) $(1/4)x = 4$
      $x = 16$

5. a) Her original weight was $133\,1/2$ pounds.
   b) The number is −24.
   c) The length of the rectangle is 40 feet.
   d) The numbers are 6 and 30 or −6 and −30.
   e) 1) $52(2) = 104$ miles
      2) $52(3\,1/3) = 173\,1/3$
      3) $52x$ miles
   f) The length of the garden is 10 ft.

Chapter Two Review Problems

1. Give a fraction to approximate the area shaded below.

2. If this set of dots represents $4/3$, find the unit set of dots.

3. On this number line, place a dot and label each of the points $5/8$ and $4/3$.

   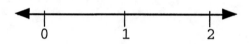

4. Write a fraction that represents the *unshaded* area of the square below.

   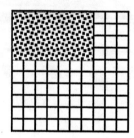

5. Because of budget reductions, a district was forced to cut $1\frac{1}{2}$ million out of a total budget of $25 million. Write the ratio of the amount of reduction to the total budget as a fraction in simplest form.

6. Solve each equation. If the answer is not an integer, give the nearest integer answer but then find the actual solution.

   a) $\dfrac{4x+1}{5} = 2$

   b) $\dfrac{4 \cdot \Box + 2}{3} = \dfrac{28}{12}$

   c) $\dfrac{6 \cdot \Box + 5}{3} = 2$

7. Use the distributive property and combine like terms to simplify these expressions.

   a) $6x^3 - \dfrac{1}{2}x^3$

   b) $3\left(\dfrac{1}{4}x + \dfrac{1}{2}\right) + \dfrac{1}{4}x$

c) $\frac{1}{2}\left(\frac{x}{4}+2\right)-2(x-4)$ d) $\frac{4x^2+6x+10}{2x}$

8. A foot is what fraction of a yard? An inch is what fraction of a yard?

9. The pitch or grade of a roof is a rise (vertical distance) of $4\frac{1}{2}$ inches for every run (horizontal distance) of $13\frac{1}{2}$ inches. Give the pitch of the roof, rise over run, in simplest fraction form.

10. Given the shading below on the unit rectangle, write the multiplication problem indicated by this shading.

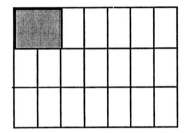

11. One standard paper size in the United States is $8\frac{1}{2}$" by 11". A student teacher needs to cut the shorter side of this paper into 4 equal strips and then cut each strip in half to make equal-sized bookmarks for one of his classes.

   a) Give the dimensions of the bookmark.

   b) Find the number of $8\frac{1}{2}$ x 11 sheets of paper that need to be cut to make at least 30 bookmarks.

12. Find the minimum length of a bolt that will go through a piece of wood that is $1\frac{3}{4}$" thick, a washer that is $\frac{1}{16}$" thick and a nut that is $\frac{3}{8}$" thick.

13. Order the following rational numbers, smallest to largest, left to right.

$$-\tfrac{3}{8},\ \tfrac{7}{16},\ -\tfrac{1}{2},\ \tfrac{0}{9},\ \tfrac{1}{3},\ \tfrac{3}{10}$$

14. Use the properties of rational numbers to estimate each of the following.

   a) $\frac{1}{4}$ of 85

   b) $\frac{2}{5}$ of $31

   c) The amount of a $\frac{1}{3}$ discount on a shirt originally priced at $26.89.

   d) $4\left(5\tfrac{1}{2}\right)$ e) $4\tfrac{2}{3}+\left(3\tfrac{1}{2}-\tfrac{2}{3}\right)$

15. On May 1, Bill weighed 175 pounds. By July 1st of that year, he had gained $1/10$ of his weight. Then by September 1st of that year, he had lost $1/10$ of his weight. What did Bill weigh on September 1st?

16. Divide $(-8a^2 - 16ab + 4b^2)$ by 32.

17. Simplify each complex fraction.

a) $\dfrac{1/5 + 5/6}{1/15}$ b) $\dfrac{x/3 + 2x/9}{x}$ c) $\dfrac{x/3 + 2x/9}{2x/9 - x/3}$

Exploration Problems

1. To estimate the circumference of the earth (the distance around a circle or sphere), Eratosthenes, a Greek mathematician, used the angle of the sun's rays on the earth to his advantage. The picture below will help depict this.

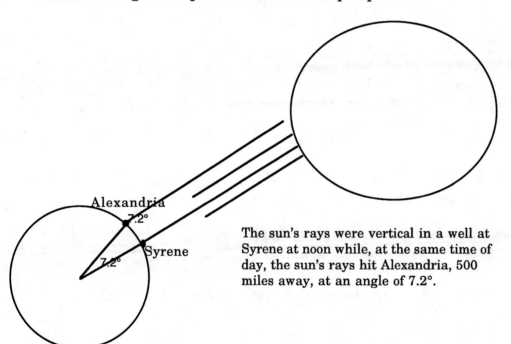

The sun's rays were vertical in a well at Syrene at noon while, at the same time of day, the sun's rays hit Alexandria, 500 miles away, at an angle of 7.2°.

Using the diagram, and the properties of similar triangles, write the proportion that Eratosthenes used to estimate the circumference of the earth and solve it to find his estimate in miles. Then use an atlas, or the Internet, to find the actual circumference of the earth, probably given at the equator since the earth is not a perfect sphere, and find the difference between that value and Eratosthenes' calculated value.

2. This right triangle contains an angle of about 53°. The three sides forming the triangle are shown. Six ratios can be formed from these three numbers.

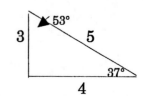

a) Write the six ratios.

b) One of the ratios is actually the "sine," a special ratio in trigonometry. Use your calculator to find the sine of 53° by pressing

   53 sin or sin 53 =

   and determine which ratio generates a value very close to this.

c) Another trigonometric ratio is "tangent." Use your calculator to find the tangent of 53° by pressing

   53 tan or tan 53 =

   and determine which of your ratios generates a value close to this.

d) The "cosine" is another trigonometric ratio. Determine which ratio represents the cosine of 53°.[1]

---
[1] *Algebra*, University of Chicago Project, 1993.

Solutions - Chapter Two Review Problems

1. $3/4$

2. 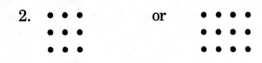 or

3. (number line showing points at $5/8$ between 0 and 1, and $4/3$ between 1 and 2)

4. $72/100 = 18/25$

5. $1\frac{1}{2}/25 = 15/250 = 3/50$

6. a) 2; $9/4$ b) 1; $5/4$ c) 0; $1/6$

7. a) $(6 - 1/2)x^3 = (5\frac{1}{2})x^3$ or $(11/2)x^3$ b) $\frac{3}{4}x + \frac{3}{2} + \frac{1}{4}x = x + \frac{3}{2}$
   c) $\frac{-15x}{8} + 9$ d) $\frac{2x^2 + 3x + 5}{x}$

8. a) $1/3$ b) $1/36$

9. $4\frac{1}{2}/13\frac{1}{2} = 1/3$

10. Problem: $1/3 \cdot 2/7 = 2/21$

11. a) Each bookmark is $2\frac{1}{8}$ inches by $5\frac{1}{2}$ inches.
    b) At least 4 sheets are needed (8 bookmarks per sheet).

12. The bolt must be at least $2\frac{3}{16}$ inches long.

13. $-1/2, -3/8, 0, 3/10, 1/3, 7/16$

14. a) $1/4 \cdot 84 = 21$ b) 12 c) $1/3(27) = \$9$ discount
    d) $4(5 + 1/2) = 22$ e) $7\frac{1}{2}$

15. $173\frac{1}{4}$ pounds

16. $\frac{-2a^2 - 4ab + b^2}{8}$

17. a) $31/2$ b) $5x/9x = 5/9$ c) 5

Chapter 2 Worksheet 1

1. Given the length ——— representing $2/3$, draw the unit length.

2. Given the length ——— representing $5/2$, draw the unit length.

3. Given the polygon ▭ representing $3/2$, draw a picture of the unit area.

4. Given the polygon ▭ representing $2/5$, draw a picture of the unit area.

5. Given the unit rectangle shown, shade $4/15$.

6. ESTIMATE the value of the following to the nearest whole:

   a) $\frac{1}{4} + \frac{2}{3}$
   b) $\frac{9}{10} - \frac{7}{8}$
   c) $\frac{2}{3} - \frac{1}{4}$
   d) $\frac{7}{8} \cdot \frac{9}{10}$
   e) $12 \div \frac{4}{7}$

7. Give a fraction that approximates the shaded area of the unit rectangle below.

8. Given this set of dots which represents $5/4$, draw the unit dot arrangement.

9. Fifteen minutes is what fraction of a hour?

10. Estimate (to the nearest half):
$$1\frac{7}{8} + \frac{4}{5} - \frac{3}{8}$$

11. Draw a diagram to show how many $\frac{1}{3}$'s there are in 2.

12. Do each of the following computations. Use some of the properties to simplify the work.

   a) $\frac{1}{3} + \left(\frac{3}{4} + \frac{2}{3}\right)$

   b) $\frac{2}{3} \cdot 18$

   c) $\frac{21}{4} \div \frac{3}{2}$

   d) $\dfrac{\frac{1}{4} - \frac{2}{3}}{\frac{1}{6}}$

13. Simplify each of the following.

   a) $\frac{a}{2} + \frac{5a}{2}$

   b) $\frac{2x}{5} + \frac{x}{2}$

   c) $\dfrac{\frac{x}{3}}{\frac{x}{4}}$

   d) $\dfrac{\frac{2a}{3} + \frac{a}{4}}{\frac{a}{2}}$

   e) $\frac{1}{2}(x^2 + 3x - 2)$

14. Put ">", "<" or "=" between each pair of rational numbers to make each statement true.

   a) $\frac{4}{5}$ $\frac{5}{6}$

   b) $\frac{14}{3}$ $\frac{9}{2}$

   c) $\frac{10}{15}$ $\frac{4}{6}$

15. Kirstin ate $\frac{1}{6}$ of a loaf of cinnamon-raisin bread. If 10 slices were left, how many slices were in the whole loaf?

Chapter 2 Worksheet 2

1. Order the following smallest to largest, left to right.

   $$\frac{3}{4}, \frac{5}{6}, \frac{5}{12}, \frac{1}{3}, -\frac{1}{2}, -\frac{2}{3}$$

2. Combine like terms using the distributive property.

   a) $3x - 5x$

   b) $\frac{1}{4}x + \frac{1}{3}x$

   c) $6(x - \frac{1}{2}x)$

   d) $2x - \frac{1}{4}x$

3. Find a replacement for each of the following windows that would make the statement true. Use the circle method.

   a) $3 \cdot \square - 4 = 11$

   b) $\frac{-5}{7} = \frac{-20}{3 \cdot \square + 1}$

   c) $\frac{2 \cdot \square + 3}{3} = 7$

   d) $\frac{2}{3} = \frac{8}{2 \cdot \square + 2}$

   e) $\frac{32}{\square - 12} = 8$

4. Use properties and shortcuts to simplify each of the following.

   a) $6(1\frac{2}{3})$

   b) $5 - 1\frac{1}{2}$

   c) $8 \div \frac{1}{3}$

   d) $1\frac{1}{4} + 2\frac{1}{3} + \frac{3}{4}$

   e) $3 \cdot 3\frac{2}{3}$

   f) $\frac{2}{5} + \left(3\frac{2}{3} + \frac{3}{5}\right)$

5. Jan has 12 chocolate cookies remaining. This is $3/7$ of the original batch. How many cookies were in the batch?

6. Perform each of the following computations and simplify.

   a) $\frac{x}{6} + \frac{y}{2}$

   b) $\frac{3a}{4} \cdot \frac{6a}{15}$

   c) $\frac{6x^2 - 24xy}{3x}$

CHAPTER THREE - Dimensional Analysis and Other Applications of Fractions, Decimals and Percents

Section 3.1 - Ratios and Unit Rates

We encounter **ratios** or mathematical comparisons frequently. One day the Detroit Tigers beat the Toronto Blue Jays 12 to 1, and the following day the Blue Jays turned the tables, winning 13 to 4. In each case the result given is a comparison between the number of runs scored by one team and the number scored by the other. Using mathematical symbolism, the ratios could be written as fractions: $12/1$ or $13/4$, or they could be written with the following notation: 12:1 and 13:4. [Recall our saying earlier that the phrase "ratio of" referred to the operation of division, and, note the presence of the word *ratio* in the term ra<u>tio</u>nal number.]

The ancient Greeks determined that there existed a ratio between two dimensions of an object which was the "most pleasing" to the eye, and they named this relationship the Golden Ratio. We'll encounter it in an exploration problem.

A ratio is a comparison between two quantities **which are measured using the same units**. For example, if we wish to compare the weight of two items using a ratio, both weights must be given in the same units—both in pounds or both in ounces or both in kilograms, etc.

When reconstituting frozen orange juice we usually mix one part (typically 1 can) of juice with three parts (3 cans) of water, so the ratio of juice to water is 1:3. This comparison of the volumes of the components is an example of a **part-to-part ratio**.

From another perspective, the juice container now holds four parts of liquid, of which $1/4$ is juice and $3/4$ is water. In this case we are comparing the volume of each component with the total volume, often called a **part-to-whole ratio**.

Small gasoline engines such as those found on lawn mowers, weed whips, etc., might require a pint of oil mixed with two gallons of gas. If the tank won't hold that much, using a ratio can help us decide how much gas and how much oil we should use. Since there are four quarts in one gallon and two pints in one quart, each gallon of gas equals 4 • 2 or 8 pints. The "recipe" called for two gallons, or 16 pints. Therefore, the ratio we are looking for is 1 pint of oil to 16 pints of gasoline. Since we now have both components measured in the same units, we can mix 1 <u>part</u> of oil with 16 <u>parts</u> of gasoline. A <u>part</u> can be any size measure we choose. We would use whatever amount is convenient.

Consider the following information from a label on "low-fat" hot dogs:

Calories per serving: 110; Calories from fat: 72.

In this case the important ratio, 72:110, compares the calories from fat to the total number of calories per serving, an important relationship in a health-conscious society. In this case, more than $1/2$ of the calories in this "low-fat" hot dog comes from fat!

> Sample Problems
>
> 1. A suit originally priced at $200 has been reduced by $40. Write in reduced fraction form the ratio of:
>   a) the amount of the reduction to the original price
>   b) the sale price to the original price
>
>   Solution:
>   a) $\$40/\$200 = 1/5$
>   b) $\$160/\$200 = 4/5$
>
> 2. Bruce sold 28 tickets to the chicken dinner sponsored by his Rotary Club. John sold 21 tickets. Elizabeth sold 27. The club sold 243 tickets before the day of the dinner.
>   a) Write a ratio which compares the combined number of tickets sold by these three members to the total number sold by the club.
>   b) Approximately what fraction of the tickets sold is this?
>
>   Solutions: a) $\dfrac{28+21+27}{243} = \dfrac{76}{243}$ b) About $1/3$
>
> 3. A mixture for hummingbird syrup is 3 parts water to 1 part sugar. What fraction of the mixture is water? sugar?
>
>   Solution: there are 4 parts total, so $3/4$ of the mixture is water and $1/4$ is sugar.

In addition to the phrase "ratio of," "per" is also included in our list of words and phrases which imply division. "Per" is used when describing a **rate**. Perhaps the most common rate we encounter is the speed at which we may *legally* drive, whether 10 miles per hour in a school driveway or 65 (or 70) miles per hour on an interstate highway. Notice how the rate is stated: **two quantities with different labels are being compared** using "per." Some other common rates you probably have encountered are: miles per gallon; students per class; cents per ounce; people per square mile. Each of these is an example of a **unit rate**, that is, if the rate were written in fraction notation the denominator would be one hour, one gallon, one class, one ounce, one square mile, respectively.

Note: We are not using "rate" and "speed" interchangeably. "Rate" is being used in a much broader sense than just to refer to the speed at which an object is traveling.

We mentioned two other unit rates when we considered the oil and gas mixture: 4 quarts per gallon and 2 pints per quart. We can convert the 2 gallons of gas to pints by using a **rate-factor** model, that is, by multiplying rates and labels as follows:

$$2 \text{ gallons} \cdot \dfrac{4 \text{ quarts}}{1 \text{ gallon}} \cdot \dfrac{2 \text{ pints}}{1 \text{ quart}} = 2 \cdot 4 \cdot 2 \text{ pints} = 16 \text{ pints}$$

In this example we've "canceled" the labels much as we "cancel" in fraction multiplication. The only label that does not cancel is "pints" which is what we are looking for. This technique is useful for solving some problems efficiently, and we'll utilize it further.

Sample Problems

1. On a typical summer day in Chicago the temperature rose from 64° F at 6 AM to 90° at noon.

   a) What was the average increase in temperature per hour?
   b) What fraction of a degree would this be per minute?

   Solution: a) (90° – 64°) ÷ 6 hours = 26° ÷ 6 hours = $4\frac{1}{3}$ degrees per hour.

   b) 6 hours • 60 min/hr = 360 minutes. 26° ÷ 360 minutes = $13/180$ of a degree per minute.

2. Onions are being sold at a farmer's market in two ways: $3 for 5 pounds or $2 for 3 pounds.

   a) What is the cost per pound for each size bag?
   b) What is the quantity purchased per dollar?
   c) Which is the better buy?

   Solutions: a) $3 ÷ 5 = $3/5$ of $1(or $.60) per pound. $2 ÷ 3 = $2/3$ of $1 (about $.67) per pound.

   b) 5 lb. ÷ $3 = $1\frac{2}{3}$ lbs. per dollar; 3 lb. ÷ $2 = $1\frac{1}{2}$ lbs. per dollar.
   c) The 5-lb. quantity is the better buy.

Proportions

If 7 children eat a total of 14 cheese sandwiches at a picnic, how many sandwiches would be needed if 3 more children joined the group and ate at the same rate? We can answer this question by using unit rates, since we easily see that 14 sandwiches for 7 children implies that we need 2 sandwiches per child. Adding 3 children means that we have a total of 10 children and must make 2•10 or 20 sandwiches. But there is another approach which works for this kind of problem. It is especially useful when the answer is not obvious. We can solve the problem by using a proportion.

As defined in chapter 2, a **proportion** is a statement of equality between two ratios or between two rates. (We utilized proportions when reducing fractions to lowest terms and when generating equivalent fractions.) Since we know that when two fractions are equal their cross-products are equal, we can solve the above problem using a proportion as follows:

$$\frac{2 \text{ sandwiches}}{1 \text{ child}} = \frac{\square \text{ sandwiches}}{10 \text{ children}}, \text{ or, } \frac{2}{1} = \frac{\square}{10}.$$

If 1 • □ = 2 • 10, the window replacement must be 20.

To estimate the fish population in a lake, fisheries research personnel capture a number of fish, tag them, and then return them to the water. After allowing a period of time for the tagged fish to mix with the rest of the population, a second sample of fish is taken. The number of fish caught and the number with tags are recorded. Experience has shown that, because of the natural mixing that occurs, the ratio between the number of fish with tags in the second sample to the size of the second sample will approximate the ratio between the number originally tagged to the fish population of the lake. A proportion can be set up as follows:

$$\frac{\text{\# tagged in second sample}}{\text{second sample size}} = \frac{\text{\# tagged originally}}{\text{total number of fish in lake}}$$

Let's assume some values for this problem. One hundred fish are originally caught, tagged, and returned to the lake to mix with the uncaught fish. A week later, 200 fish are caught, and 40 of them have tags. Substituting these numbers in the above proportion, we have

$$\frac{40 \text{ tagged fish in second sample}}{200 \text{ fish in second sample}} = \frac{100 \text{ originally tagged}}{\text{total number of fish in lake}}$$

Using notation we've seen previously, $\frac{40}{200} = \frac{100}{\square}$

What number would we use in the window to make the statement true? If the fractions are equal, then the cross products are equal. $40 \cdot \square = 100 \cdot 200$, and the missing number, by trial-and-error, equals 500. Of course, in real life studies, the value would not necessarily work out to be a whole number, and we would round the result appropriately.

Replacing the window with a variable, n, to represent the total number of fish in the lake, we have a more formal algebraic proportion:

$$\frac{40}{200} = \frac{100}{n}$$

and we can use the cross-product technique to solve for n.

We know that
$$40n = 100 \cdot 200$$
or
$$40n = 20000.$$

Above we used trial-and-error, asking ourselves what number times 40 would give 20000, but we need to find a more efficient approach.

Recall from chapter 2 that any number times its reciprocal equals one. Since our goal is to solve for $1n$, we can use the properties of equality and multiply both sides of our equation, $40n = 20000$, by 1/40.

$$\frac{1}{40}(40n) = \frac{1}{40}(20000)$$

and $1n$ or $n = 500$

Since dividing by 40 gives the same result as multiplying by $\frac{1}{40}$, we could also approach this problem by dividing both sides of our equation by 40.

$$\frac{40n}{40} = \frac{20000}{40}$$

and again, $\quad n = 500$

Therefore, we conclude that there are about 500 fish in the lake.

Sample Problems

1. A recipe for buttermilk pancakes calls for 3 cups of flour and 2 eggs. To make a larger batch using 3 eggs, how much flour will be needed?

   Solution:
   Let n represent the number of cups of flour needed.

   Then $\quad \dfrac{3 \text{ cups flour}}{2 \text{ eggs}} = \dfrac{n \text{ cups flour}}{3 \text{ eggs}} \quad$ or $\quad \dfrac{3 \text{ cups flour}}{n \text{ cups flour}} = \dfrac{2 \text{ eggs}}{3 \text{ eggs}}$

   and so $\quad 2n = 3(3) \quad$ using either proportion
   $\quad\quad\quad\quad 2n = 9$
   $\quad\quad\quad (\tfrac{1}{2})2n = (\tfrac{1}{2})9 \quad (\tfrac{1}{2}$ is the reciprocal of 2)
   and $\quad\quad n = \tfrac{9}{2}$ or $4\tfrac{1}{2}$

   Therefore, $4\tfrac{1}{2}$ cups of flour will be needed.

2. On a map of the United States, the map legend for the state of Arizona shows that 1 inch represents 27 miles. If Phoenix and Tucson are about 4 inches apart on the map, what is the approximate distance, in miles, between them?

   Solution: Let x represent the number of miles between the cities.

   $$\frac{1 \text{ inch}}{27 \text{ miles}} = \frac{4 \text{ inches}}{x \text{ miles}} \quad \text{or} \quad \frac{1 \text{ inch}}{4 \text{ inches}} = \frac{27 \text{ miles}}{x \text{ miles}}$$

   Either proportion is correct. The cross-product will be the same in each case. It is important to notice that more than one arrangement of terms is possible when setting up a proportion. We'll consider this fact further when we look at some geometric figures. Solving, we have

   $\quad\quad\quad\quad 1x = 27(4)$
   and $\quad\quad x = 108$

   Therefore, Phoenix and Tucson are about 108 miles apart.

Note that in solving each of these examples we began by defining the variable used. This step is crucial, not only to insure the correct placement of the variable in the equation, but also to guarantee that the solution answers the question asked.

In the examples studied so far, we have encountered **direct** variation. In other words, when one component of a ratio increased in value, the other also increased. Recall that the more children we fed the more sandwiches we needed.

In some situations, however, an increase in one component leads to a decrease in another. Consider a 200 mile automobile trip. The faster one drives, the less time the trip takes. Thus, for a given distance, the average speed and the time required for the trip are said to vary **inversely**. Similarly, the more teeth found on a gear, the fewer turns the gear makes in a given amount of time. If a gear with 42 teeth rotates 2 times per minute (speed is 2 rpm), a 21 tooth gear connected to it will turn twice as fast: 4 rpm. Notice that the product of number of teeth and revolutions per minute is 84 for each gear!

Consider another example. If a gallon of ice cream lasts a family of four an average of three days, and assuming the ice cream is eaten at the same rate, how long will a gallon last if two of the family members go to a convention for a week?

If the number of people eating is cut in half, the ice cream should last the remaining family members twice as long. Therefore, with two people eating it the ice cream should last 6 days. Once again, notice that the products are equal:

4 people • 3 days = 2 people • 6 days.

Similarity

Look at the two rectangles shown.

Does the relationship between the longer and the shorter sides of each of them appear to be the same? Since the longer side of the left hand rectangle looks about 4 times the shorter side, while the longer side of the right hand rectangle looks only about one and a half times the shorter side, the relationship is not the same.

Now consider these rectangles:

In <u>both</u> rectangles the longer side appears to be about double the shorter side. In other words, the ratio between the longer side and the shorter side of each rectangle is about 2:1 and the rectangles have the same shape. This is one way to illustrate that these rectangles are **similar**.

If any two figures are similar, the ratio of <u>any</u> **two corresponding sides** is <u>always</u> the same. Examine similar rectangles TUVS and WXYZ on the top of the next page.

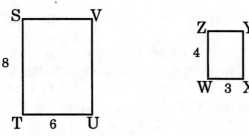

The ratio of side ST to side ZW is 8:4 or 2:1; the ratio of side TU to side WX is 6:3 or 2:1. Of course, as shown previously, <u>within</u> each figure, comparing length to width, we have equal ratios; that is,

8:6 equals 4:3.

In the triangles below,

 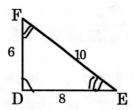

the ratio of the vertical side, CA, of the smaller triangle to the vertical side, FD, of the larger triangle, is 3:6 or 1:2. Comparing the horizontal sides and the third sides of the smaller and larger figures, 4:8 and 5:10, respectively, the same ratio, 1:2, holds. **Similar figures** have another important characteristic: **corresponding angles are equal**. As indicated on the triangles above, angles A and D are corresponding and equal in measure, angles B and E are corresponding and equal, and angles C and F are corresponding and equal. We name similar figures by naming their corresponding angles in the same order, that is, triangle *ABC* is similar to triangle *DEF*.

Note: **Equal corresponding angles *alone* do not assure similarity; corresponding sides must also be proportional.** Squares and other rectangles all contain 4 right angles, yet the figures may have different shapes.

We can use the ratios described for similar figures to solve some commonly encountered problems. For example, to estimate the height of a flagpole, the length of its shadow can be compared with the length of the shadow of a yardstick (known height of 3 feet) at the same time of day, and a proportion can be written and solved.

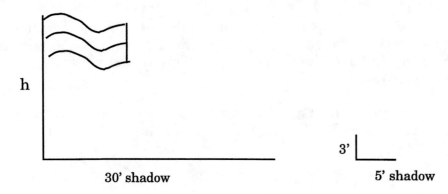

Let *h* represent the height of the flagpole.
Then

and
$$\frac{h \text{ feet}}{30 \text{ feet}} = \frac{3 \text{ feet}}{5 \text{ feet}}$$
$$5h = 30(3)$$
$$5h = 90$$
$$(1/5)5h = (1/5)90$$
$$1h = 18$$

Therefore, the flagpole is about 18 feet tall.

Sample Problems

1. If the figures shown below are similar, find the length of sides x and y.

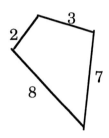
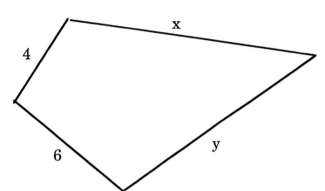

Solution:

$$\frac{2}{4} = \frac{8}{x} \quad \text{or} \quad \frac{2}{8} = \frac{4}{x}$$

$$2x = 32$$
$$x = 16$$

(Note the figures are in different positions and be careful to identify the positions which actually correspond.)

Side x is 16 units long.
And,

$$\frac{y}{7} = \frac{6}{3} \quad \text{or} \quad \frac{y}{6} = \frac{7}{3}$$

$$3y = 42$$
$$y = 14$$

Side y is 14 units in length.

2. For the similar triangles shown, name the pairs of corresponding sides and the pairs of corresponding angles.

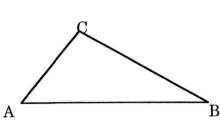
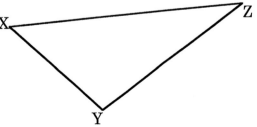

Solution: AC & XY   ∠A & ∠X
            BC & ZY   ∠C & ∠Y
            AB & XZ   ∠B & ∠Z

Problem Set 3.1

1. One student finished a test in 55 minutes. Another student finished in 1 hour and 5 minutes. Find the ratio of these two times.

2. Solve each of the following for the unknown.

   a) $\dfrac{3}{5} = \dfrac{x}{15}$ b) $\dfrac{2}{7} = \dfrac{x+3}{14}$ c) $\dfrac{1}{4} = \dfrac{x-3}{x}$

3. A teacher has 30 compositions to read. In 40 minutes, she read 5. Assuming that she continues to read at the same rate, how many minutes will it take her to read all 30 compositions? Solve by two methods: unit rate and by using a variable in a proportion. How many hours will it take?

4. An intermediate algebra class at Mountain Junior College has 18 men and 14 women.
   a) Give the ratio of men to women in the class.
   b) Give the ratio of women to men in the class.
   c) What fraction of the class is men? Women?

5. Josh is attending a university whose academic year consists of 2 semesters, each 16 weeks long (including exam week). He has budgeted $1120 for incidental expenses for the 32 weeks he's on campus. After 3 weeks in school, he has spent $114 for incidentals. Assuming he continues his spending at the same rate, will his budget for incidentals be adequate? If not, how much more will he need to complete the year?

6. If a college composition class of 30 students at Mountain Junior College is $3/5$ men, how many men are in the class? How many women are in this class?

7. Two gears are shown in the sketch. Gear A has 30 teeth and

   gear B has 15 teeth. Answer the following questions.

   a) How many revolutions and in what direction will gear B turn if A is turned 8 revolutions counterclockwise?
   b) What is the ratio of the gears?[1]

8. A group of college students plan to bicycle across Iowa, from Sioux City to Dubuque in early June on their 15-speed mountain bikes in 5 days. This distance is approximately 295 miles. Is this a reasonable goal? Justify your answer.

9. With respect to energy use, it is claimed in a train company advertisement that railroads are twice as efficient as cars and four times as efficient as jet airplanes in terms of travel. If the statement is valid, what is the relative efficiency of cars and jet airplanes?[2]

---

[1] *Sourcebook of Applications*, NCTM, 1980, page 37.
[2] Ibid, 44.

10. A painter has made light blue by mixing 5 parts white and 1 part blue paint. She has 2 liters of a darker blue which is one-half white and one-half blue. How many liters of white paint should be added to the darker blue to make it the lighter blue?

11. A machine weighing 35 tons is to be installed on a concrete base with an area of 4 square feet. The weight of the machine is to be uniformly distributed over the bed and the concrete is capable of supporting 125 pounds per square inch. Is the base strong enough to support the machine?[3] Justify your answer.

12. To estimate the height of a tree, a person 65 inches tall stands near the tree. The person's shadow is measured to be 15 inches. At the same time, the shadow of the tree is measured to be $7\frac{1}{2}$ feet, what is the approximate height of the tree, in feet?

13. To estimate the population of brook trout in a stream in northern Michigan, the Department of Natural Resources uses the following procedure. Biologists and technicians wade in the stream for a given distance, say $\frac{1}{2}$ mile, carrying electrodes that produce a current in the water, mildly stunning, or shocking, the fish. While the brook trout are stunned, they are captured, counted, a specific fin is clipped, and the fish are returned to the water. Other fish are merely released, not clipped. The next day, the researchers return to the same portion of the stream, again wading in the water with electrodes, and count the number of brook trout with fin clips and the total number of brook trout they find that day. Use this "mark, capture, and recapture method" to estimate the number of fish in one $\frac{1}{2}$ mile area of a stream where the DNR fin-clipped 125 brook trout the first day, and then counted 76 brook trout the next day with the specific fin-clip out of a total of 600 brook trout caught the second day.

14. In the similar figures ABC and EDC below, find the missing values, x and y. Note that side AB corresponds to side DE, side BC corresponds to side DC, and side AC corresponds to side EC. The corresponding angles are also shown.

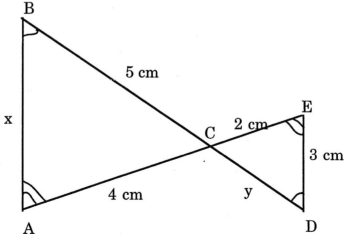

---

[3] Ibid, 50.

15. In a blueprint for a new house, $1/4$ inch represents 1 foot. If the dimensions of a ranch-style house on the blueprint are 7 inches by $12\frac{1}{2}$ inches, find the area of the first floor.

16. Surveyors determine the grade of a road by taking the ratio of the vertical change of a road bed for a given horizontal distance (sometimes called rise over run).

    Find the grade of the road pictured below.

    $2\frac{1}{5}$ ft. [diagram of a long thin wedge]
    50 ft.

17. In the recipe for hummingbird syrup (ratio of water to sugar, 3:1), find the amount of water and sugar needed to make <u>approximately</u> 8 cups of solution. (Note that the volume will not be exactly 8 cups because sugar dissolves in water.)

18. Solve using the rate-factor model.
    If a faucet drips 8 times in 20 seconds,

    a) how many drips are there in an hour?

    b) If it takes approximately 304 drips from this faucet to make $1/4$ cup of water, how many gallons of water are wasted in a 24-hour day?
    (one gallon = 16 cups)

19. On an Alaska state map, $4/5$ inches is approximately 120 miles. If the distance between Glennallen and Anchorage is $4\frac{1}{8}$ inches on the map, approximately how far apart are the cities, in miles?

20. If the ratio of gas to oil in a weed whip is 16:1, explain why you should not use 16 pints of gas to 1 gallon of oil in this weed whip.

Exploration Problems

1. One way to determine whether or not a person is at risk for developing heart disease is to calculate the ratio of the person's waist measurement to hip measurement. For men, if the ratio is greater than 1, the risk increases. For women, if the ratio is greater than $4/5$, the risk increases. Given data on the following people:

Name	Waist Measure	Hip Measure
Sam	33	33
George	34	36
Sue	34	36
Beth	25	32
Bob	38	36

Determine which of these people is at risk for heart disease. Of those at risk, determine their individual maximum waist measurement so that they would no longer be at risk.

2. Using horizontal and vertical lines, draw a rectangle of any size on grid paper and find the length and width. Find the perimeter and the area of this rectangle. If you were to draw another rectangle twice as wide and twice as long, what is the ratio of the perimeter of the larger rectangle to the perimeter of the first rectangle? What is the ratio of their areas?

3. To make an office handicap-accessible, a ramp must be constructed at the front door. The grade of this ramp cannot exceed $2/100$. Given a certain office door is $2\frac{1}{2}$ feet above street level, what is the minimum horizontal distance between the door and the street?

Solutions - Problem Set 3.1

1. 55:65 or 11:13

2. a) 9   b) 1   c) 4

3. 240 minutes or 240 ÷ 60 = 4 hours to read all compositions

4. a) 18:14 or 9:7
   b) 7:9
   c) $9/16$; $7/16$

5. Josh will not have enough money; he'll need $96 more.

6. 18 men; 12 women.

7. a) 16 clockwise;   b) A:B = 2:1

8. Opinions will vary. Riders must average 295 ÷ 5 or 59 miles per day.

9. Cars are twice as efficient as jets.

10. 4 liters

11. The base is strong enough because the machine will exert a pressure of about $121\frac{1}{2}$ pounds per square inch, <u>or</u>, the base will support 72000 pounds.

12. The tree is 390" or $32\frac{1}{2}$ ft. tall.

13. There are about 987 brook trout in the stream.

14. x = 6 units; y = $2\frac{1}{2}$ units

15. Area of the floor is 28' x 50' = 1400 sq. feet.

16. The grade is $11/250$.

17. Need approximately 2 cups of sugar and 6 cups water for 8 cups syrup.

18. a) 1440 drips per hour
    b) $28\frac{2}{5}$ cups or $1\frac{4}{5}$ gallons of water are wasted.

19. About $618\frac{3}{4}$ miles

20. 16:1 implies that the parts are the same. Using 16 pints of gas to 1 gallon of oil would not be appropriate, and, in fact, would probably ruin the weed whip.

Section 3.2 - Relationship Among Fractions, Decimals, and Percents

The relationship among common fractions, decimals, and percent is important, and we need to know how to interpret information given in any of these formats. Fortunately, our monetary system provides an easy basis for our initial work.

Think of some of the coins we use and compare their values to a dollar. A quarter is $1/4$ of a dollar, and its value can be represented in decimal form as $.25. Thus, we have the same quantity represented as a common fraction and as a decimal. Now recall that "per" means "divided by" and "cent" is derived from the Latin "centum" meaning one hundred. Therefore, **percent** means **divided by 100**. Since our dollar is worth 100 cents, a quarter is 25 cents or 25 hundredths or 25 percent of a dollar. Similarly, $1/2$ of a dollar is $.50 or $50/100$ or 50 percent of a whole dollar.

The coin analogy is of limited use, so let's look at some visual representations of the relationship among fractions, decimals, and percent.

In each of the 10 x 10 squares below there are 100 small squares, and each small inner square represents $1/100$ or 1 % of the area. **Keep that fact in mind as you study the figures shown**.

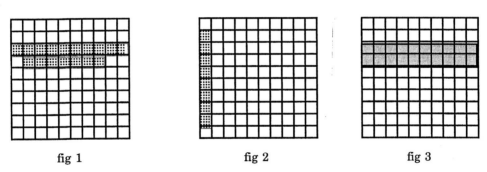

fig 1   fig 2   fig 3

In **fig 1**, 17 small squares are shaded, representing 17%. We write this as $17/100$ or, recalling the decimal place value system, as .17. (There will be more on place value later in this section.) **Fig 2** shows 8% or $8/100$ shaded (the <u>reduced</u> fraction is $2/25$); the equivalent decimal is .08. In **fig 3**, 20% is shaded. Notice that, although we could use $20/100$ as a fraction depicting this amount, we can also see that <u>2 of 10 rows</u> or <u>1 of 5 rows</u> has been shaded, a visual confirmation that $20/100 = 2/10 = 1/5$. The decimal value is written as .20 or as .2.

Sample Problems

1. a) Draw a 10 x 10 square and shade 40 squares.
   b) Then write the percent, a decimal value, and the reduced fraction value for the amount shaded.
   c) How many <u>rows</u> are there in the 10x10 square?
   d) How many of them have been shaded?
   e) What does this tell you about the relationship between .40 and .4?

151

Solutions: a)

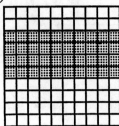

b) $^{40}/_{100}$ or 40% has been shaded. This is equivalent to .40. In fraction form we have $^{40}/_{100} = ^{2}/_{5}$.

c) There are 10 rows in the square.

d) 4 rows have been shaded, so $^{4}/_{10}$ or .4 is shaded.

e) .4 is equivalent to .40.

2. Draw a 10 x 10 square and shade $^{1}/_{2}$ of a small square. Write the amount shaded as a percent.

Solution:

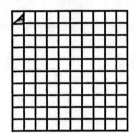

$^{1/2}/_{100}$ or $^{1}/_{2}$% or .5% has been shaded.

Note: Since each small square is $^{1}/_{100}$ of the outer square, we have shaded $^{1}/_{2}$ of $^{1}/_{100}$ or $^{1}/_{2}$ of 1% or $^{1}/_{200}$ of the original square. Try dividing 1 by 200 on your calculator. The result should be .005, the decimal equivalent of $^{1}/_{2}$%. The calculator offers an easy way to convert a common fraction to its decimal form.

3. If we were to shade $^{1}/_{3}$ of a 10 x 10 square, how many small squares would be shaded?

Solution: In this case, the answer is not a whole number of squares since $100 \div 3 = 33^{1}/_{3}$. Therefore, we would have to shade 33 whole squares, representing 33%, and then $^{1}/_{3}$ of another square, representing $^{1}/_{3}$%, so the percent equivalent of $^{1}/_{3}$ is $33^{1}/_{3}$ %.

Terminating and Repeating Decimals

Use your calculator to divide 1 by 3. The result, .33333..., is called a **repeating decimal**. Now change $2/3$ to its decimal equivalent using your calculator. Note whether your calculator shows .6666..., for the answer or whether it rounds off, displaying .66666667. Either is considered correct. The difference is the result of how different calculators are programmed: some round the digit in the last location of display and others leave the number as is. It is important to note both represent the decimal equivalent of $2/3$, .6666..., with the 6 repeating.

Since it is clumsy to write repeating digits, we customarily use a line segment, or bar, above the digit or block of digits that repeats: $.\overline{3}$ or $.\overline{6}$ for the two cases of repeating decimals we've encountered so far.

Change $1/11$ to its decimal form using your calculator. The result is again a repeating decimal, but in this case there are two digits that repeat. We write the decimal value $.\overline{09}$, with the bar over the pair of digits that repeat.

The decimals we dealt with earlier in this section, such as .25 and .08, had a finite number of digits to the right of the decimal point. This means that zeros could be added to the right of .4, for example, without changing the value of the decimal. (We can use equivalent fractions, $4/10 = 40/100$, etc., to check this.) Decimals of this type are called **terminating** decimals. As noted previously, converting a terminating decimal to a percent is generally not a difficult task.

Use a calculator again and find the decimal equivalent for $2/11$. The result is a repeating decimal, .181818... or $.\overline{18}$. Writing the percent equivalent for a repeating decimal is not as easy as writing one for a terminating decimal. However, the percent equivalent of a fraction such as $2/11$ can be found easily using a proportion. Since percent is the comparison of a value with 100, what we are looking for in each case is the numerator of a fraction equivalent to the given fraction but whose denominator is 100.

Let n represent the numerator we are seeking.
then
$$\frac{n}{100} = \frac{2}{11}$$

and $\qquad 11n = 200 \quad$ (using cross-products)

$$n = \frac{200}{11} \text{ or } 18\frac{2}{11}$$

so
$$\frac{2}{11} = 18\frac{2}{11}\%$$

Consider the fraction $1/6$. Changing $1/6$ to a decimal using a calculator gives the result, .166666... or $.1\overline{6}$.

Using the proportion method again, we can find the exact percent equivalent of $1/6$ as follows:

$$\frac{n}{100} = \frac{1}{6}$$

then

$$6n = 100$$

$$n = \frac{100}{6} \text{ or } 16\frac{2}{3}\%$$

so

$$\frac{1}{6} = 16\frac{2}{3}\%$$

<u>Sample Problems</u>

1. Find the decimal and exact percent equivalents for $3/7$.
   Solution:
   $$3 \div 7 = .428571428571... \text{ or } .\overline{428571}$$

   and, $$\frac{n}{100} = \frac{3}{7}$$

   $$7n = 300$$

   $$n = \frac{300}{7} \text{ or } 42\frac{6}{7}$$

   so
   $$\frac{3}{7} = 42\frac{6}{7}\%$$

2. Use a calculator to write decimal equivalents for the following fractions, then write corresponding percent values.

   a) $27/50$ b) $2/25$ c) $3/200$

   Solution: a) $27 \div 50 = .54$; since .54 equals 54 hundredths, it also equals 54%.
   b) $2 \div 25 = .08$ and 8 hundredths equals 8%.
   c) $3 \div 200 = 1.5 \div 100 = .015 = 1.5\%$

We can see from these examples that whenever the decimal equivalent of a fraction is a repeating decimal, the percent equivalent will contain a mixed number, the fraction portion of which is a <u>repeating decimal</u>.

Ordering Decimals

We know how to compare the values of common fractions. To compare the values of decimal fractions most easily, we'll arrange them vertically, lining up decimal points, and then compare digits in the corresponding place value slots, <u>left to right</u>. You may find it helpful to add zeros to the right of non-zero digits, or to write more of a repeating decimal, in order to assure that the decimal parts extend the same number of places to the right of the decimal point in all the numbers. For example, to order .707, .077, .$\bar{7}$, and .77, we can rewrite the decimals in a column, adding zeros when necessary until each decimal is carried out at least to the thousandths place.

$$.707$$
$$.077$$
$$.777...$$
$$.770$$

Comparing digits in the tenths, hundredths, and thousandths place, respectively, results in

$$.077 < .707 < .770 < .\bar{7}$$

If it is necessary to compare a common fraction with a decimal, first change one of the values to the other form, and then use the appropriate method of comparison.

<u>Sample Problem</u>

Write the following decimal fractions in order, smallest to largest. Use "<" between unequal values, "=" if two values are equivalent.

$$3/10, .43, .034, .3, .\overline{34}, 33\tfrac{1}{3}\%$$

Solution: .3000
.4300
.0340
.3000
.3434...
.3333...

Therefore: $.034 < 3/10 = .3 < 33\tfrac{1}{3}\% < .\overline{34} < .43$

Size Changes

Many copy machines permit us to enlarge or reduce our originals. On one model the range allowed is from 64% to 141%. If the text on an original document is 6 inches by 9 inches, a reduction to 67% (to approximately $2/3$ of the original size) would measure about $(2/3)6"$ and $(2/3)9"$, or 4 inches by 6 inches. Similarly, a size increase to 115% (or 1.15 times the original size) would result in a document measuring (1.15)6" by (1.15)9" or 6.9 inches by 10.35 inches.

Note that instead of using 115%, we used 1.15 in the calculation. **This is because percent is only a notation. We must use the fraction or decimal equivalent of a percent value when doing calculations.**

<u>Sample Problems</u>

1. A picture is to be enlarged by 25%.

   a) The enlargement will be what percent of the original picture?

b) If the picture is originally 9 inches by 12 inches, what will be the size of the enlargement?

   Solution:
   a) The original picture can be thought of as the "unit" or 100%. Therefore, the enlargement will be 100 + 25 or 125%.

   b) Using the decimal equivalent of 125%, 1.25, we can find the dimensions of the enlargement by (1.25)9 = 11.25 inches and (1.25)12 = 15 inches.
   Or, using a proportion with n as the length of one side of the enlargement we have,

   $$\frac{125}{100} = \frac{n}{9}$$ (in the proportion, the fraction equivalent of 125% is used)

   $$100n = 1125$$

   $$n = 11.25$$

   and

   $$\frac{125}{100} = \frac{n}{12}$$

   $$100n = 1500$$

   $$n = 15$$

   Note that using proportions may be easier than taking 25% of 9, for example, and adding the result to 9.

2. A jacket priced at n dollars is marked 25% off. What is the sale price of the jacket in terms of n?

   Solution:
   $$n - .25n = (1 - .25)n = .75n$$

Estimation and Mental Arithmetic

The relationship among fractions, decimals and percents can be extremely useful in mental computation work and in estimation.

One of the most useful estimation and mental arithmetic tools is the 10% method. Since 10% = $\frac{10}{100}$ = $\frac{1}{10}$ = .1 and multiplying any number by .1 moves the decimal point one place to the left, calculating 10% of any number becomes an easy task.

For example, 10% of 32 = 3.2 and 10% of $24.20 = $2.42.

Also, since 5% is $\frac{1}{2}$(10%) and 20% = 2(10%), we can easily find 5% or 20% of a number by using the appropriate multiplier and the 10% value.

From the example above, since 10% of 32 = 3.2, then 5% of 32 would be

$$\tfrac{1}{2}(10\% \text{ of } 32) = \tfrac{1}{2}(3.2) = 1.6.$$

Also,
$$20\% \text{ of } \$24.20 = 2(10\% \text{ of } \$24.20) = 2(2.42) = \$4.84.$$

To estimate the dollars and cents representing a 15% tip on a restaurant bill, we use this method, the relationship between 5% and 10% as described above, and the distributive property. If the numbers are difficult to work with, we can always round them off before doing the computation.

For example, if a restaurant bill (before tax) is $24.89, it would be reasonable to use $25 as our total. Calculating the amount of a 15% tip,

$$(10\% + 5\%)\$25 = 10\% \cdot 25 + 5\% \cdot 25$$
$$= .1 \cdot 25 + \tfrac{1}{2}(.1 \cdot 25)$$
$$= \$2.50 + \tfrac{1}{2}(2.50)$$
$$= \$2.50 + \$1.25 = \$3.75$$

Since we rounded the actual bill up to $25, this amount really represents a little more than a 15% tip!

Similarly, if a sweater originally selling for $21.99 was marked down 20%, we could estimate the amount of savings on this sweater by calculating

$$2(10\% \text{ of } \$22) = 2(.1 \cdot 22) = 2(\$2.20) = \$4.40$$

These equivalencies can be used in a slightly different way to estimate the amount of discount offered on a pair of slacks that are 25% off. Here it is easier to think of 25% as $\tfrac{1}{4}$, especially if the original price is a number compatible with 4. If these slacks were originally priced at $43.99, our estimation of the amount of discount becomes

$$\tfrac{1}{4} \text{ of } \$44 = \$11 \text{ since we can use the shortcut for multiplication.}$$

Some well-used fraction and percent equivalencies are:

$\tfrac{1}{2} = 50\%$	$\tfrac{1}{4} = 25\%$	$\tfrac{3}{4} = 75\%$
$\tfrac{1}{8} = 12.5\%$	$\tfrac{1}{3} = 33\tfrac{1}{3}\%$	$\tfrac{2}{3} = 66\tfrac{2}{3}\%$
$\tfrac{1}{5} = 20\%$	$\tfrac{2}{5} = 2(\tfrac{1}{5} = 20\%) = 40\%$	$\tfrac{3}{5} = 3(20\%) = 60\%$

Sample Problems

1. If a bottle of one brand of fruit juice contains 15 ounces, and a new bottle is marketed with a label that reads 20% more, how much juice is the customer really getting in the new bottle?

Solution:

Using $\frac{1}{5}$ instead of 20%, $\frac{1}{5}$ of 15 = 3 ounces (not too difficult to do in your head!) and the amount of fruit juice in this new bottle is

$$15 + 3 = 18 \text{ ounces.}$$

2. If a picture is to be enlarged to 150%, and its original size is 10" by 12", what will the new size be?
Solution:
$$100\% + 50\% = 1 + \tfrac{1}{2} = \tfrac{3}{2},$$

so a picture that is originally 10" by 12" will be enlarged to

$$\tfrac{3}{2}(10") \text{ by } \tfrac{3}{2}(12") \text{ or } 15" \text{ by } 18".$$

3. Which is larger: 20% of 32 or 10% of 61?

Solution:
  10% of 32 = 3.2 so
  20%(32) = 2(3.2) = 6.4. (note that we calculated 10 % first)
 In comparison,
  10% of 61 = 6.1
   so 20% of 32 is larger.

4. Many states have sales tax on nonedible (to humans) grocery items. <u>Estimate</u> the amount of tax in any state where the sales tax is 6% on the total purchase of the following items.

   1 package of paper towels for $.99;
   5-pound box of dog biscuits for $2.89;
   2 packages of trash bags at $1.49 a package.

Solution:
 Estimation of total: $1 + $3 + 2(1.50) = $1 + $3 + $3 = $7 and since <u>1%</u> of <u>7</u> = <u>.07</u>, then 6% = 6(1%)
 and 6(.07) = $.42 is the approximate amount of tax on these items.

5. <u>Estimate</u> each of the following:

a) the product of 6.23 and 4.12   b) the quotient of 36.3 and 2.2

   Solutions: a) Multiply 6 by 4 and we get 24.
        b) Divide 36 by 2 and the estimate is 18.

Problem Set 3.2

1. Shade 70% of a 100-grid. Explain how you can use the grid to show that 70% = .70 = .7.

2. Using a 10 x 10 square, illustrate the equivalence between 60% and $3/5$.

3. For problems involving these circles, remember that $C = 2\pi r$ and $A = \pi r^2$. Leave the ratios in "pi" form initially and then simplify as appropriate.

   a) Write the ratio of the radius of the smaller circle to that of the larger circle.
   b) Write the ratio of the circumference of the smaller circle to that of the larger circle.
   c) Write the ratio of the area of the smaller circle to that of the larger circle.

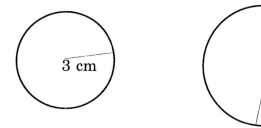

4. In a college classroom of 36 students there are 16 females. What percent of the class is male?

5. Find the decimal and percent equivalents for each of the following fractions. (Use proportions as needed to find the percents.)

   a) $5/4$ b) $4/9$ c) $1/8$ d) $11/3$ e) $13/10$ f) $1/24$

6. Put the expressions in order from smallest to largest. Use "<" between unequal values, "=" between equivalent expressions.
   a) $2/3$, $.\overline{6}$, .66, .07, .667, .6,
   b) $7/6$, 1.17, $-7/6$, −1.17, 1.16, 0, 116%

7. U. S. Navy specifications for phosphorus bronze call for 85% copper by weight, 7% tin, .06% iron, .2% lead, .3% phosphorus, and the remainder zinc. The zinc content is what percent of the total?

8. If an American made car given normal use depreciates 25% of its original cost in the first year and 15% of the remaining value in the second year, which of the following cars is more reasonably priced?

   a) A one-year-old sedan priced at $12517.50 (price new: $16687.50)
   b) A two-year-old sedan priced at $9900 (price new: $15735)

9. The sides of a triangle are 5 inches, 6 inches, and 8 inches, respectively. What will be the length of each side if the triangle is enlarged on a copy machine set for 110%?

10. A photograph measuring 9 cm by 12 cm was reduced by 10%. What are the dimensions of the smaller photo?

11. A drawing 10 inches in height has been enlarged to 12.5 inches. What size setting was used on a copy machine to produce the enlargement?

12. The Italian dinner was marvelous and the waitperson attentive. The bill was $21.20 including tax. You are feeling generous. Explain how you would use estimation to calculate mentally a tip of approximately 20%.

13. Estimate the better buy if an 18-ounce box of old-fashioned oatmeal costs $1.59 while a 42-ounce box costs $2.79. Check your conclusion with your calculator.

14. Given the rectangle in the coordinate system below, increase each of the y-coordinates by 100%. Draw the new rectangle on the coordinate system. Label the vertices. How does the size of the new rectangle compare with the size of the original?

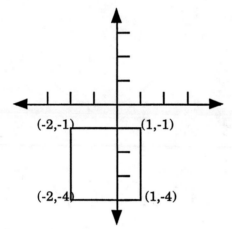

15. In each of the following, estimate the answer:

   a) $.3 \bullet 61.2$ b) $588 \div .5$ c) $88.14 + 62.3 - 15.84$

16. A local clothing store buys a certain brand of jeans from a distributor for x dollars a pair. If the store then marks each pair up 200%, find the cost of the jeans to the consumer in terms of x.

17. Compare 48% of 25 and and 25% of 48. Which is easier to compute? Justify your answer.

18. If an item is marked 20% off, what percentage of the original price are you actually paying?

Exploration Problems

1. Choose several common fractions, convert to decimal form using your calculator, and record the results. What appears to determine whether the decimal repeats or terminates? Can you find a pattern or rule which could be used to determine whether a decimal will terminate or repeat without doing the division? Choose more examples if necessary in order to demonstrate your conclusion.

2. A photograph is enlarged on a copy machine set for 120%. By how much is the area of the photograph increased?

3. Is there a limit to the length of a repeating string for a repeating decimal? Explain your answer.

4. Explain how you might estimate the amount of cash to take to the store if you wanted to buy 5 bags of chips at $1.49 each and 8 bottles of pop at 99¢ each including deposit and were returning 9 bottles for 10¢ refunds.

Solutions - Problem Set 3.2

1.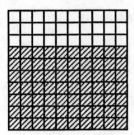

   70 squares out of 100 squares are shaded, representing $70/100 = .70$. But 7 rows out of 10 rows are also shaded and so $.70 = 7/10 = .7$.

2.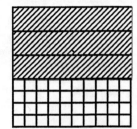

   60 squares of 100 are shaded; 3 double rows of 5 double rows are shaded.

3. a) 3:4   b) $2\pi(3) : 2\pi(4) = 3:4$ c) $\pi(3)^2 : \pi(4)^2 = 9:16$

4. $55\tfrac{5}{9}\%$

5. Use a calculator to divide for decimal equivalents.

   a) 1.25; 125%   b) .444...; $44\tfrac{4}{9}\%$ c) .125; 12.5%   d) 3.666...; $366\tfrac{2}{3}\%$

   e) 1.3; 130%   f) .041666...; $4\tfrac{1}{6}\%$

6. a) $.07 < .6 < .66 < .\overline{6} = 2/3 < .667$ b) $-1.17 < -7/6 < 0 < 1.16 = 116\% < 7/6 < 1.17$

7. 100% − other amounts = 7.44% zinc.

8. a) is worth $12515.63; b) is worth $10031.66. The two-year-old car is the better buy.

9. 5.5", 6.6", 8.8"

10. 8.1 cm; 10.8 cm

11. 125%

12. 10% is a little over $2, so 20% would be twice that or a little over $4.

13. Answers will vary, but $1.60 ÷ 20 > $2.80 ÷ 40 so the larger box is a better buy.

14. New vertices are at (–2,–2), (–2,–8), (1,–2), (1,–8). Rectangle is twice as high, so area is doubled.

15. Answers may vary but reasonable answers are a) 18, b) 1200, and c) 135.

16. $x + 2x = 3x$ (since 200% = 2). Thus, each pair of jeans will cost the consumer three times what the local store paid for them.

17. Both equal 12. Using $\frac{1}{4}$ in place of 25%, it is easier to take $\frac{1}{4}$ of 48. The reason they are the same value is shown below.
$$48 \cdot .25 = \frac{48}{1} \cdot \frac{25}{100} = \frac{48}{100} \cdot \frac{25}{1} = .48 \cdot 25$$

18. 80%

Section 3.3 - Rational Numbers in Problem Solving

"Unemployment has risen two tenths of a percent over the last 6 weeks." "Precipitation is 12% below average for the first 9 months of the year." "18 month CD's are now paying 5.76% annual yield." We've all seen or heard this kind of information. Percent notation is used so commonly that all of us need to understand information and calculation involving percent.

We have looked at the relationships which exist among fractions, decimals, and percent, have used percents in estimation problems and now we need to extend our study to additional practical situations that frequently involve percents. For example, a retail price will be based upon the wholesale price increased by a markup of a given percent; a worker negotiating for a raise is likely to speak in terms of a percent increase in salary; a college president facing a budget deficit may well ask for wage concessions in the form of a percent reduction.

More specifically, consider an appliance store that has announced a 13% price reduction on all items for one day only in honor of 13 years in business. A refrigerator originally selling for $879 must be marked down. Recall that "of" is frequently interpreted to mean multiplication in a mathematical sense, so the employee figuring the amount of discount must find 13% of $879, or (13%)879.

This case illustrates the commonly used formula:

a percent times the whole equals the part.

Utilizing the decimal equivalent for 13%:

$$(.13)879 = \$114.27$$

We have calculated the amount we <u>save</u>. To find the <u>sale price</u>, (what we'll have to <u>pay</u>), we must subtract the discount from the original price as follows:

$$\$879 - \$114.27 = \$764.73.$$

The problem can be approached another way. Usually the ads tell us how much we are going to <u>save</u>. Rarely do they mention how much we are going to <u>spend</u>. But if we think of the original price as representing 100%, then subtracting a 13% discount would suggest we will have to pay 87% of the original ticketed price.

$$100\% - 13\% = 87\%$$

Let's see whether that idea works. Calculating 87% of $879,

$$(.87)879 = \$764.73, \text{ we get the same sale price.}$$

We can extend this approach to a general case for which we can write an appropriate algebraic equation. If we let n represent the original price, the sale price, s, after the 13% discount has been taken, can be found by

$$s = n - 13\% \text{ of } n$$
or
$$s = 1n - .13n$$

$$\text{and} \qquad s = (1 - .13)n \quad \text{or} \quad s = .87n$$

On the right side, we've collected like terms using the distributive property, a technique employed before. This method will work for finding the sale price when we're given any percent reduction.

As another example, Cathy and Andy Wilson have found that they can save a considerable amount on the cost of homeowners' insurance if they only insure their house for $60,000, which is 75% of its current replacement or market value. How much is their home worth? "Of" means multiplication here, but since the current replacement value is missing, we need to take a different approach. One of the easiest is to set up a proportion.

Think of the total value of the house as being 100%. We can set up a proportion to show that 75% compared to 100% will equal $60000 compared to the market value.

Let M represent the market value.
Then
$$\frac{75}{100} = \frac{60000}{M}$$
$$75M = 6000000$$
$$M = 80000$$

The Wilsons' house is worth $80000.

An additional example which can be solved easily with a proportion is the following: An elementary school saw its enrollment drop from 473 to 298 with the closing of a nearby air force base. What percent of the student body has been lost?

First, we need to ask ourselves how many students have been lost.

$$473 - 298 = 175 \text{ students lost}$$

Now we can set up a proportion in which the original student population represents 100% and the number of students lost corresponds to the unknown percent of loss. Thus:

If S represents the percent of students lost to the school, then
$$\frac{S}{100} = \frac{175}{473}$$
$$473S = 17500$$
$$S = 36.99 \text{ or about } 37$$

Therefore the school lost about 37% of its enrollment.

Note: The proportion method of solving percent problems has an advantage over other methods because there is no need to shift the decimal point when given a percent or when solving for a missing percent.

Sample Problems

1. The winter population of a town is 963. As soon as Memorial Day arrives, the summer folks return, and the population swells by $33\frac{1}{3}\%$. What's the summer population?

   Solution: We first need the <u>amount</u> of increase which is $33\frac{1}{3}\%$ of 963. Rather than using the decimal equivalent of $33\frac{1}{3}\%$, let's use the fraction equivalent, $\frac{1}{3}$.

   $$(\tfrac{1}{3})963 = 321 \text{ summer people}$$

   Therefore, the summer population is $963 + 321 = 1284$ people.
   Finally, since $1 + \frac{1}{3} = \frac{4}{3}$, we could calculate
   $$\tfrac{4}{3}(963) \text{ which also gives us } 1284.$$

2. A bicycle originally priced at $162 has been reduced to $139. What is the percent reduction?

   Solution: A proportion can be used to solve this problem. We need to remember that the percent reduction requires a comparison of the <u>amount</u> of the reduction and the <u>original cost</u>.
   The amount of the reduction is $\$162 - \$139 = 23$

   If P is the percent reduction,

   then
   $$\frac{P}{100} = \frac{23}{162}$$

   $$162P = 2300$$
   $$P = 14.2$$

   Therefore, the reduction is about 14.2%.

3. Ms. Watson has been granted a salary increase of 7% and is now earning $45207.50 per year. What was her salary before the raise?

   Solution: There are occasions when a percent greater than 100 makes a great deal of sense, and this problem illustrates that fact. (Certainly a team can't win more than 100% of the games it plays, but a salary after a raise is more than 100% of the previous salary.) An equation similar to the one used for the discount problem or a proportion may be used. Compare the two solutions and choose the approach you prefer. Either is correct.
   Let S represent the previous salary.

   $$\text{Then } S + .07S = 45{,}207.50$$
   $$1.07S = 45{,}207.50$$
   $$S = 42250$$
   OR

166

Since the old salary was 100% and the new one is 7% greater, the new salary can be represented by 107%, and:

$$\frac{107}{100} = \frac{45207.50}{S}$$

$$107S = 4520750$$

$$S = 42250$$

Ms. Watson's salary was $42,250 before her raise.

Applications in Geometry

In section 1.5, we did much work with perimeter and area of geometric figures with integer dimensions. We will now extend this work and look at some practical applications involving rational numbers.

A room with dimensions $16\frac{1}{4}$' by $20\frac{1}{2}$' has a perimeter of

$$P = 2\left(16\frac{1}{4}\right) + 2\left(20\frac{1}{2}\right)$$

$$= 2\left(16 + \frac{1}{4}\right) + 2\left(20 + \frac{1}{2}\right)$$

$$= 32 + \frac{1}{2} + 40 + 1$$

$$= 73\frac{1}{2} \text{ feet}$$

(Use decimal equivalents, key this into your calculator and generate the answer of 73.5 as a check.)

The area of the floor of this room is:

$$A = \left(16\frac{1}{4}\right)\left(20\frac{1}{2}\right) \text{ or } (16.25)(20.5))$$

$$= 333\frac{1}{8} \text{ or } 333.125 \text{ sq. ft.}$$

To carpet this room, we would have to buy $333\frac{1}{8}$ sq. ft. of carpeting. However, carpeting is usually sold by the square yard. To find the amount and cost of carpeting this room, we need to find a "conversion factor" to change the square feet to square yards. Using the diagram below may help to picture the process.

Since 3 feet = 1 yard, the diagram above represents 3 ft. • 3 ft. = 9 sq. feet or 1 yd ^{2}.

Thus, $\dfrac{333.125 \text{ sq. ft.}}{9 \text{ sq. ft}} = 37.01$ sq. yards of carpeting

The same conversion can be done using a proportion as follows:

Letting x represent the amount of carpet required,

then $\qquad \dfrac{333.125 \text{ sq. ft}}{x} = \dfrac{9 \text{ sq. ft.}}{1 \text{ sq. yd.}}$

$$9x = 333.125$$
$$x = 37.01 \text{ sq. yds.}$$

If the carpeting is priced at $18.99 a square yard, the total cost of the carpeting alone, excluding labor, pad, etc., (rounded to the nearest cent) is:

$$37.01 \cdot \$18.99 = \$702.82$$

Sample Problems

1. A room is $12\frac{1}{2}'$ x 16' and 8' high. There is one 4' x 3' window on the south wall of the room and one 3' x 7' doorway on the north wall.

   a) Find the total amount of wall and ceiling space (in square feet) that will require paint in this room.

   b) If one gallon of paint covers approximately 400 square feet, find the amount of paint needed to coat all the walls and the ceiling of this room.

   c) If a gallon of paint costs $16.49, find the cost of painting this room.

   Solutions:

   a) ceiling: $12.5 \cdot 16 = \mathbf{200}$ sq. ft.
      walls: $2(12.5 \cdot 8) = \mathbf{200}$ sq. ft. and $2(16 \cdot 8) = \mathbf{256}$ sq. ft.
      window area: $4 \cdot 3 = \mathbf{12}$ sq. ft.
      doorway: $3 \cdot 7 = \mathbf{21}$ sq. ft.
      $\qquad 200 + 200 + 256 - 12 - 21 = 623$ sq. ft. needs painting.

   b) $623 \div 400 = 1.56$ gallons, so we'll need 2 gallons.

   c) $\$16.49 \cdot 2 = \$32.98$ for paint.

Applications and Their Graphs

In chapter one we plotted points on the Cartesian coordinate system and looked at geometric figures and their locations, positions, and sizes. Now we will consider other information which can be pictured on a graph, frequently utilizing only the first quadrant. Remember that all values are *positive* in the first quadrant. This fact is important in many practical applications. For example, the graph below shows the salary a worker would earn for various lengths of time worked, at a rate of $6.25 per hour. The amount earned will, of course, increase as the number of hours worked increases. Notice that the number of hours worked is shown along the x-axis (horizontal) while the amount earned is shown along the y-axis. We sometimes say the amount earned *depends upon* or *is a function of* the number of hours worked.

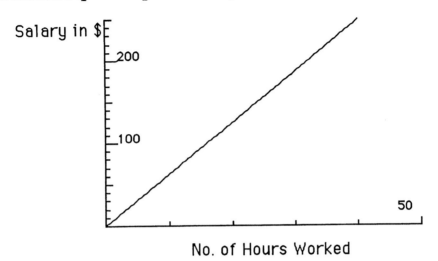

To determine *from the graph* the amount earned, for example, for 32 hours of work, carefully draw a vertical line from "32" on the x-axis up to the graphed line and then from that point on the line draw a horizontal line to the y-axis. (See graph below.) You can read the "pay" from the y-axis, or at least obtain a reasonable estimate.
It appears the pay for 32 hours is $200.

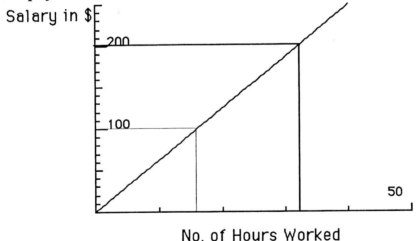

Likewise, if you know how much someone earned, you can start from that amount on the y-axis, move horizontally to the graphed line, then vertically down to the x-axis and determine the number of hours worked. Note that, from the graph above, a paycheck of $100 means someone worked about 16 hours.

We can write an algebraic equation which represents the graph drawn above. Let S represent the salary earned and h the number of hours worked. Then, since a worker receives $6.25 for each hour worked, we can write the following equation:

$$S = 6.25h$$

Sample Problems

1. The graph below shows the relationship between the original price of an item at a hardware store and the end of the season 40% off sale price.

   a) From the graph, find the sale price for an item which originally cost $50.
   b) From the graph, *estimate* the original price of an item on sale for $55.
   c) Using "depends upon" or "function of" terminology, tell what the graph illustrates.
   d) Write the formula for the sale price in terms of the original price.

   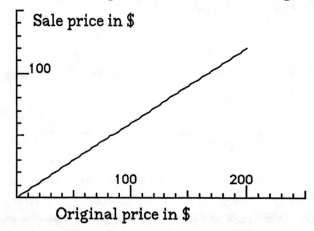

   Solution:

   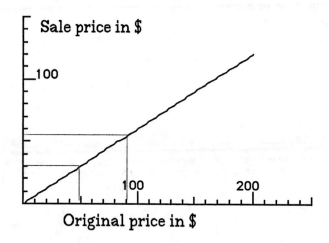

   a) $30
   b) approximately $90
   c) The sale price depends upon or is a function of the original price.
   d) If S represents the sale price and P represents the original price, $S = .6P$

2. The graph below shows how the number of hours spent driving between two parks varies depending upon the average speed of the car making the trip.

   a) At an average speed of 35 miles per hour (lots of sightseeing time taken), how long will the trip take?

   b) What speed must be averaged in order to make the trip in 8 hours?

c) If the graph is extended the line would meet the x-axis. Should it be continued that far? Why or why not?

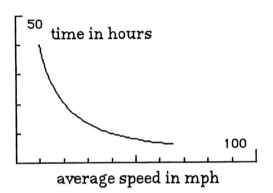

Solution: a) approximately 11 hours
            b) approximately 50 mph
            c) The graph should not be continued because the time would then be 0 hours which does not make sense. Also, legal speed limit should be remembered!

Problem Set 3.3

1. Find each of the following. Use fraction equivalents as appropriate to make the computations easier.

   a) 25% of 60 b) 130% of 20 c) 4 is what percent of 6?

2. One week in January a midwestern college closed for 2 days due to severe cold weather and wind chill. What percent of the work week was lost?

3. The regular Highway 41 route from Ahmeek to the Cliffview site in Keweenaw County, MI is 7.4 miles. Cliff Drive, a shorter route, is approximately 6.8 miles long. What percent of the distance is saved by taking Cliff Drive?

4. U. S. Navy specifications for phosphorus bronze call for 85% copper by weight, 7% tin, .06% iron, .2% lead, .3% phosphorus, and the remainder zinc. How many pounds of each element are needed to make a ton (2000 pounds) of bronze?[1]

5. A car manufacturer recommends that during the summer a car radiator should contain 60% coolant. If the radiator holds a total of 5 gallons, how much coolant should be used? How much water?

6. Beth is working in an upscale clothing store where she earns $200 base salary per week plus a 15% commission on all sales. What will be her before tax earnings for a 2-week period in which she sold merchandise worth $6000?

7. An item is marked 20% off. It is now selling for $62 and this includes the 20% discount. Find the original price.

8. To determine retail price, a merchant marks up all clothing 100% of its wholesale price when it arrives at the store. At the end of the season unsold items are marked down 40%. If the wholesale price of a summer blazer is $75,
   a)   What is the retail price?
   b)   What is the end-of-season sale price?
   c)   The final sale price is what percent of the wholesale price?

9. An "extra-strong for the UP" antifreeze solution is 85% antifreeze, 15% water. If your car radiator requires 4 gallons of solution, how much pure antifreeze does it hold? How much water?

10. Given a choice between closing down a mine or taking a pay cut, miners agree to a one-year reduction in salary of $1\frac{1}{2}$%. If Max earns $37,577.75 per year after the pay cut, how much was he earning before?

11. The sale price for a pair of shoes is $56. If the shoes were originally marked $72, find the percent of discount using a proportion.

---
[1] *A Sourcebook of Applications of School Mathematics,* NCTM, 1980, page 55.

12. In 1985 the number of AIDS cases reported in a midwestern city was 192. Of these, 12 of the victims were children 17 or younger, 141 were adult males, and 39 were adult females.

    a) What percent of the victims were children?
    b) 5 years after the 1985 report, new data showed 44 children, 210 adult males, and 66 adult females out of 320 victims. Has the percentage of male AIDS victims increased or decreased? By how much?

13. In a Michigan store you saw a shirt you liked for $28.75 plus tax. Michigan sales tax is 6%. Your son Bob saw the same shirt selling for $27.99 in a Chicago store, but there the sales tax is 8.75 %. Where is the shirt the better buy?

14. A page contains printing measuring 12 cm by 15 cm. It was produced by reduction on a copy machine set for 80%. What were the original dimensions of the printed area?

15. After typing a birthday greeting to your 95 year old aunt, you enlarge the message for easier reading. If the printed area is 4 inches in width and 5 inches in height, what copy machine enlargement should you choose to make the printed area $6\frac{1}{2}$ inches high?

16. Baseball batting averages are calculated much like percentages, but the ratio is formed based on a comparison with 1000 instead of 100. Thus a lifetime average of .325 means the batter got approximately 325 hits for every 1000 times at bat. If a batter gets 37 hits in 141 times, what is his batting average? How many hits will he need to get in his next 50 trips to the plate in order to maintain this average?

17. Jim's garden measures 20 yards by 15 yards. At the garden supply store he finds organic fertilizer labeled "100 square feet per bag." How many bags will Jim need? If he wants to keep the deer out of his garden this year, how much will he have to spend for fencing which is on sale at $19.95 per 100 feet?

18. The following graph shows the circumference of a circle as a function of the length of the radius.
    a) From the graph, approximate the circumference of a circle with a radius of 5 cm.
    b) From the graph, approximate the radius of a circle having a circumference of 200 cm.

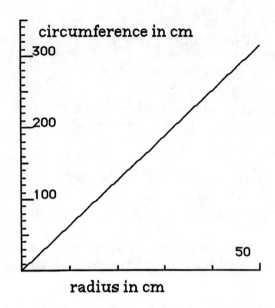

19. The graph below shows the relationship between the daily sales at a small concession stand and the amount of profit.
    a) How much must be sold to realize a $20 profit?
    b) Explain why this graph is drawn using more than just the first quadrant?

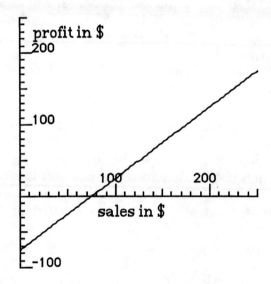

20. According to the U.S. Census Bureau, in 1980, out of a total population of 226,545,805, there were 22,599,676 persons 65 or older. In 1990, the total population was 248,709,843, and 31,241,831 were 65 or older. If the 65 and older group continues to grow at the same rate, predict the percent of the total population that will be in this age group in the year 2000.

Exploration Problems

1. Kris, a Bantam hockey player from the Upper Peninsula of Michigan is traveling with his team to Canada for a hockey tournament. At the border one of the chaperones asked Kris to exchange $750 of US currency into Canadian money. Kris received $959.25.
   a) What is the US to Canadian exchange rate in %?
   b) Upon return, Kris converted the remaining $119 Canadian money back into US funds, receiving $86.87. Find the Canadian to US exchange rate in %.
   c) At these rates, if a traveler were to exchange $100 US money for Canadian funds and then immediately convert the money received back into US currency, how much money would he receive? Explain the transaction.

2. If the baseball player in problem 16 hopes to improve his average to .275 by the time he has been to the plate 200 times in the same season, how many hits will he need to get? Would it be possible for him to improve his average to more than .300 by then? What's the best batting average he could attain?

3. A few years ago cement truck drivers struck for 46 days. Assuming these drivers made $13.50 an hour and worked 260 eight-hour days per year before the strike, what percentage increase is needed in yearly income to make up for the lost wages within one year after the end of the strike?[2]

---
[2] *A Sourcebook of Applications*, NCTM, 1980, page 85.

Solutions - Problem Set 3.3

1. a) 15   b) 26   c) $66\frac{2}{3}\%$

2. $\frac{2}{5} = 40\%$

3. approx. 8.1%

4. 1700 lbs., 140 lbs., 1.2 lbs., 4 lbs., 6 lbs., and 148.8 lbs respectively.

5. 3 gallons of coolant; 2 gallons of water

6. $1300

7. $77.50

8. a) $150
   b) $90
   c) 120%

9. 3.4 gallons of antifreeze, .6 gallons of water

10. $38,150

11. 22.2%

12. a) 6.25%
    b) decreased by 7.8%

13. It's a better buy in Chicago.

14. 15 cm x 18.75 cm

15. 130%

16. .262 He will need to get 13 hits.

17. 27 bags. He'll probably have to buy 3 rolls for $59.85.

18. a) between 31 and 32 cm
    b) about 32 cm

19. a) $90 - $95
    b) Some sales must be made to cover expenses before a profit can be realized. Thus, there would be a loss shown until the expenses are covered, and a loss is represented by a negative value on the "profit" or y-axis.

20. About 15%.

Chapter Three Review

1. Use a 100-grid or draw a diagram to show that the area occupied by 20% is the same as the area occupied by $\frac{1}{5}$.

2. <u>Estimate</u> each of the following:

    a) the cost of 10.5 gallons of gasoline at $2.12 a gallon

    b) the amount of sales tax at a rate of 6% on a pair of shoes selling for $44.99

3. Complete the chart below.

Decimal	Fraction	Percent
.32	____	____
____	$1\frac{2}{3}$	____
____	____	1.5%

4. Six cans of brand A cat food were on sale, 6 cans for $1.89. Brand B of cat food was on sale, 5 cans for $1.49. Which brand is the better buy?

5. Solve for the unknown in each of the following.

    a) $\frac{3}{4} = \frac{x}{5}$ b) $\frac{3}{2} = \frac{x+1}{5}$ c) $\frac{4}{x-5} = \frac{1}{2}$

6. A car is traveling at an average speed of 52 miles per hour. At this rate, how long will it take to make the trip from Houghton, MI to Marquette, MI, a distance of approximately 100 miles?

7. In one math classroom, the ratio of males to females is 18:14. If this ratio is representative of the entire school population and there are a total of 7,852 students, find the number of males and females on this campus. (Round each calculation to the nearest whole person.)

8. The wholesale price of a guitar is x dollars. If Maki's Music Store marks up the price of the guitar 125%, write the expression for the selling price of the guitar in terms of x.

9. Translate each of the following into a mathematical sentence and solve for the variable. Write your answer in a complete sentence.

    a) 22% of a number is 88. Find the number.

    b) A tie was originally selling for $19.99 but was marked down to $16.99. What percent decrease in price (to the nearest tenth of a percent) is $16.99?

10. To fit into a folder a 9" x 12" drawing needed to be reduced in size by 25%. Find the new size of the picture after the reduction.

11. The following diagram represents cans placed in a box.

    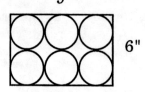

    If the area of the bottom of each can be found by the formula $A = 3.14\, r^2$, where r represents the radius of each can as shown on the diagram, find the percent of the area of the bottom of the box occupied by the bottom of the cans. What percent of the space is wasted?

12. Triangle ABE below is similar to triangle ACD where side AB corresponds to side AC, AE corresponds to AD and BE corresponds to CD. Use the lengths given on the diagram (in inches) to find the length of AB and the length of AD.

    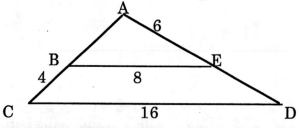

13. If R is a rectangle pictured as

    then draw a diagram representing

        a) 150% of R   b) 75% of R

14. A television set has been marked down to $299. This price includes a 25% discount. Find the original price of the television set.

15. Write a real-life word problem that involves a percent mark-up.

16. On the advice of a financial advisor, a teacher invested $5000 in a high-stake account. The teacher made no withdrawals or deposits in that account and in five years, the amount had increased to $8,200. What was the average annual percent return on this investment?

17. The diagram on the top of the next page represents a grassy area 7 feet by 10 feet surrounded by a uniform walk 3 feet wide.

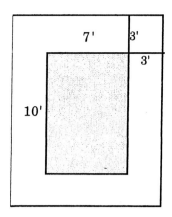

a) Find the outside perimeter of this grassy area and walk.

b) Find the area occupied by the uniform walk.

18. Draw a graph to show the relationship between pounds of birdseed sold (at $.69 per pound) and the cost of the birdseed, excluding tax.

Exploration Problems

1. The Greeks applied the Golden Ratio to much of their architecture. The rectangular front of the Parthenon is an example of this ratio. The rectangle is drawn such that if a square with sides equal to the width is cut off from the rectangle, then the remaining rectangle will be similar to the original. Write a proportion relating the dimensions of the new rectangle to those of the original.

2. Use the Internet to search for prices of various items such as cars, heating oil, bananas, etc., in 1960 and then again in 1999. Find the percent increase (or decrease) in the prices of these items in this time span.

Solutions - Chapter Three Review Problems

1. 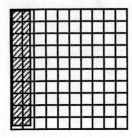 $20/100$ squares or 1 of every 5 double columns is shaded.
   Thus, $20\% = 1/5$.

2. a) $10(1 + .12) = 10 + 1.20 = \$11.20$ for the gasoline
   b) 1% of 45 = .45 so 6% = 6(.45) = \$2.70 tax

3.
Dec.	Frac.	Percent
.32	$32/100$	32%
$1.\overline{6}$	$1\,2/3$	$166\,2/3\,\%$
.015	$15/1000$	1.5%

4. Brand B is cheaper at $.298 cents per can.

5. a) $3\,3/4$ b) $1\,3/2$ c) 13

6. Approx. 1.92 hours or about 1 hr. 55 min.

7. 4417 males and 3435 females

8. $x + 1.25x$ or $2.25x$

9. a) $.22 \times n = 88$, $n = 400$ b) $3/19.99 \times 100\%$ = approx. 15%

10. $6\,3/4"$ by 9"

11. 78.5% is occupied by the cans and 21.5% of the area is wasted.

12. AB = 4 inches, AD = 12 inches

13. a) 150% of R:    b) 75% of R:

14. The original price of the TV was $398.67.

15. Answers will vary.

16. 64% return, 12.8% average annual return

17. a) P = 58'   b) Area of walk is 138 sq. ft.

18. The graph looks like:

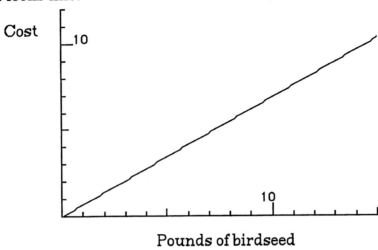

Chapter 3 Worksheet 1

1. To estimate the number of fish in a pond, 150 fish are tagged and then returned to the pond. Later, a sample of 100 fish shows 6 tagged. Using a proportion, find the approximate number of fish in the pond. Use T to represent the total number of fish.

2. On a map, the legend says that one-half inch represents 50 miles in real world measurements. If two cities are $2\frac{3}{4}$ inches apart on the map, what is the real world distance between them? Write two proportions you could use to solve this problem. One of the proportions should use ratios; the other should use rates. Remember to identify the variable you are using.

3. Complete the following chart of fraction, decimal, percent equivalents.

fraction	decimal	percent
$\frac{7}{8}$	_____	_____
_____	.07	_____
_____	_____	45%
$\frac{5}{6}$	_____	_____
_____	1.5	_____

4. Order the following, smallest to largest, left to right.

$.44, \;.\overline{4}, \;.4, \;\frac{9}{20}, \;44.5\%, \;.044$

5. If the sides of a triangle are 3 inches, 4 inches, and 5 inches respectively, find the length of the sides of a new triangle if it is produced using a copy machine set for 120%.

6. After a raise your hourly wage of $6.40 increased to $6.72. Use a proportion to find the percent of your raise.

7. A coat cost $74.88 including 4% sales tax. Find its cost before the tax was added using a proportion.

8. A salad dressing mixture contains oil and vinegar in the ratio of 2 parts oil to 1 part vinegar. Find the amount of oil and the amount of vinegar in 12 ounces of this mixture.

Chapter 3 Worksheet 2

1. A map legend is keyed so that ½ inch represents 50 km.
   If 1 km = .625 mile, find the distance, in miles, between two cities that are $1\frac{1}{4}$ inches apart on this map.

2. A waiter served a table of 4 at a gourmet restaurant and the total bill was $72.35, including 6% tax. What was the amount of the bill before the tax was added?

3. A car salesman receives a weekly salary of $300 plus a 7% commission on all his sales. If his sales for one week totaled $14,215, what was his salary that week?

4. A TV set marked down to $299.99 included a 25% discount. What was the original price of this TV set?

5. For gears that are connected, the product of the number of teeth and the corresponding rpm (revolutions per minute) for each gear is the same. A gear which has 21 teeth and turns at a speed of 900 rpm is connected to another gear which turns at a speed of 270 rpm. Does the second gear have more teeth or fewer teeth than the first? How many teeth does it have?

6. To check the accuracy of the speedometer in a car, math professors on their way to a conference clocked their car traveling a 1-mile section of I-75 between Mackinaw City and Detroit. The driver kept a fairly constant speed at 65 mph, as least as indicated on the speedometer. If the professors clocked a time of 52 seconds for that 1-mile section, what was their calculated speed, in miles per hour? What is the difference between the speedometer reading and the calculated speed? Discuss the implications of this result.

7. Calculate the number of square yards of carpeting needed to cover a living room floor that is $14\frac{1}{4}$ feet by $17\frac{1}{2}$ feet.

8. The graph below shows the relationship between the amount of time in a taxi cab and the cost of the ride. Use the graph to answer the following questions.

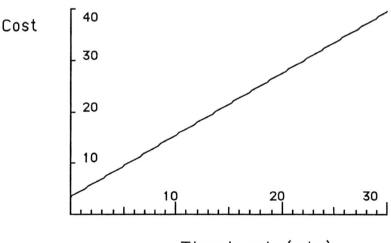

Time in cab (min.)

   a) Find the approximate cost of a 20 minute taxi ride.
   b) If the cost of the ride was $15, what was the length of the ride?

CHAPTER FOUR – Probability and Statistics

Section 4.1 - Concepts of Probability

In one toss of a fair coin, what is the chance of getting a "head" on the top face? What is the chance of getting a "tail"?

Our intuition tells us that since there are two possible outcomes in a coin toss, specifically, a head or a tail, the chance of getting either one of them is one out of two, or $\frac{1}{2}$. What we've considered here is the "likelihood" of getting a head (or tail) on any one toss of a fair coin. This likelihood of an event happening is the informal definition of **probability**.

As another example, when tossing a fair die and observing the number on the top face, there are six possible outcomes: 1, 2, 3, 4, 5, and 6. Since there is a total of 6 possible outcomes and each number occurs only once, the chance of any one of these numbers occurring is $1/6$. However, if the question is, "What is the chance of getting an even number in the toss of a fair die?", there are now 3 ways to get an even number (by tossing a 2, 4 or 6) out of a total of 6 possible outcomes and so here the probability is

$$\frac{3}{6} \text{ or } \frac{1}{2}$$

We define the probability of any event A as

$$P(A) = \frac{\text{number of ways event A can occur}}{\text{total number of outcomes}}$$

The notation P(A) does not mean P times A, but instead means the probability of event A occurring.

Let's use this definition to find other probabilities. If a bag contains two red, two blue, and one white marble, and one marble is selected at random from the bag, what is the probability of selecting a red marble? A white marble? A red or blue marble?

First, since there are two red marbles out of a total of 5 marbles in the bag, the probability or chance of drawing a red marble is

$$P(\text{red}) = \frac{2}{5}.$$

Similarly, since there is one white marble out of a total of five marbles, the probability of selecting a white marble at random is

$$P(\text{white}) = \frac{1}{5}.$$

Finally, since selecting a red marble <u>or</u> blue marble is considered "success", and there are two of each in the bag, the probability of drawing a red or blue marble is

$$P(\text{red or blue}) = \frac{2+2}{5} \text{ or } \frac{4}{5}$$

Many times probability is calculated from data that has been gathered through experiment or accumulated over a long period of time. For example, meteorologists calculate the chance or probability of precipitation each day based on a computer model and long-term accumulated records. For a specific day, they look at conditions such as wind direction and velocity, temperature and relative humidity. Based on their model and records, they then use the number of days when the weather conditions were similar to the day in question as a basis for comparison and consider the number of <u>those</u> days that precipitation actually occurred. This ratio, often in percent form, does represent the probability of precipitation that day.

When calculated probabilities are based on data from observation or experiment, they are called experimental probabilities.

If we tossed two fair dice and added the numbers on the top faces, what possible sums could occur? Realizing that each die has the numbers 1, 2, 3, 4, 5, and 6, it is not difficult to calculate all possible sums,

$$1 + 1 = 2,\ 1 + 2 = 3,\ 1 + 3 = 4,\ \ldots,\ \text{up through } 6 + 6 = 12.$$

The possible sums are then

$$2,\ 3,\ 4,\ 5,\ 6,\ 7,\ 8,\ 9,\ 10,\ 11 \text{ and } 12.$$

Are these sums equally likely to occur? Or is there a better chance of tossing one sum than another?

The easiest way to answer this question is to generate the **sample space** of all possible sums and the different ways these sums can occur. For clarity, think of using two different-colored dice, say red and blue, so that the <u>ordered pair</u> (1,2) represents a 1 on the red die and a 2 on the blue, and the ordered pair (2,1) represents a 2 on the red die and a 1 on the blue. In our sample space, these represent different outcomes.

In table form:

Possible Sums:

2	3	4	5	6	7	8	9	10	11	12	
(1,1)	(1,2)	(1,3)	(1,4)	(1,5)	(1,6)	(2,6)	(3,6)	(4,6)	(5,6)	(6,6)	ways to generate sums
	(2,1)	(3,1)	(4,1)	(5,1)	(6,1)	(6,2)	(6,3)	(6,4)	(6,5)		
		(2,2)	(2,3)	(2,4)	(2,5)	(3,5)	(4,5)	(5,5)			
			(3,2)	(4,2)	(5,2)	(5,3)	(5,4)				
				(3,3)	(3,4)	(4,4)					
					(4,3)						

From this table, it is apparent that all sums are <u>not</u> equally likely to occur. The sum of 7 occurs, theoretically, most often, and the sums of 2 and 12 occur least often. [However, each of these 36 individual ordered pairs has an equal chance of being tossed.] Also note the pattern in the table. The number of ways consecutive sums can be generated increases by one through the sum of seven, and then decreases by one each time.

We can calculate what we call the **theoretical** probabilities, based on an analysis of all 36 possible outcomes, each of which occurs exactly one time. (When actually doing the experiment of tossing the dice, for example, you may get the pair (3,4) more than one time. This would then be used in the calculation of the **experimental** probabilities. The exploration problems at the end of this section will illustrate this.)

To find the probability of tossing a sum of 7, we use the definition of probability:

$$P(\text{sum of 7}) = \frac{\text{number of ways 7 can occur}}{\text{total number of outcomes}} = \frac{6}{36} \text{ or } \frac{1}{6}$$

Also, to find the probability of tossing a sum of 2,

$$P(\text{sum of 2}) = \frac{1}{36} \text{ since a sum of 2 can occur only one way, (1,1).}$$

Let's find the probability of tossing a sum that is even. From our table of all possible outcomes, the sums of 2, 4, 6, 8, 10, and 12 are even. We need to count the number of ways each of those sums can occur, add them together—and again compare to the total of 36. Thus,

$$P(\text{even sum}) = \frac{1+3+5+5+3+1}{36} = \frac{18}{36} \text{ or } \frac{1}{2}.$$

Counting the number of ways to generate a sum that is odd we have:

$$2 + 4 + 6 + 4 + 2 = 18,$$

and comparing to the total number of outcomes we get:

$$P(\text{odd sum}) = \frac{2+4+6+4+2}{36} = \frac{18}{36} \text{ or } \frac{1}{2}. \quad \text{Does this result surprise you?}$$

Looking at a different set of restrictions, what if we want to know the probability of the sum being even <u>and</u> greater than 4?

The sums selected for consideration here must satisfy both requirements, <u>even</u> and <u>greater than 4</u>. Those sums are 6, 8, 10 and 12. Adding the number of ways these sums can occur:

$$P(\text{sum is even } and \text{ greater than 4}) = \frac{5+5+3+1}{36} = \frac{14}{36} \text{ or } \frac{7}{18}.$$

Note that from the addition of rational numbers, the above expression can also written as

$$\frac{5+5+3+1}{36} = \frac{5}{36} + \frac{5}{36} + \frac{3}{36} + \frac{1}{36}$$

$$= P(\text{sum of 6}) + P(\text{sum of 8}) + P(\text{sum of 10}) + P(\text{sum of 12})$$

Finally, let's find P(even sum <u>or</u> sum greater than 4). In this case, <u>either</u> an even sum <u>or</u> a sum greater than 4 constitutes the event, so the number of ways the sums of 2, 4, 6, 8, 10, 12, 5, 7, 9, and 11 can occur must be counted.

By adding the number of ways each of these sums can occur,

$$P(\text{even sum } or \text{ sum greater than 4}) = \frac{1+3+5+5+3+1+4+6+4+2}{36}$$

$$= \frac{34}{36} \text{ or } \frac{17}{18}$$

$$= P(2) + P(4) + \ldots + P(11)$$

[Notice that only the ways to generate the sum of 3 are excluded since 3 is the only number that does not satisfy either condition.]

We can make an important conclusion from our work here:

> **In general, to find the probability of an event consisting of several different outcomes, we add the number of ways each of those outcomes can occur and compare this to the total number of outcomes.**

In some situations, like the one above, we can find the probability of an event by calculating the probability of each outcome comprising the event and then adding the probabilities together. This only works, however, if the outcomes are what we call **mutually exclusive**—that is, outcomes that cannot occur at the same time.

As an example, consider the experiment of selecting one card from a standard deck of 52 cards. The standard deck consists of 4 suits: hearts and diamonds, both red, and spades and clubs, both black in color. Each suit consists of 13 cards, an ace, those numbered 2 through 10, and the face cards: jack, queen, and king. In the selection of one card at random, if we want to find the probability that the card is a

    a) king,

we take the ratio of the number of kings in the deck to the total number of cards, so: $\quad P(\text{king}) = \dfrac{4}{52} \text{ or } \dfrac{1}{13}.$

To find the probability of selecting a

    b) king *or* an ace, $\quad P(\text{king or ace}) = \dfrac{4+4}{52} = \dfrac{8}{52} \text{ or } \dfrac{2}{13}$

because there are 4 kings and 4 aces and one *or* the other satisfies the requirement.

We can also write this as $\dfrac{4}{52} + \dfrac{4}{52} = P(\text{king}) + P(\text{ace}) = \dfrac{8}{52}$

since the outcomes of king and ace are **mutually exclusive**. A single card can be a king *or* an ace but cannot be both.

However, to find the probability of selecting a

    c) heart or a king,

we look at the number of ways to get a heart <u>or</u> a king

$$P(\text{heart or a king}) = \frac{13 + 3}{52} = \frac{16}{52} \text{ or } \frac{4}{13} \neq P(\text{heart}) \neq P(\text{king}).$$

The reason we didn't add 13 and 4 is because the king of hearts was already included in the "heart" count and we don't want to count it twice. This is the type of problem where the probabilities of the outcomes comprising the event cannot just be added because the outcomes of heart and king have something in common. In other words, the selection of a heart is not mutually exclusive from the selection of a king since there is one card that is the "king of hearts."

Finally, the probability of selecting

d) a king and a queen,

$$P(\text{king and queen}) = \frac{0}{52} = 0$$

since there isn't one card that is both a king and a queen.

It is crucial at this point to realize that we have been working with probabilities that were not based on experiment or sampling. We instead considered the sample space only and calculated probabilities based on that list of all possible outcomes. For example, in the toss of a coin, we discussed that it is equally likely to get a head or a tail. So, theoretically, if we tossed the coin two times, we should get one head and one tail. And if we tossed the coin 50 times, we should get 25 heads and 25 tails (since each should happen $1/2$ of the time). In reality, however, this is unlikely to happen. Chance, the toss itself, and other factors can influence the results. If we tossed a coin 1000 times, we again will probably not get 500 heads and 500 tails, but we may get a ratio closer to $1/2$ heads and $1/2$ tails. And if we toss that coin 100,000 times, we may get a ratio even closer to $1/2$ for each. The conclusion that can be made here is that

Theoretical probability is what we *expect* will happen in the long run as experiments are repeated an *infinite* number of times. That is when we expect the experimental probability to equal the theoretical probability.

Sample Problems

1. From the table of all possible outcomes for the toss of two dice, find these theoretical probabilities:

   a) P(sum of 4 or sum of 10)

   b) P(sum of 1)

   c) P(odd sum less than 5)

   Solutions:

   a) $P(\text{sum of 4 or sum of 10}) = \frac{3 + 3}{36} = \frac{6}{36} \text{ or } \frac{1}{6} = P(\text{sum of 4}) + P(\text{sum of 10})$

   b) $P(\text{sum of 1}) = \frac{0}{36} = 0$

c) P(odd sum less than 5) = $\frac{2}{36}$ or $\frac{1}{18}$ (Only a sum of 3 needs to be considered.)

2. There are six marbles in a bag: 3 red, 2 green, and 1 blue. If a marble is selected at random from the bag, find the probability that it is

   a) blue

   b) blue or red

   c) red, blue, or green

Solutions:

   a) P(blue) = $\frac{1}{6}$

   b) P(blue or red) = $\frac{1+3}{6}$ = $\frac{4}{6}$ or $\frac{2}{3}$ = P(blue) + P(red)

   c) P(red, blue, or green) = $\frac{3+2+1}{6}$ = $\frac{6}{6}$ = 1 = P(red) + P(blue) + P(green)

   In other words, we are sure to get a red, blue, or green marble from this bag.

3. Can you devise an event where the probability of that event will be 0? 1?

   Possible solutions:

   for P = 0: The event of tossing a 7 on one toss of a fair die;

   for P = 1: Precipitation in the form of snow sometime between November and March in Fargo, North Dakota.

   Therefore, the probability of events certain not to happen is 0 and the probability of events certain to happen is 1.

Geometric Probability

The outside east side wall of a 3-story dormitory is 120' feet long and 32' high. If each story contains 10, 3' x 4' windows, we can find the ratio of window space to wall space by calculating the area of all the windows and comparing that to the area of the wall.

Area (of windows) = (10)(3 • 4) = 120 sq. ft. for each story
120 • 3 stories = 360 sq. ft.

Area (of east wall) = 120 • 32 = 3840 sq. ft.

Thus the ratio: $\frac{360}{3840}$ = $\frac{9}{96}$ = $\frac{3}{32}$ represents the fraction of the total wall space occupied by windows.

From a different perspective, if students are playing football in the grassy area next to the east wall, and a student throws the football that randomly hits the wall, this ratio also represents the probability that the ball will hit a window!

Sample Problems

1. The area of a circle can be found by the formula, $A = \pi r^2$, where r represents the radius of the circle and $\pi \approx 3.14$.

   A simple dartboard has the following configuration and dimensions:

   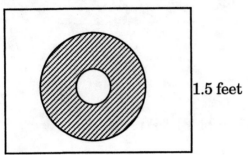

   radius outer circle = .5 feet

   radius of inner circle = .125 feet

   Given that a person shoots a dart and hits the board,

   a) what is the probability that the person hits the bull's eye?

   b) what is the probability that the dart hits the shaded portion?

   Solutions:

   a) Area of bull's eye = $3.14 (.125)^2$
   = .05 sq. ft. (to the nearest hundredth)

   Area of dartboard = $2 \cdot 1.5 = 3$ sq. ft.

   Probability dart hits bull's eye = $\dfrac{.05}{3}$ = .016 or $\approx$ .02

   b) Area of shaded portion = $3.14(.5)^2 - 3.14(.125)^2 \approx .736$ sq. ft.

   Probability dart hits shaded region = $\dfrac{.736}{3}$ = .245

Problem Set 4.1

1. A bag contains 7 slips of paper, each containing the name of a different day of the week. If one slip is selected at random, list all the possible outcomes of this experiment.

2. An experiment consists of a combination of tossing a coin, observing the top face, and then tossing a fair die and noting its top face.

   a) Write the sample space for this experiment.

   b) Find the probability that the coin landed on heads.

   c) Find the probability that the die had a number larger than 4 on its top face.

   d) Find the probability that the die had an even number on the top face and the coin was a head.

3. A bag contains 2 red, 3 green, and 2 yellow balls.

   i) If 1 ball is drawn at random, what is the probability that it is

      a) green   b) green or yellow   c) green and red

   ii) How many red balls must be added to make the P(red) = $1/2$?

4. Given the spinner at right with colored regions as indicated, if a person takes a spin, what is the probability that the spinner will land on

   a) a green region

   b) a yellow region

   c) a blue region

   d) a blue or yellow region

   e) a blue or green region

   f) a blue or red region

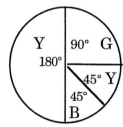

5. Two dice are tossed and the sum of the numbers on the top faces is recorded. Find the probability that the sum is

   a) 8   b) greater than 11

   c) at least 6   d) greater than 10 and less than 2

   e) 10 or less than 4   f) at most 12

6. A bag contains 3 red, 4 green, and 5 yellow marbles.

   If one marble is selected at random, find the probability that the marble is

   a) red   b) green or yellow

   c) not red and not yellow

7. A card is selected at random from a standard deck of 52 cards. Find the probability that the card is

   a) a "2"

   b) a red "10" or a club

   c) a jack or a queen

   d) a king and a heart

   e) a king or a heart

   f) a number larger than 3

8. Data values can be gathered from an experiment or sampling and a **frequency distribution** is often made of these data values. A frequency distribution is a table that lists the data points or categories with the number of times each of those points or categories occurred in the sampling. Different types of graphs can then be made where the data values are usually plotted on the horizontal axis, and the number of times those data values occurred are plotted on the vertical axis.

   Below is a bar graph of data on eye color gathered in a university statistics class of 40 students.

   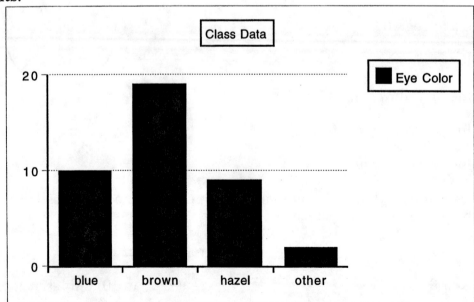

   The **relative frequency** of blue eyes, for example, can be found by taking the ratio of the number of times blue eyes occurred to the total number of students in the class. The relative frequency of blue eyes in this class is the same value as the probability of selecting, at random, a person with blue eyes from this class. Use the information above to answer the following questions.

  a) If a person is selected at random from this statistics class, what is the probability that the person has blue eyes?

  b) If a person is selected at random from this class, what is the probability that the person has hazel-colored eyes?

  c) If a person is selected at random from this class, what is the probability that the person has blue or brown eyes?

9. In a recent election, 96 voters favored the school millage increase and 73 favored the bond issue to build a new fire station. 127 ballots were cast. If 64 voters favored both the millage and the bond issue, what is the probability that a randomly selected voter supported one but not both of the propositions?

10. Given **figure 1** below with the outer square measuring 3' on a side, the inner square 1' on a side:

  a) What is the probability that a dart which hits the board will land in the inner square?

  b) What is the probability that the dart will land in the shaded portion?

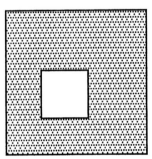

figure 1

11. Marty and Chris were playing tennis near a fenced yard with a small wading pool as shown in **figure 2**. If Marty randomly hits a ball into the yard, what is the probability, expressed as a percent, that it will land in the wading pool?

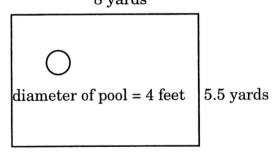

figure 2

Exploration Problems

1. Toss two dice; add the top faces. (We have already found all possible sums in this section.) Do this experiment 30 times and record your data in a table similar to the one below, making an "X" each time that sum results, beginning on the lowest part of the table and recording upward.

   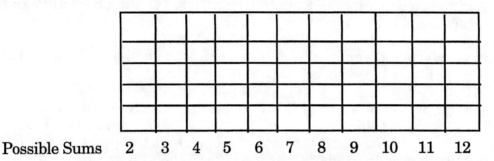

   Possible Sums   2   3   4   5   6   7   8   9   10   11   12

   Then *based on your data from a total of 30 trials*, find

   a) the number of even sums that occurred

   b) the number of odd sums that occurred

   c) the P(even sum)

   d) the P(odd sum)

   e) the sum(s) that occurred most often

   Compare your results to the *theoretical* results in this section. Also pool the class data and compare the experimental results to what we called the theoretical results. If there are differences, explain possible reasons for them.

2. Toss two dice and multiply the numbers on the top faces. Do this experiment 30 times. List all possible products that could result when two dice are tossed and put these at the bottom of a table similar to that constructed in exploration problem 1. Record the results of your 30 trials. Then find

   a) the number of even products that occurred and the number of odd products that occurred

   b) the P(even product)

   c) the P(odd product)

   d) the product(s) that occurred most often.

   e) Compare your part *b* and *c* results. If they are not at all alike, give a possible explanation and justify that explanation.

3. In the election described in problem 9 above, there was also a referendum to determine the feelings of the community regarding the installation of a traffic signal at an intersection where several serious accidents had recently occurred. 103 of the voters favored the new signal. If the only other information available is that 96 favored the school millage and 73 favored the bond issue, answer the following questions:

   a) Of the 127 voters, what is the greatest number that *could* have favored *all three* issues?

   b) What is the least number that *could* have favored all three?

Solutions - Problem Set 4.1

1. {Monday, Tuesday, Wednesday, Thursday, Friday, Saturday, Sunday}

2. a) Sample Space: {H1, H2, H3, H4, H5, H6, T1, T2, T3, T4, T5, T6}
   b) P(heads) = $6/12$ or $1/2$
   c) P(number larger than 4) = $4/12$ or $1/3$
   d) $3/12$ or $1/4$

3. i) a) $3/7$
      b) $5/7$
      c) $0/7 = 0$
   ii) Add 3 red balls.

4. a) P(green) = $1/4$ b) P(yellow) = $5/8$
   c) P(blue) = $1/8$ d) P(blue or yellow) = $3/4$
   e) P(blue or green) = $3/8$ f) P(blue or red) = $1/8$

5. a) $5/36$ b) $1/36$
   c) $26/36$ or $13/18$ d) 0
   e) $6/36$ or $1/6$ f) $36/36 = 1$

6. a) P(red) = $3/12$ or $1/4$ b) P(green or yellow) = $9/12$ or $3/4$
   c) $4/12 = 1/3$

7. a) P(2) = $4/52$ or $1/13$ b) P(red 10 or club) = $15/52$
   c) P(J or Q) = $8/52$ or $2/13$ d) P(king and heart) = $1/52$
   e) P(king or heart) = $16/52$ or $4/13$ f) P(number > 3) = $28/52$ or $7/13$

8. a) P(blue eyes) = $10/40$ or $1/4$
   b) P(hazel eyes) = $9/40$
   c) P(blue or brown) = $29/40$

9. $41/127$

10. a) $1/9$
    b) $8/9$

11. about 3%

Section 4.2 - Tree Diagrams and Compound Events

In the toss of two fair coins, we can list all possible outcomes using ordered pairs: {(H,H), (H,T), (T,H), (T,T)}. Let's examine another convenient way to generate the sample space. The outcomes from the toss of the first coin can be shown as

Given that we have the first coin accounted for, the outcomes possible on the second coin can also be shown as

Putting these together and forming the **compound event** of observing the results on the first *and* second coins, we have the **tree diagram** below.

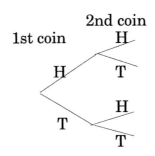

diagram 1

The possible outcomes of the second coin branch out from both outcomes of the first coin because whether the first result is Head or Tail, a Head or Tail can still occur on the second coin. (We say that the event of tossing a Head or Tail on second coin in this case is **independent** of the outcome on the first coin.)

Reading the diagram, the branch HH represents a Head on the first coin and a Head on the second, and the branch TH represents a Tail on the first coin and a Head on the second, etc.

Following along the branches of the tree, we can list the sample space as

$$S = \{HH, HT, TH, TT\}$$

Here there are <u>four</u> possible, *equally likely*, outcomes and we can use this information to find probabilities. For example,

P(2 heads) = P(2H) = $\frac{1}{4}$ since there is only *one* way out of *four* to get 2 heads, and P(exactly 1 head) = P(1H) = $2/4$ since there are *two* ways, HT and TH, out of four possible outcomes that result in exactly 1 head.

Let's redraw the diagram and label the probabilities along each branch,

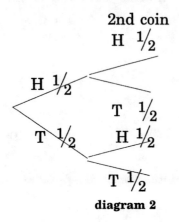

diagram 2

Comparing our earlier results for P(2H) = $1/4$ and the probabilities along that branch of the tree diagram, which are $1/2$ and $1/2$, we can actually generate the probability of getting a head on the 1st coin and a head on the 2nd by multiplying the numbers along the branch and we have

$$P(2H) = 1/2 \cdot 1/2 = 1/4$$

It is important to reflect on our results for P(exactly 1 head) in this context. The two events, HT and TH, satisfy the statement "exactly 1 head". We find the probability of each, and again multiply the numbers along the appropriate branches. However, since both are outcomes satisfying the requirement "exactly 1 head", we need to add these computed probabilities as we did with the dice toss sums. So

$$P(\text{exactly 1 H}) =$$
$$P(HT) + P(TH) = 1/2 \cdot 1/2 + 1/2 \cdot 1/2$$
$$= 1/4 + 1/4 = 2/4$$

and this agrees with our previous result.

Returning to the bag of marbles containing 3 red, 2 blue, and 1 green marble, let's change the experiment so that we select 1 marble, observe its color, replace that marble, and then draw another. This is called **sampling with replacement** and the tree diagram with probabilities labeled along the branches can be shown as:

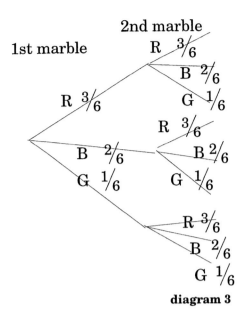
diagram 3

To find the probability of selecting 2 red marbles, we multiply the numbers along the only branch satisfying the requirement, R and R, and we have

$$P(R \text{ and } R) = \tfrac{3}{6} \cdot \tfrac{3}{6} \text{ or } \tfrac{1}{2} \cdot \tfrac{1}{2} = \tfrac{1}{4}$$

(Note that there are a total of 9 outcomes or branches here – RR, RB, etc. – and $P(R \text{ and } R)$ does not equal $\tfrac{1}{9}$ since these outcomes are <u>not</u> equally likely to occur. Can you see why?)

The probability of selecting a red and then a blue marble,

$P(R_1 \text{ and } B_2)$, notation indicating red is drawn first and blue, second,
$$= \tfrac{3}{6} \cdot \tfrac{2}{6} = \tfrac{6}{36} \text{ or } \tfrac{1}{6}$$

But to find the probability of selecting a red and a blue marble, not necessarily in that order,

$P(R \text{ and } B) = P(R_1 \text{ and } B_2)$ *or* $P(B_1 \text{ and } R_2)$ since both branches constitute a red and a blue marble and we have
$$P(R \text{ and } B) = \tfrac{3}{6} \cdot \tfrac{2}{6} \text{ or } \tfrac{2}{6} \cdot \tfrac{3}{6}$$
$$= \tfrac{6}{36} + \tfrac{6}{36} = \tfrac{12}{36} \text{ or } \tfrac{1}{3}$$

Notice that when the order of the selection was not important, we, in fact, doubled our chances of getting a red and a blue marble (since $\tfrac{1}{6} \cdot 2 = \tfrac{1}{3}$).

It is also important to realize that we used two operations when calculating probabilities in this compound event. It should be apparent from several examples above that generally

in probability, "And" indicates the operation of multiplication, and "Or" indicates the operation of addition.

Sample Problems

1. In the toss of two fair coins, find the probability of getting

   a) exactly zero heads

   b) at least 1 head

   Solutions:

   a) From the sample space, {HT, TH, TT, HH}, TT is the only outcome out of 4 where there are no heads so
   $$P(TT) = P(0\ H) = 1/4 \quad (\text{or } 1/2 \cdot 1/2 = 1/4)$$

   b) *at least* 1 head in this case means 1 or 2 heads so

   $$P(\text{at least 1 H}) = P(TH) + P(HT) + P(HH)$$
   since all satisfy the requirement that *at least* one head was tossed.
   Thus, P(at least 1 H)
   $$= \frac{1}{2} \cdot \frac{1}{2} + \frac{1}{2} \cdot \frac{1}{2} + \frac{1}{2} \cdot \frac{1}{2} = \frac{1}{4} + \frac{1}{4} + \frac{1}{4} = \frac{3}{4}$$

   [Note: Another way to approach this problem is to use the statement, P(at least 1 H) + P(0H) = 1, since these events (at least 1 H and 0 heads) comprise the whole sample space and the P(Sample Space) = 1. Thus, we could find P(at least 1 H) by taking
   $$1 - P(0H) = 1 - 1/4 = 3/4.$$

   The events *at least 1 H* and *0H* are called **complements**.]

2. Use **diagram 3** to help find the following probabilities when sampling two marbles with replacement.

   a) $P(R_1 \text{ and } G_2)$

   b) P(B and G)

   Solutions:

   a) $P(R_1 \text{ and } G_2) = \frac{3}{6} \cdot \frac{1}{6} = \frac{3}{36} = \frac{1}{12}$

   b) $P(B \text{ and } G) = P(B_1 \text{ and } G_2) + P(G_1 \text{ and } B_2)$
   $$= \frac{2}{6} \cdot \frac{1}{6} + \frac{1}{6} \cdot \frac{2}{6} = \frac{2}{36} + \frac{2}{36} = \frac{4}{36} = \frac{1}{9}$$

Finally, let us consider the same bag of marbles, containing 3 red, 2 blue, and 1 green, and sample two marbles, one after another, *without* replacing the first before the second is drawn. This is called **sampling without replacement**.

To find the probability of selecting two red marbles in this context, we must realize that the second selection will depend on what happened in the first selection.
That is, if a red marble is drawn first and kept out of the bag, the probability of selecting another red marble is now different. Specifically, the probability of selecting two red marbles is

$$P(R \text{ and } R) = \frac{3}{6} \cdot \frac{2}{5} = \frac{6}{30} = \frac{1}{5}$$

The $2/5$ component is the probability of selecting a second red marble given that a red marble was drawn first. If a red marble was drawn first, then there are only 2 red marbles left out of a total of 5 marbles now in the bag!

Completing the tree diagram and labeling the probabilities we have

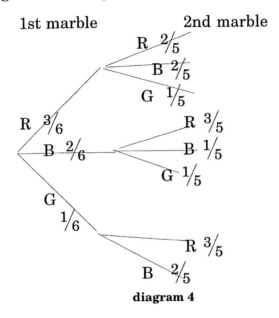
diagram 4

The tree diagram can also be used to find the probability of selecting a green marble on the second draw, under the condition that the first marble drawn was blue. In this case, we limit our discussion to the branch indicated below.

1st marble   2nd marble
  B 2/6
         G 1/5

Note that the probability we are looking for, $P(G_2 \text{ given } B_1)$, is $1/5$, the second part of the branch only, and, in this case, the probabilities are not multiplied.

What if we want to know the probability of selecting a green marble on the second draw given that a green marble was drawn first? Again by limiting our discussion to this branch,

1st marble   2nd marble
  G 1/6   ?

it is easy to see that there is no branch indicating we could select a second green marble. So: $P(G_2 \text{ given } G_1) = 0$

Sample Problems

1. Use the tree **diagram 4** for sampling two marbles **without replacement** and find the probability of drawing

    a) a blue and then a red marble

    b) two green marbles

    c) a blue marble on the second draw, given a blue marble was drawn first.

    d) a blue marble is drawn second.

    Solutions:

    a) $P(B_1 \text{ and } R_2) = \dfrac{2}{6} \cdot \dfrac{3}{5} = \dfrac{6}{30} = \dfrac{1}{5}$

    b) $P(G \text{ and } G) = \dfrac{1}{6} \cdot \dfrac{0}{5} = \dfrac{0}{30} = 0$ which agrees with our result above.

    c) $P(B_2 \text{ given } B_1) = 1/5$

    d) $P(R_1 \text{ and } B_2) + P(B_1 \text{ and } B_2) \text{ and } P(G_1 \text{ and } B_2)$—all must be considered.
    Thus, P(B drawn 2nd) =
    $$\dfrac{3}{6} \cdot \dfrac{2}{5} + \dfrac{2}{6} \cdot \dfrac{1}{5} + \dfrac{1}{6} \cdot \dfrac{2}{5} = \dfrac{6}{30} + \dfrac{2}{30} + \dfrac{2}{30} = \dfrac{10}{30} = \dfrac{1}{3}$$

2. Find the probability of drawing two aces from a standard deck of 52 cards if

    a) the first card is *replaced* before the second is drawn;

    b) the first card is *not replaced* before the second is drawn.

    Solutions:

    a) $P(\text{Ace and Ace}) = \dfrac{4}{52} \cdot \dfrac{4}{52} = \dfrac{16}{2704} = \dfrac{1}{169}$

    b) $P(\text{Ace and Ace}) = \dfrac{4}{52} \cdot \dfrac{3}{51} = \dfrac{12}{2652} = \dfrac{1}{221}$

Problem Set 4.2

1. a) Use a tree diagram to construct the sample space of all outcomes possible when three fair coins are tossed. List the outcomes.

   b) When three fair coins are tossed, find the probability of getting

   1) exactly 3 heads

   2) at most 1 head

   3) 3 tails

   4) at least 1 tail

2. A bag of marbles contains 3 red, 4 green, and 5 yellow. If it is helpful, use tree diagrams for parts i) and ii).

   i) If 2 marbles are drawn at random **with replacement,** find the probability that

   a) both marbles are green

   b) the first is yellow and the second green

   c) one is red and one is green

   d) the second is green

   ii) From the same bag, if 2 marbles are drawn at random **without replacement,** find the probability that

   a) both marbles are yellow

   b) the first is red and the second green

   c) one is green and one is yellow

   d) the second is red given the first was yellow

3. A bag contains 10 equal size pieces of paper: four are labeled with a "4", three are labeled with a "3", two are labeled with a "2", and one is labeled "1". Two pieces of paper are drawn from this bag, but the first is replaced before the second is drawn. Find the probability that

   a) both pieces are marked with a "3"

   b) the first is labeled "4" and the second is labeled "1"

   c) one is labeled with a "2" and the other is labeled with a "3"

4. An unorganized drawer of 14 socks contains socks that are navy, brown or black. If you reach into the drawer and draw socks at random, how many socks would you have to draw to be sure that you have 1 pair of the same color?

5. The diagram below is another way to depict the sampling of two marbles from a total of 3W and 1B marble, without replacement. Pairs are indicated by connecting lines. Write the sample space of all possible outcomes (with respect to color). Then find the P(2W) and P(W and B).

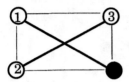

6. Two cards are drawn at random, **with replacement**, from a standard deck. Find the probabililty that

   a) both cards are kings

   b) a king is drawn first and then a "10" is drawn

   c) a jack is drawn, given that an ace was drawn first

   d) If the cards were drawn at random, **without replacement**, find the probability that

   i) both cards are queens

   ii) a jack is drawn, given that an ace was drawn first.

   e) Compare your answers to parts c) and ii) above. Give a possible explanation for your results.

7. Two boxes contain letters as described below. The experiment is to select a letter from box 1, put it into box 2, and then select a letter from box 2. Make a tree diagram and label the branches with letters and probabilities to help you answer the following questions.

   |MATH| |SAT|

   Box 1   Box 2

   a) What is the probability that the letter drawn from box 1 is a "T"?

   b) What is the probability that the letter drawn from box 2 is an "A", given that an "A" was drawn from box 1?

   c) What is the probability that a "T" was drawn from box 1 and a "S" was drawn from box 2?

   d) What is the probability that the letter drawn from box 2 was a "T"?

Exploration Problem

A bag contains 2 yellow marbles and 1 green marble. A game between two players, David and Hakeem, consists of drawing two marbles (without replacement). If the two marbles are the same color, David wins; if they are different colors, Hakeem wins.

a) Determine whether or not the game is fair.

b) If the game is not fair, how would you make it fair without changing the rules?

Solutions - Problem Set 4.2

1. a) [Tree diagram showing three coin flips with outcomes: HHH, HHT, HTH, HTT, THH, THT, TTH, TTT]

   b) 1) $\frac{1}{8}$ 2) $\frac{4}{8} = \frac{1}{2}$ 3) $\frac{1}{8}$ 4) $1 - P(0T) = 1 - \frac{1}{8} = \frac{7}{8}$

2. i) a) $\frac{4}{12} \cdot \frac{4}{12} = \frac{1}{3} \cdot \frac{1}{3} = \frac{1}{9}$ b) $\frac{5}{12} \cdot \frac{4}{12} = \frac{20}{144} = \frac{5}{36}$

   c) $\frac{3}{12} \cdot \frac{4}{12} + \frac{4}{12} \cdot \frac{3}{12} = \frac{24}{144} = \frac{1}{6}$ d) $\frac{4}{12}$ or $\frac{1}{3}$

   ii) a) $\frac{5}{12} \cdot \frac{4}{11} = \frac{20}{132} = \frac{5}{33}$ b) $\frac{3}{12} \cdot \frac{4}{11} = \frac{12}{132} = \frac{1}{11}$

   c) $\frac{4}{12} \cdot \frac{5}{11} + \frac{5}{12} \cdot \frac{4}{11} = \frac{40}{132} = \frac{10}{33}$ d) $\frac{3}{11}$

3. a) $\frac{3}{10} \cdot \frac{3}{10} = \frac{9}{100}$ b) $\frac{4}{10} \cdot \frac{1}{10} = \frac{4}{100} = \frac{1}{25}$ c) $\frac{2}{10} \cdot \frac{3}{10} + \frac{3}{10} \cdot \frac{2}{10} = \frac{12}{100} = \frac{3}{25}$

4. 4 socks to be sure that there is one pair of the same color.

5. Sample Space: $\{(W_1W_2), (W_2W_3), (W_1W_3), (W_1B), (W_2B), (W_3B)\}$
   and $P(2W) = \frac{1}{2}$ and $P(W \text{ and } B) = \frac{1}{2}$

6. a) $\frac{4}{52} \cdot \frac{4}{52} = \frac{16}{2704} = \frac{1}{169}$ b) $\frac{4}{52} \cdot \frac{4}{52} = \frac{16}{2704} = \frac{1}{169}$

   c) $\frac{4}{52}$

   d) i) $\frac{4}{52} \cdot \frac{3}{51} = \frac{12}{2652} = \frac{1}{221}$

   ii) $\frac{4}{51}$

   e) $\frac{4}{52} < \frac{4}{51}$ so the problem of selecting a jack the 2nd time is larger—fewer cards and the same number of jacks.

7.

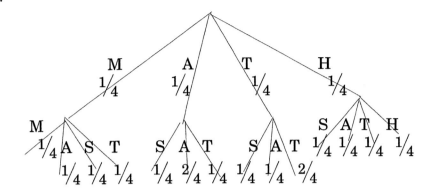

a) $P(T) = 1/4$

b) $P(A_2 \text{ given } A_1) = 2/4 = 1/2$

c) $P(T_1 \text{ and } S_2) =$
$$\frac{1}{4} \cdot \frac{1}{4} = \frac{1}{16}$$

d) $P(T_2) = P(M_1 \text{ and } T_2) + P(A_1 \text{ and } T_2) + P(T_1 \text{ and } T_2) + P(H_1 \text{ and } T_2) =$
$$\frac{1}{4} \cdot \frac{1}{4} + \frac{1}{4} \cdot \frac{1}{4} + \frac{1}{4} \cdot \frac{2}{4} + \frac{1}{4} \cdot \frac{1}{4} = \frac{5}{16}$$

Section 4.3 - Statistics

The test scores on the first test in a college algebra class at Northeastern University were as follows:

66, 80, 58, 65, 81, 71, 75, 72, 78, 70, 85, 87, 91, 98, 95, 45, 75, 78, 80, 68, 72, 73, 70, 68, 65, 66, 90, 90, 94, 70, 62, 95, 94, 78, 86, 87, 91, 88, 89, 92.

Data in general, which is the result of gathering information, is classified as **quantitative**, having a numerical value, or **qualitative**, categorized by a quality or characteristic such as color. The above test score data is quantitative.

When data is written in a nonorganized list, it is not particularly useful because important information (such as trends and averages) is not readily apparent.

One method of "picturing" the data is through a **line plot**. A horizontal line with a numerical scale appropriate for the data is drawn and each data point is plotted accordingly above the line. For the test score data above the line plot is:

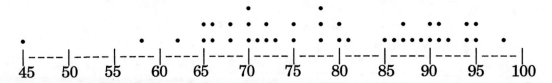

This graph gives us an idea of where the "clustering" is in the set of test scores. In addition, we can easily see the lowest and highest scores on the test in this class as well as the value that occurs most often.

Another graph that can give us much information is the **stem-and-leaf plot.**
Since the test score data is two-digit, we will let the ten's digit be the <u>stem</u> and the one's digit of each score be the <u>leaf</u>. We first write, vertically and in chronological order, the ten's digits of the data as follows:

Stem	Leaf
4	
5	
6	
7	
8	
9	
10	

(Note that the "stems" are in ascending order; descending order is also appropriate.)

We then record the one's digits from each test score. The first five scores listed in the data set are shown as "leaves" on the next diagram.

```
Stem  |  Leaf
-----------------------
  4
  5      8
  6      6 5
  7
  8      0 1
  9
 10
```

where 6 | 65 represents test scores of 66 and 65, respectively. Finally, we will put the remaining "leaves" in. The stem-and-leaf plot is:

```
Stem  |  Leaf
-----------------------
  4      5
  5      8
  6      6 5 8 8 5 6 2
  7      1 5 2 8 0 5 8 2 3 0 0 8
  8      0 1 5 7 0 6 7 8 9
  9      1 8 5 0 0 4 5 4 1 2
 10
```

An alternative to this is the *ordered* stem-and-leaf plot where the leaves are put in chronological order, left to right as shown below.

```
Stem  |  Leaf
-----------------------
  4      5
  5      8
  6      2 5 5 6 6 8 8
  7      0 0 0 1 2 2 3 5 5 8 8 8
  8      0 0 1 5 6 7 7 8 9
  9      0 0 1 1 2 4 4 5 5 8
```

There are several advantages with this type of graph: 1) Each data point is listed; 2) it can be read as a horizontal bar graph in the sense that the stem values that occur most often are apparent; and 3) the organization of the data is conducive to finding certain measures of clustering, as we shall see in the next section.

The **vertical bar graph** is another option for representing the same data. From the next graph, it is again easy to see that scores in the 70's occurred most often. However, in a bar graph, individual values are typically lost since the values are grouped as 40's, 50's, etc., and the vertical axis only indicates how many scores are in each category. Bar graphs are easy to read, however, and qualitative data are often represented in this form.

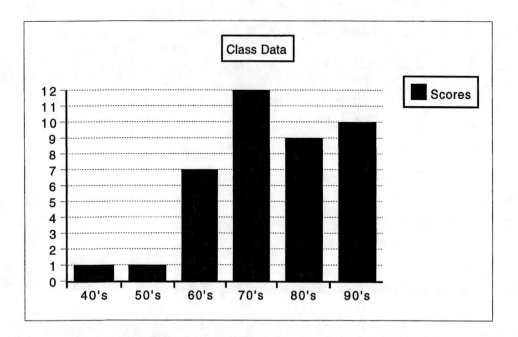

Finally, another graph that is particularly useful with qualitative data, especially when the parts are compared to the whole unit, is a **circle graph**. For example, if summer weather data is collected for a period of 30 days in Sault Ste. Marie, MI and the days classified as 1) rainy; 2) sunny; 3) partly sunny; and 4) cloudy, the following table might be generated.

No. of rainy days	6
No. of sunny days	7
No. of partly sunny days	9
No. of cloudy days	8

So that we can label the circle graph correctly, we must find the percent of the 30 days that rain occurred. As we did in chapter 2, we find the fraction of the 30 days that rain occurred and multiply by 100% as follows:

$$\frac{6}{30} \cdot 100\% = 20\%$$

Similarly, the percent of the 30 days that were sunny is

$$\frac{7}{30} \cdot 100\% = 23.3\%$$

and the percent of the 30 days that are partly sunny is

$$\frac{9}{30} \cdot 100\% = 30\%$$

and finally, the percent of the days that were cloudy is

$$\frac{8}{30} \cdot 100\% = 26.7\%$$

(Note that each percent was calculated to the nearest tenth and that the percents add to 100.)

The circle, 360°, represents the entire 30 days, and the portion of the circle each of the classifications will occupy can be found by multiplying each fraction (or percentage) above by 360. Thus,

$$\frac{6}{30} \cdot 360 = 72° \text{ is accounted for by the "rainy days" classification,}$$

$$\frac{7}{30} \cdot 360 = 84° \text{ by the "sunny days,"}$$

$$\frac{9}{30} \cdot 360 = 108° \text{ by the "partly sunny days,"}$$

and

$$\frac{8}{30} \cdot 360 = 96° \text{ is accounted for by the "cloudy days."}$$

(All measurements are rounded to the nearest degree.)
(Also be sure the degrees add to 360!)

To construct a circle graph, we use a compass to draw the circle and then use a protractor, which is a tool for measuring angles, to mark off the number of degrees needed for each category.

We begin by drawing a horizontal line (from the center of the circle to the circle itself). This serves as the base line for the measurement of the first angle. Complete the first angle by drawing the second ray and then use that ray as the base line for the next angle. Proceed in this manner until all sections are accounted for. The finished circle graph, complete with percentage labels, is shown below at the right.

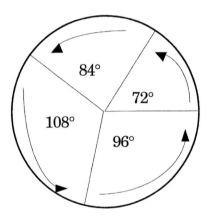
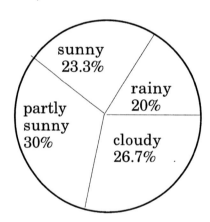

Sample Problems

1. Five professors were selected at random and asked to measure, to the nearest tenth of a centimeter, the lengths of the pencils they had in or on their desks.

   a) Make a line plot and a stem-and-leaf plot of the data recorded below. Hint: For the stem-and-leaf plot, use the whole numbers as the stems and the decimal parts as the leaves.

Professor	Pencil Lengths
#1	16.0, 15.3, 12.5, 10.2
#2	12.5, 14.5, 9.0, 9.2
#3	8.2, 16.2, 13.2, 17.0, 14.5, 14.1
#4	10.4, 13.2, 9.8, 10.5
#5	9.5, 15.2

b) To the nearest centimeter, what is most common pencil size on the desks of these professors?

Solutions:

a) Line plot:

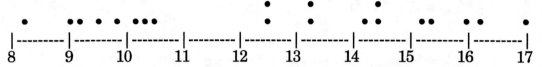

Stem-and-leaf plot (ordered):

Stem	Leaf
8	2
9	0 2 5 8
10	2 4 5
11	
12	5 5
13	2 2
14	1 5 5
15	2 3
16	0 2
17	0

b) A pencil size of 9 centimeters is the most common.

2. Use the data given below from the eye color survey of 40 students, problem #8 in problem set 4.1, to construct a circle graph.

Blue eyes, 10   Brown, 19
Hazel, 9   Other, 2

Solution:
Blue: Brown: Hazel: Other:
$\frac{10}{40} = 25\%$ $\frac{19}{40} = 47.5\%$ $\frac{9}{40} = 22.5\%$ $\frac{2}{40} = 5\%$

Thus, 25% of 360 = 90° for blue eye category; 47.5% of 360 = 171° for brown; 22.5% of 360 = 81° for hazel eyes and 5% of 360 = 18° for other eye colors.

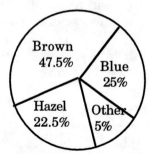

(Note that we used percentages again here but sometimes circle graphs are labeled with fractional equivalents instead.)

3. Make a bar graph of the following data on leaders in the history of the NBA for most field goal shots made (through 1994-95 regular season).[1]

    Kareem Abdul-Jabbar   15,837
    Wilt Chamberlain   12,681
    Elvin Hayes   10,976

Solution:

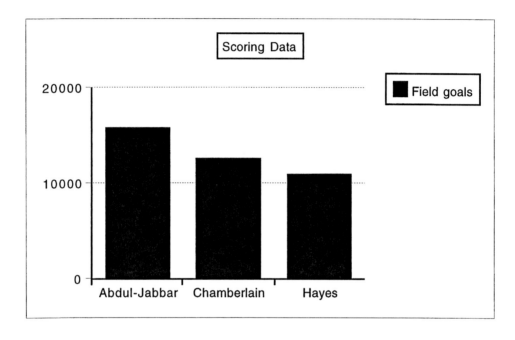

[1] Data as reported in the Marquette *Mining Journal*, Tuesday, May 16, 1995.

Problem Set 4.3

1. The heights (to the nearest whole inch) of 14 members of a college girls basketball team are: 66, 70, 70, 71, 71, 68, 69, 71, 72, 69, 72, 73, 70, 74.

   a) Draw a line plot of this data.

   b) Do you think this data is representative of the heights of all female students on this campus? Justify your answer.

2. Use this line plot to answer the following questions.

   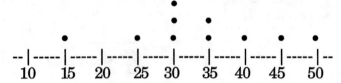

   a) What percent of the data lies *below* 25?

   b) What percent of the data lies *between* 25 and 35, endpoints not included?

   c) What data value occurred most often?

3. The following is the monthly budget of expenditures for a family of 4 in a midwestern town.

   >Clothing: $230
   >Food: $460
   >Electricity, water: $120
   >House Phone: $55
   >Car Phone: $25
   >House payment: $625
   >   (incl. insurances,taxes)
   >Miscellaneous: $300

   a) If all monthly income is budgeted as above, what is the total monthly income for this family?

   b) Draw a bar graph with these categories on the horizontal axis and the expenditures on the vertical axis.

   c) Find the percent of the budget represented by each category (to the nearest tenth of a percent).

   d) Draw a circle graph of the budget information and label accordingly.

4. Given the **double bar graph** on the top of the next page that relates the average salaries (in thousands) of men and women in the same rank at universities A, B, C, D, and E in Minnesota.[1] From the graph answer the following questions. (Series 1 represents salaries for men and Series 2 represents salaries for women.)

---

[1] Information reported in *Academe*, March/April, 1995.

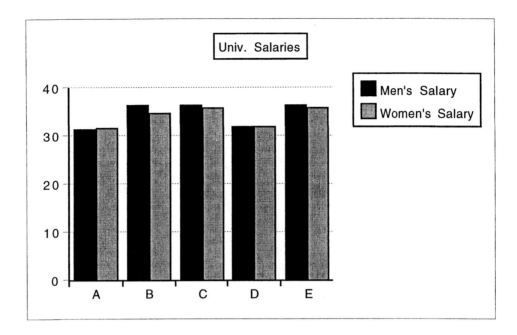

a) At which university, if any, are the salaries of men and women the same?

b) At which university does the greatest gap in salaries exist? Estimate the amount of difference in salaries.

c) Identify the university at which the average women's salary is larger than the average men's salary at this rank.

5. The table below gives the frequency distribution for the normal annual precipitation (in inches) for 29 cities of the U.S.[2]

Normal Annual Precipitation	Frequency
0 to less than 12	5
12 to less than 24	6
24 to less than 36	8
36 to less than 48	6
48 to less than 60	4

a) How many cities have less than 48 inches of precipitation annually?

b) How many cities have 36 or more inches of precipitation each year ?

c) What percent of the cities have less than 24 inches of precipitation annually?

6. The following data are the first quiz scores from a Beginning Algebra class at Gitchee Gumee College.

75 42 89 68 45 90 75 89 66 87 94 45 75 62 98 100 83 86 57 97

a) Make an ordered stem and leaf plot of the data.

b) Find the score that occurred most often, if one exists.

[2] Data from *Introductory Statistics*, P.Mann, 1992, Wiley.

7. The following circle graph indicates the portion of a $12.5 million school budget devoted to Administration, Instruction, Support Personnel, and Physical Plant. Find the dollar amounts devoted to each of these categories.

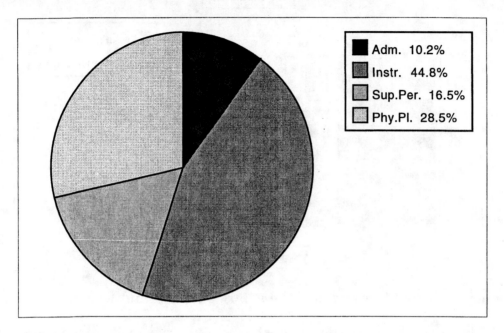

8. Toss one die 60 times, each time observing the top face and recording your results. Then make a bar graph of your data, with the possible outcomes (1,2,3,4,5,6) on the horizontal axis and the frequency of occurrence on the vertical axis. Finally, from your data, answer the following questions.

   a) What number occurred most often?

   b) What number occurred least often?

   c) How many times would you expect each number to occur? Explain your reasoning.

Exploration Problem

A **simulation** is an experiment that can be used to "simulate" a real-life context. The context here is that a cereal company has decided to package six different-colored pens in one of its cereal products for one year. Each box will contain only one pen. Assuming, of course, that consumers will keep buying the cereal until they have one pen of each color, devise an experiment to simulate this situation and find the average number of boxes a consumer must purchase to obtain one pen of each color.

Solutions - Problem Set 4.3

1. a)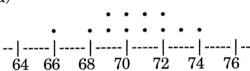

   b) No, because these girls are often recruited for height and probably, as a group, are taller than the girls on this campus, in general.

2. a) $1/10 = 10\%$ b) $3/10 = 30\%$ c) 30

3. a) $1815

   b)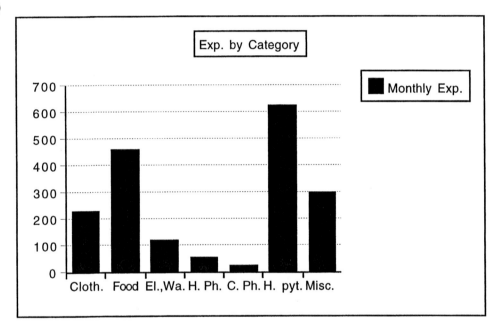

   c) Clothing: 12.7%   (45.6°)
   Food: 25.3%   (91.2°)
   Elect.,Wt.:   6.6%   (23.8°)
   H.Phone: 3.0%   (10.9°)
   C.Phone: 1.4%   (5.0°)
   H. Pyt.:   34.4%   (124.0°)
   Misc.:   16.5%   (59.5°)
   Totals: 99.9%   360°
       (Due to rounding, called the "round-off error")

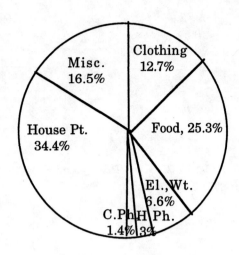

4. a) At University D   b) At B, about $2500   c) University A

5. a) 25 cities   b) 10 cities   c) $11/29 = 37.9\%$

6. a)
Stem	Leaf
4	2 5 5
5	7
6	2 6 8
7	5 5 5
8	3 6 7 9 9
9	0 4 7 8
10	0

    b) 75

7. The amounts are:
    Adm.: $1,275,000
    Instr.:   5,600,000
    Sup.Per.:   2,062,500
    Phy.Pl.:   3,562,500

8. a) and b): Results will vary.
    c) You would expect to get each number $1/6$ of the time. Thus, you'd expect each number to occur $1/6 (60)$ or 10 times.

Section 4.4 - Measures of Clustering and Box Plots

Measures of Clustering

After an exam, a question commonly asked is "What was the class average on this exam?" We probably have an intuitive sense that the average is the "middle." However, the **average** or **arithmetic mean**, or **mean**, really is the "weighted" average since data points that occur more than once "weight" the data in a sense. The mean is calculated by adding all the data values together and dividing by the total number of data points.

The notation, Σx, called *summation notation*, is the mathematical way of representing the sum of all the data values, and, in statistics, n usually is the total number of data points. The sample mean, symbolized by $\bar{x}$, is thus

$$\bar{x} = \frac{\Sigma x}{n}$$

To find the mean using any calculator, you can add the individual data pieces and divide the total by the number of data values. However, the calculation can be simplified with a scientific calculator by using the "memory plus" key, usually indicated on the calculator as Σ+ or M+. After the last data value is entered, use the "recall," usually the RM or MR key, and divide by the n, the number of data values entered. For the test score data on the previous page, we have

$$\bar{x} = \frac{66 + 80 + 58 + \ldots + 92}{40} = \frac{3138}{40} = 78.45$$

Notice that we did not round to the nearest integer and that the mean is, in this case, not one of the original data values.

Most scientific calculators have a statistics mode. Once in that mode, entering the data is all that is required, and various values, such as the mean, can be generated by pressing that specific key. Graphing calculators operate differently. Refer to your calculator manual for instructions.

Another measure of "clustering" is the data value that occurs most frequently in a data set. This value is called the **mode**. Sometimes, there is no mode, and other times there may be more than one value that occurs most often. It is conventional to identify one or two modes, but we rarely go beyond that. If there are two modes, the data set is said to be <u>bimodal</u>. In our test score data, the modes are 70 and 78 so the data set is bimodal.

<u>Sample Problem</u>

1. Given this data set, find the mode, if it exists, and use the M+ or Σ+ key on a scientific calculator (or use a graphing calculator) to find the mean.

   Data: 2.2, 3.1, 6.2, 5.1, 4.5, 8, 5.6, 7.2

   Solution: There is no mode here.
   The mean, $\bar{x}$, is 5.2375 or 5.24 (to the nearest hundredth).

The last indicator of "clustering" is the **median**. <u>**When the values are written in ascending, or descending, order, this is the point exactly in the middle of the data set.**</u> As a simple example, consider the data set:

$$0, 20, 2, 2, 5, 9, 7$$

First, write the numbers in chronological order:

$$0, 2, 2, 5, 7, 9, 20$$

It is not difficult to see that the data point "5" is located halfway between the lowest and highest values. However, if the data set is large, finding the median may be more difficult. To ease the process, there is a formula that we can use to find the *position* of the median. For n again representing the total number of data values, the *position* of the median is found by calculating

$$\frac{n+1}{2}$$

In our data set above, there are seven values so the median is the

$$\frac{n+1}{2} = \frac{8}{2} \text{ or 4th number (from either end).}$$

Therefore, in this example, the median is **5**.

What happens if there is an even number of data values? Let's look at the data set:

$$0, 3, 4, 2, 10, 3, 20, 12$$

Ordering these values: 0, 2, 3, 3, 4, 10, 12, 20 and counting, we have 8 data points. Using the formula for position, the median is at the

$$\frac{8+1}{2} = 4.5 \text{ position}$$

This means that the median is located halfway between the fourth and fifth number, in this case between 3 and 4, and we find the median by taking the <u>average</u> of these two data values. Therefore, the median is

$$\frac{3+4}{2} = 3.5$$

[Note again that the mean and the median aren't always actually data values in the set.]

It is very important to understand the context in which each of these measures is used to describe data. Let's consider these sample data sets listed in chronological order for convenience:

    Data Set 1: 1, 1, 2, 3, 3, 6, 8, 20
    Data Set 2: 1, 2, 2, 2, 4, 5, 5, 6

Some of the statistics of Data Set 1 are:

>Bimodal, 1 and 3
>Mean = 5.5
>Median = 3

For Data Set 2 we have:

>Mode = 2
>Mean = 3.375
>Median = 3

Note that only the median is the same for both sets. Drawing a line plot for both data sets using the same numerical scale helps the differences to "stand out."

Data Set 1:
```
          •   •
      • • •       •   •                                                 •
  -- | --- | --- | --- | --- | --- | --- | --- | --- | --- | --- | --- | --- | --- | --- | --- | --- | --- | --- | --- | --- |
     0   1   2   3   4   5   6   7   8   9  10  11  12  13  14  15  16  17  18  19  20
```

Data Set 2:
```
              •
              •       •
          • •     • • •
  -- | --- | --- | --- | --- | --- | --- | --- | --- | --- | --- | --- | --- | --- | --- | --- | --- | --- | --- | --- | --- |
     0   1   2   3   4   5   6   7   8   9  10  11  12  13  14  15  16  17  18  19  20
```

The values in Data Set 2 are more clustered. There is also a smaller gap between the low and high data value (a statistic called the **range**.) It is even more interesting to compare the means. The individual data values of both sets are not terribly different, except for the value of 20 in the first set. In fact, it is this data point that makes the mean of Data Set 1 larger than that of Data Set 2. Because each value is entered into the calculation of the mean, the "20" carries more weight and skews the mean to the right. Challenge: Change Data Set 1 to: 1, 1, 2, 3, 3, 6, 8, 8. How do you think this will affect the mean? Does it affect the median?

Sample Problems

1. A sample of 15 students was selected at random from a large sociology class. The students were asked: How many television sets do you have at your permanent residence? The responses were:
   >3, 1, 2, 2, 2, 4, 3, 0, 1, 1, 1, 3, 3, 2, 2

   a) Find the mean number of television sets at the permanent residences of this sample of students.

   b) Find the mode.

   c) Find the median number of television sets.

   d) Find the range in number of television sets.

Solution: a) $\bar{x} = \dfrac{3+1+2+\ldots+2}{15} = \dfrac{30}{15} = 2$

b) mode is 2

c) data in order: 0, 1, 1, 1, 1, 2, 2, 2, 2, 2, 3, 3, 3, 3, 4
so the position of the median is the

$\dfrac{15+1}{2} = 8^{th}$ number and the median = 2.

d) the range is $4 - 0 = 4$.

2. If the data set, 1,1,2,3,5,6,10,20, represents a sample of scores on a 20-point sociology quiz, which statistic, the mean or the median, "best" describes the class performance on this quiz? Why?

Solution: $\bar{x} = 6$ and the median is 5. The median seems to describe the class performance better because the mean was inflated by the scores of 10 and 20 when, in fact, most of the students scored lower than those numbers.

Box and Whisker Plots

Scores on standardized tests are sometimes returned to a school district by class and grade level so that teachers have information on their particular classes, as well as on how the students compared to other classes and to a developed "norm." This information is often given in graph form and a form that is being used more frequently now is the **box and whisker plot**, commonly called a **box plot**. For this type of graph we need the median, the lower extreme and upper extreme values, and the first and third quartiles.

We will construct a box plot for the test score data given at the beginning of section 4.3 and find each of the values needed as we proceed. The *ordered stem and leaf plot* for the data that we developed previously will help us to find the median. The ordered plot is reproduced below.

```
Stem  |  Leaf
------------------------
  4   |  5
  5   |  8
  6   |  2 5 5 6 6 8 8
  7   |  0 0 0 1 2 2 3 5 5 8 8 8
  8   |  0 0 1 5 6 7 7 8 9
  9   |  0 0 1 1 2 4 4 5 5 8
```

Since there were 40 test scores, the *position* of the median is

$$\dfrac{n+1}{2} = \dfrac{41}{2} = 20.5$$

Since the data are already in sequential order, we only need to count down to the 20th number and take the average of the 20th and 21st number.

The *median* is thus the average of 78 and 78 which equals 78.

The **lower extreme** is merely the lowest data value and the **upper extreme** is the highest data value.

In this case, the *lower extreme* is 45 and the *upper extreme* is 98.

The last two quantities needed are the first and third quartiles. If we draw a line through the median value, we now have half of the data above and half below this value. Counting the number of data points in each, and using the formula for the position of the median, we can generate the "median" of each half of the data. For the lower half this means we have generated the point $1/2$ of $1/2$ or 1/4 of the way from the beginning of the list. This is called the **first quartile**. For the second half, we have generated $1/2$ of $1/2$ or $1/4$ beyond the original median, or half-way point. Since $1/2 + 1/4 = 3/4$, we have now found the **third quartile**. The median is at the 20.5th position, and therefore there are 20 data values on each side of the median.

Thus,

1st quartile is at position $\dfrac{20 + 1}{2} = 10.5$

and, taking the average of 70 and 70, the *1st quartile*, designated as Q_1, is 70.

The *3rd quartile*, Q_3, is also in the 10.5 position, but *above* the median, so we need to take the average of 89 and 90 which is 89.5.

To make the graph, we first draw a number line labeled sequentially with values appropriate to the data.

```
--|---------|---------|---------|---------|---------|---------|---------|---------|---------|---------|
  45       50        55        60        65        70        75        80        85        90        95       100
```

In summary, the statistics we have found are:
> median is 78;
> $Q_1 = 70$;
> $Q_3 = 89.5$;
> L.E. = 45;
> U.E. = 98.

Below the line we then place a short vertical line at the median, and at both the first and third quartiles. Letting the latter represent the ends of a box, we connect these "sides" with horizontal lines as "lengths."

```
--|---------|---------|---------|---------|---------|---------|---------|---------|---------|---------|
  45       50        55        60        65        70        75        80        85        90        95       100
                                                   [=================|===========]
```

Finally, we make a point at the lower extreme, a point at the upper extreme and connect these points with horizontal lines to the box. These lines form the "whiskers." The box plot of the test score data is shown below.

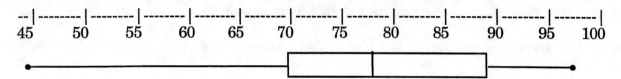

Note that box plots are extremely useful when comparing data sets. For example, let's examine the two box plots below of test score data from 4th graders on a state-mandated mathematics test.

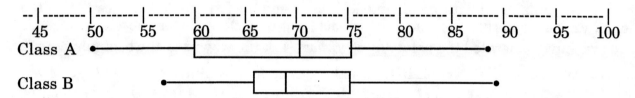

From these graphs we can determine that Class A had a higher median score but less clustering of scores between Q_1 and Q_3 (can you tell why?). Class A did have a lower extreme than class B. Class B had a better "top" score, had more clustering of scores and had a "better" lower score.

Sample Problems

1. In surveying a hospital's records for 15 consecutive days in July, statisticians found the following number of calls requesting an ambulance:

   3, 1, 0, 2, 5, 1, 2, 3, 0, 1, 0, 2, 2, 5, 3

   a) Find the median number of calls requesting an ambulance from this hospital in those 15 days.

   b) Find the first and third quartiles for this data.

   c) Give the lowest and highest number of calls requesting an ambulance.

   d) Draw a box plot for this data.

   Solution:

   a) 0,0,0,1,1,1,2,2,2,2,3,3,3,5,5 so the median number of calls per day is 2 (the 8th number).

   b) Separating the data with a line through the median, we have

   0, 0, 0, 1, 1, 1, 2, 2, 2, 2, 3, 3, 3, 5, 5

   so the 1st quartile is at position $\dfrac{7+1}{2} = 4$ and

thus, $Q_1 = 1$. The third quartile is 4 places to the right of the median and so $Q_3 = 3$.

c) The low and high values (the extremes) are 0 and 5, respectively.

d) The box and whisker plot:

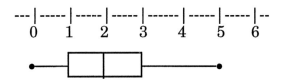

2. Shown below is a box plot of ambulance data from another town of comparable size during the same 15 days in July. Compare this information with that given in problem 1 above and write out your observations.

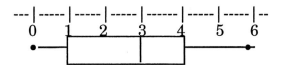

Solution:
The median number of calls per day is higher and the data is less clustered here. Also, although the lower extremes are the same, the upper extreme is greater for this town (that is, this hospital received as many as 6 calls in one day).

Problem Set 4.4

1. The following data represents a sample of heights (in inches) of students enrolled one fall in an elementary statistics course.

   58, 70, 72, 65, 64, 73, 72, 57, 66, 66, 68, 71

   a) Find the mean height. 12

   b) Find the median height. middle

   c) Give the modal height, if one exists. 66

2. Given the following data on the ages of 38 past U.S. presidents at the time of their death.

President	Age at Death	President	Age at Death
G. Washington	67	J. Garfield	49
J. Adams	90	C. Arthur	57
T. Jefferson	83	G. Cleveland	71
J. Madison	85	B. Harrison	67
J. Monroe	73	W. McKinley	58
J. Q. Adams	80	T. Roosevelt	60
A. Jackson	78	W. Taft	72
M. VanBuren	79	W. Wilson	67
W. Harrison	68	W. Harding	57
J. Tyler	71	C. Coolidge	60
J. Polk	53	H. Hoover	90
Z. Taylor	65	F. Roosevelt	63
M. Fillmore	74	H. Truman	88
F. Pierce	64	D. Eisenhower	78
J. Buchanan	77	J. Kennedy	46
A. Lincoln	56	L. Johnson	64
A. Johnson	66	R. Nixon	81
U. Grant	64	R. Reagan	93
R. Hayes	70	G. Ford	93

   a) Construct an ordered stem-and-leaf plot.

   b) Use this to find the median age at the time of the presidents' deaths.

   c) Find the first and third quartiles.

   d) Construct a box plot of this data.

   e) Find the percent of the data which lies between
      i)   the lower extreme and the first quartile.
      ii) the first quartile and the median.
      iii) above the third quartile.
      iv) between the first and third quartiles.
      Do these values differ with different data sets? Justify your reasoning.

3. Terrill received his phone bill which included several long distance calls. The length of each call is given in minutes.

   2, 1, 4, 1, 13, 39, 1, 1, 34, 3, 8, 1, 10, 28, 7, 3, 3, 3, 12, 10, 2, 23, 10, 1, 38, 50, 13, 13, 6, 33, 2.

   Find the mean length of his long distance calls, the median length, the mode if it exists, and the range. Which statistic best describes the length of time Terrill tends to spend on a long distance call? Justify your answer.

4. The age data for 18 students in a night-class of Cultural Geography are grouped in the frequency distribution given:

Age	Frequency (no. of students)
18	3
19	2
20	3
21	4
22	0
23	2
26	1
35	1
36	2

   a) Find the mean, median, and mode for this data and label each.

   b) Find the first and third quartiles, the upper and lower extremes, and then make a box plot of the age data.

5. An employee group of a company is bargaining for higher wages. They have the salaries of all employees, including the top administrators, accessible to them. This employee group is hoping to elicit support from the public sector and the local newspaper is willing to print an article on their bid for higher salaries. Which statistic, the mean or the median, of the salaries of all employees should they include in their press release? Why?

6. Search the Internet for graphs representing various categories of data. Based on your search,

   a) What type of graph seems to be used most frequently?

   b) Find two box plots based on comparable data and the same numerical scale. Compare the data sets by writing your observations.

Exploration Problems

1. Give an example of two data sets, each containing at least 4 pieces of data, with the same mean but different medians.

2. Give an example of two data sets, each containing at least 5 pieces of data, with the same median but different means.

3. As a representative of the employee group described in problem 5 above, it is your responsibility to write the article described. Write the article. Then write a rebuttal from the employer's point of view.

Solutions - Problem Set 4.4

1. a) $\bar{x} = 66.8$ inches   b) $\tilde{x}$ is 67   c) There are two modes, 66 & 72

2. a)   Stem   Leaf
   ```
   ----------------------
     4          6 9
     5          3 6 7 7 8
     6          0 0 3 4 4 4 5 6 7 7 7 8
     7          0 1 1 2 3 4 7 8 8 9
     8          0 1 3 5 8
     9          0 0 3 3
   ```

   b) $(68 + 70) / 2 = 69$ c) $Q_1 = 63$, $Q_3 = 79$

   d)
   ```
        --|-----|-----|-----|-----|-----|-----
          40   50   60   70   80   90
   ```

   e) i) 25%   ii) 25%   iii) 25%   iv) 50%   These percentages are not dependent on the data set. For example, 50% of the data <u>always</u> lies between the first and third quartiles.

3. mean = $375 \div 31 \approx 12.1$ minutes
   median = 7 minutes
   mode = 1 minute
   range = $50 - 1 = 49$ minutes
   Median value is the better description; there are only a few very lengthy calls and these tend to skew the mean.

4. a) The mean age is 23.1 years, the median is 21 years and the mode is 21 years.

   b) $Q_1 = 19$ and $Q_3 = 23$, the lower and upper extremes are 18 and 36, resp. The box plot is
   ```
        --|---|---|---|---|---|---|---|---|---|---|--
          18 20 22 24 26 28 30 32 34 36 38
   ```

5. The median salary because this statistic has a lower value. The mean salary is inflated by the salaries of the few top administrators and it would not be in the employees' best interest to advertise the higher figure.

6. a) and b): Answers will vary.

Chapter Four Review

1. A marble is selected at random from a box containing 2 red, 2 blue, and 3 green marbles. Find the probability that the marble is:

    a) green

    b) red or blue

    c) green and blue

2. Given a unit rectangle with regions marked as shown. If a dart is thrown at random and hits the rectangle, what is the probability that it lands on region

    a) "1"

    b) "3" or "4"

    c) not marked with a "2"?

    d) "1", "2", "3", or "4"

    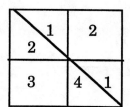

3. One red, 3 yellow, and 1 green chip are placed in a bag. A chip is selected at random, the color noted, then *replaced* into the bag and another chip drawn. Find the probability that:

    a) both chips are yellow.

    b) the first chip is green and the second chip is yellow.

    c) a green chip and red chip are selected.

    d) a red chip is selected on the second draw.

4. Draw the tree diagram and label the branches to depict the experiment of selecting two ping-pong balls, **without replacement**, from a box containing 2 gold, 3 purple, and 5 white balls. Then find the probability of selecting

    a) 2 gold balls

    b) at least one white ball

    c) a gold ball on the second draw, given a purple was drawn first.

5. List the sample space for all possible outcomes when 4 coins are tossed. Then find the probability of tossing

    a) exactly 2 heads   b) exactly 3 tails

6. Given the rectangle with regions marked A and B as described below.

   a) Which region appears to occupy the most area? Justify your answer without doing any calculations.

   b) If a region was selected at random, what is the probability of selecting a region marked "A"?

7. a) One card is drawn at random from a standard deck of 52 cards. Find the probability that the card

   i) is a "2"

   ii) has a number on it

   iii) is a queen or a heart

   b) Two cards are drawn at random, **with replacement**. Find the probability that

   i) a king and a "5" are drawn, in that order

   ii) a heart and a club are drawn

   c) Two cards are drawn at random, **without replacement**. Find the probability that

   i) a king and a "5" are drawn, in that order

   ii) two diamonds are drawn

8. Two boxes containing the words "PROBABILITY" and "STATISTICS", respectively are shown below.

   Box 1   Box 2

   a) If box 1 is selected and three letters are drawn **without replacement** and recorded in order, what is the probability that the letters spell "BAT"?

   b) If two letters are drawn from box 1, **without replacement**, recorded in order, and then two letters are drawn from box 2, **without replacement**, and recorded in order, what is the probability that the outcome is "BATS"?

9. A survey was taken in a college introductory statistics course of 30 students to gather information on the number of pairs of shoes owned by each student.

The information from the survey of these students is plotted below.

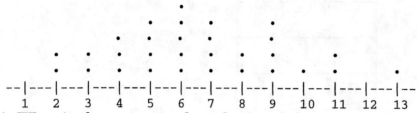

a) What is the mean number of pairs of shoes owned by these students?

b) What is the mode?

c) What is the median?

d) What is the range in number of pairs of shoes?

e) Find the lower and upper extreme values, the first and third quartiles and then draw a box and whisker plot of this data.

10. The following stem and leaf plot shows the pre-tax prices, to the nearest dollar, of several brand-name humidifiers (tabletop models).[1] Find the median and mean of this data. (Note that the ten's digit of each price is the stem.)

Stem	Leaf
2	3 5 7 7 9
3	7 8
4	3 5
5	3 5 9
6	0 4 4
7	4
8	
9	
10	5

11. This circle graph shows the percent of students in the given age categories at a midwestern university. If there are 8676 students attending this university, find:

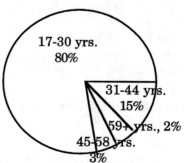

a) the number that are in the age bracket 17 - 30 years

b) the number that are 59 and older.

---
[1] from *Consumer Reports 1995 Buying Guide*, page 210.

12. Give an example of a data set with a median larger than the mean.

13. The following data represents the final score distribution in an intermediate algebra class of 24 students during the winter semester.

    63, 81, 82, 75, 78, 78, 86, 77, 93, 58, 69, 73, 72, 71, 68, 70, 76, 85, 86, 91, 92, 98, 89, 65

    a) Make a bar graph showing the frequency of A's, B's, etc. where the grading scale is

    90 - 100   A
    80 - 89   B
    70 - 79   C
    60 - 69   D
    59 and below   F

    b) Based on the grading scale above, make a bar graph showing the relative frequency (percent) of A's, B's, etc.

    c) Compare this graph with that from part a) above and write out your observations.

Exploration Problems

1. On a circular dartboard with radius = .5 feet, if the probability of landing in the bull's eye is .15, what is the area of the bull's eye circle?

2. You are making 100 cookies and have exactly 150 chocolate chips. Devise a way to estimate the probability that a cookie selected at random will contain no chocolate chips.

Solutions - Chapter 4 Review

1. a) 3/7   b) 4/7   c) 0

2. a) 1/4   b) 3/8   c) 5/8   d) 1

3. a) 9/25   b) 3/25   c) 2(1/25) = 2/25   d) 1/5

4.

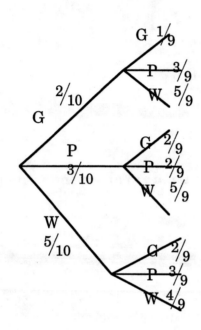

a) P(G and G) = 2/90 or 1/45

b) P(G_1 and W_2) + P(P_1 and W_2) + P(W_1 and G_2) + P(W_1 and P_2) + P(W_1 and W_2) =

$$\frac{10}{90} + \frac{15}{90} + \frac{10}{90} + \frac{15}{90} + \frac{20}{90} = \frac{70}{90} = \frac{7}{9}$$

c) 2/9

5. {HHHH,HHHT,HHTH,HHTT,HTHH,HTHT,HTTH,HTTT, TTTT,TTTH,TTHT,TTHH,THTT,THTH,THHT,THHH}
   a) 6/16 or 3/8   b) 4/16 or 1/4

6. a) In rows 2 and 3, there are equal parts designated to A and B. In row 1, however, more area is designated by "A". Thus, "A" seems to occupy more area.

   b) 10/18 or 5/9

7. a) i) 4/52 or 1/13   ii) 36/52 or 9/13   iii) 16/52 or 4/13

232

b) i) $\dfrac{4}{52} \cdot \dfrac{4}{52} = \dfrac{1}{169}$ ii) $\left(\dfrac{13}{52} \cdot \dfrac{13}{52} = \dfrac{1}{16}\right)2 = \dfrac{2}{16} = \dfrac{1}{8}$

c) i) $\dfrac{4}{52} \cdot \dfrac{4}{51} = \dfrac{16}{2652} = \dfrac{4}{663}$ ii) $\dfrac{13}{52} \cdot \dfrac{12}{51} = \dfrac{156}{2652} = \dfrac{1}{17}$

8. a) $\dfrac{2}{11} \cdot \dfrac{1}{10} \cdot \dfrac{1}{9} = \dfrac{2}{990} = \dfrac{1}{495}$ b) $\dfrac{2}{11} \cdot \dfrac{1}{10} \cdot \dfrac{3}{10} \cdot \dfrac{3}{9} = \dfrac{18}{9900} = \dfrac{1}{550}$

9. a) $\bar{x}$ = 6.6 pairs b) mode = 6 pairs c) median = 6 pairs

   d) range is 13 - 2 = 11 pairs e) Lower extreme = 2

   Upper extreme = 13, Q_1 = 5, Q_3 = 9

   --|---|---|---|---|---|---|---|---|---|---|---|--
    1 2 3 4 5 6 7 8 9 10 11 12 13

   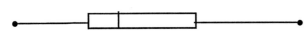

10. Median = $45   Mean = $48.71 (rounded to the nearest cent)

11. a) approx. 6941 students   b) approx. 174 students

12. One example:
    1, 1, 1, 3, 3, 3, 4
    Here the median is 3 and the mean is 2.29.

13. a)
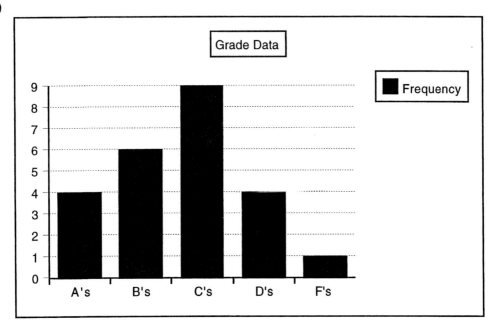

b)

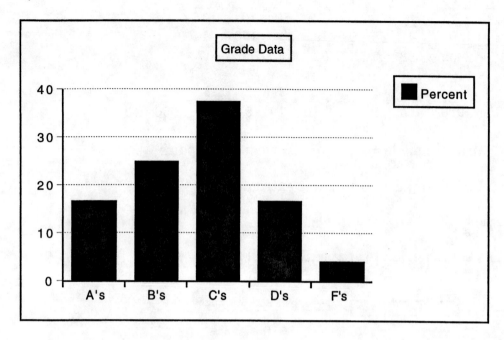

c) The graphs have the same shape, only the vertical scale has changed to reflect the values of the percentages.

Chapter 4 Worksheet 1

1. An experiment consists of drawing 1 slip of paper from a jar containing 12 slips of paper, each with a different month of the year written on it.
   a) Write out the sample space for this experiment.
   b) If 1 slip is drawn, what is the probability that the name of the month begins with "J"?
   c) If 1 slip is drawn, what is the probability that the name of the month has exactly 4 letters?

2. If one of the first 25 natural numbers (1,2,3,...25) is chosen at random, and each has an equally likely chance of being chosen, find the probability that
   a) the number is even.
   b) the number is at least 10.
   c) the number is less than 3 or greater than 20.
   d) the number is even and greater than 13.

3. A card is drawn from a standard deck of 52 cards. Find the probability that the card is:
   a) a heart
   b) a club or an ace
   c) a "10" or a king
   d) a club and an ace
   e) not a diamond
   f) not a heart or a king

4. A golf bag contains two red tees, four blue tees, and five white tees.
   a) If one tee is drawn at random from the golf bag, what is the probability that the tee is
      i) red
      ii) red or white

   b) If two tees are drawn at random, **with replacement**, from the golf bag, draw a tree diagram to show the sample space. Label the probabilities along the branches. Then answer the questions below.

      i)   What is the probability that both tees are blue?
      ii) What is the probability that the first one is blue and the second one white?
      iii) What is the probability that one is blue and one is white, not necessarily in that order?

   c) Suppose two tees are drawn at random, **without replacement**, from the same golf bag (containing 2 red, 4 blue, and 5 white tees).
      Draw a tree diagram to show the sample space for this experiment. Label the probabilities along the branches and then answer the questions below.
      i)   What is the probability that both tees are red?
      ii) What is the probability that a white tee is drawn first and then a red tee is drawn?
      iii) What is the probability that a white tee and a red tee are drawn, not necessarily in that order?
      iv) What is the probability that a white tee is drawn second?

v) What is the probability that a white tee is drawn second, given that a red tee was drawn first?

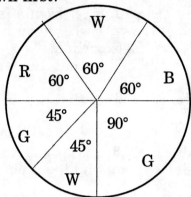

5. Given the spinner shown with red, white, blue, and green regions. If the spinner is spun once, find the probability that it will land in:
   a) a white region
   b) a green region
   c) a purple region
   d) a blue or green region
   e) a region which is neither blue nor green

Chapter 4 Worksheet 2

You have only the face cards of all suits, jacks, queens, kings, from which to choose.
1. One card is drawn at random. Find the following probabilities.

   a) P(queen)
   b) P(red king)
   c) P(red king or black jack)
   d) P(king or jack)
   e) P(king and queen)

2. 1 red, 4 yellow, and 3 blue tiles have been stored in a box. A tile is selected at random, **replaced**, and a second tile is drawn. Find the following probabilities.

   a) $P(R_1 \text{ and } B_2)$
   b) P(Y and B)

3. From the same set of tiles used in problem 2, a tile is drawn, **not replaced**, and a second tile is drawn.

   a) Make a tree diagram to represent this situation, including the probabilities for each branch.
   b) Find the probability of drawing 2 yellow tiles.

4. A pie is tossed at the 5' x 7' rectangular target shown. If it hits the rectangle, what is the probability that it hits one of the identical triangles?

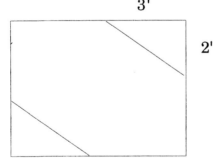

5. Design a geometric probability problem so that the P(success) = .25.

6. Two boxes contain letters as shown below.

   | OWOSSO | | SNOW |

   Box 1   Box 2

   a) If a box is chosen at random and then a letter is drawn, find the probability of drawing a "W".

   b) If a letter is drawn from box 1 and put into box 2 and then a letter is drawn from box 2, what is the probability that the last letter drawn is a "W"?

   c) If 2 letters are drawn from box 1, without replacement, and the order recorded, and a 3rd letter is drawn from box 2, what is the probability the outcome is "SOS"?

Chapter 4 Worksheet 3

1. Given the following data which is a sample of quiz scores on a 10-point quiz ($\frac{1}{2}$ points counted) in a beginning algebra class:

   1.5, 2, 10, 9, 8.5, 3.5, 2.5, 1, 2.5, 4, 8, 7, 7.5, 2, 1.5, 9

   a) Construct a line plot of the data.

   b) Construct a stem-and-leaf plot of the data; let the stem be numbers 1 - 10 and the decimal parts be the leaves.

   c) Find the mean, median and mode of the data.

   d) Which statistic, the mean or the median, best represents the data? Why?

   e) Find the first and third quartiles and make a box and whisker plot.

2. Given the following circle graph. If the total budget for this organization is $1,425,600, find the dollar amount allocated to each category.

   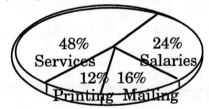

3. A university must absorb $1.28 million in budget cuts for the next academic year. The cuts have been allocated by division as follows:

University Advancement:	$ 10,240
University Relations:	$ 53,760
President:	$ 72,960
Student Affairs:	$ 90,880
Finance and Administration:	$227,840
Academic Affairs:	$824,320

   Construct a circle graph and a bar graph to represent this data.

CHAPTER FIVE - Linear Equations and Inequalities

Section 5.1 – Formal Equation Solving

When we first solved equations using the cover-up method, we circled the processes that needed to be done according to the order of operations. Our goal was to keep working until we had the term containing the unknown in the innermost circle. Then, we often performed one more operation to solve for the unknown or variable itself. When we used the pan balance method, we crossed out equal numbers of boxes and marbles that occurred on both sides to eliminate extra information and get a better picture of the content of the boxes. Let's examine both processes more carefully now.

In an equation such as $2x + 1 = 7$, we circled the "2x" term and realized that this had to equal 6 because $6 + 1 = 7$. We showed this as

$$\overset{6}{\boxed{(2x)}} + 1 = 7.$$

Thus, $\qquad 2x = 6$ and $x = 3$.

On the pan balance we represented the same equation as

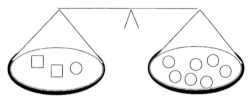

and crossed out equal quantities from both sides

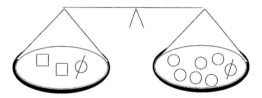

which led to $2 \cdot \square = 6$ (Note that $2x = 6$ as written above and $2 \cdot \square = 6$ mean the same thing!) and again

$$\square = 3$$

Another way of analyzing these models is through the comparison of the *symbolic* processes that correlate with specific steps. On the pan balance we subtracted 1 marble from each side to maintain the balance. We can show this as

$$2x + 1 - 1 = 7 - 1,$$

but, using the property, **a – b = a + (– b)**, we can also write

$$2x + 1 + (-1) = 7 + (-1).$$

Since $1 + (-1) = 0$ by the additive inverse property, our equation simplifies mathematically to

$$2x + 0 = 7 + (-1) \quad \text{or} \quad 2x = 6$$

[Note this latter statement shows up in all 3 methods!]

From the pan balance, we then had to consider two boxes balancing 6 marbles and use this information to determine the number of marbles in each box. Our approach with the cover-up method at this point was to think, "what times 2 is 6?" Of course, these two statements really say the same thing. The relationship between division and multiplication is what is important here. We can divide the marbles up equally between the 2 boxes by doing the operation, $6 \div 2$. Or, we can determine "what times 2 is 6" by using division.

Rewriting the statement

$$2x = 6 \text{ as}$$

$$\frac{2x}{2} = \frac{6}{2}$$

(which is the division of <u>both</u> sides by 2, again to maintain the balance)

gives us $\quad x = 3$, and this agrees with our result above.

[Check: $2x + 1 = 7$; $2(3) + 1 = 7$]

Now consider the problem

$$\frac{1}{3}x = 4$$

A reasonable interpretation of this equation would be: Four marbles are contained in $1/3$ of a box. Because our goal is to solve for $1x$, how many marbles are in one whole box?

We can again use the relationship between multiplication and division to solve for x. We isolate x by dividing both sides of the equation by $1/3$. However, since dividing by $1/3$ gives the same result as multiplying by its reciprocal, 3, it would definitely be easier to use the reciprocal here. Thus, we have

$$3\left(\frac{1}{3}\right)x = 3(4)$$

and $\quad x = 12$

[Check: $1/3(x) = 4$; $1/3(12) = 4$]

As another example to illustrate the mathematical process, consider the equation

$$4x - 6 = -2$$

Using the cover-up method, we would circle the "$4x$" term and write

$$\overset{4}{(4x)} - 6 = -2.$$

so $x = 1$.

Using our "new", or **formal**, method, we can employ the additive inverse to move the "–6" and isolate the "4x" term as shown below.

$$4x - 6 + 6 = -2 + 6$$ [We add the <u>opposite</u> of –6, which is +6, to both sides because 4x – 6 is the same as 4x + (–6).]

Then $4x + 0 = 4$

and $4x = 4$.

(Note how this relates to the circled quantity on the bottom of the previous page.)

Finally, to solve for x, we only need to use the division process and we have

$$\frac{4x}{4} = \frac{4}{4}$$

and $x = 1$ [Check: 4(1) – 6 = –2. This solution also agrees with our earlier result.]

<u>Sample Problem</u>

Show how both the cover-up (circle) method and the formal method can be used to solve for n in the equation:

$$4n - 5 = 7$$

Solution:

Cover-up method   **Formal** method

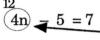
$\overset{12}{(4n)} - 5 = 7$ $4n - 5 + 5 = 7 + 5$ (using add. inv. of –5)

$4n = 12$ ⟵⟶ $4n = 12$

and $n = 3$

$$\frac{4n}{4} = \frac{12}{4}$$

so $n = 3$

Several important points must be made here.

1) The formal method of solving an equation uses the **additive and multiplicative properties of equality** that we discussed in section 1.3. That is, **an equation's balance will be maintained when the same quantity is added to (or subtracted from) both sides or when both sides are multiplied (or divided) by the same quantity.**

2) The goal of solving any equation is **first to isolate the term containing the variable before attempting to solve for the variable itself.** We did this in all methods that we have discussed.

In the examples we have done so far in this section, the variable appeared in only one term of each equation. However, in equations like

$$3x - 1 = 2x - 5$$

the variable appears in more than one term. (Recall that <u>terms</u> are separated by plus and minus and equal signs). Previously, when the variable appeared on both sides of the equals sign, we used a table to help find the value of x that made the statement true. For some of these problems we used the pan balance, but then our constant terms were always positive. If one or more terms in an equation is negative, that model is not as useful. The cover-up method is not helpful here, either.

The <u>formal</u> method will take any "guesswork" out of this problem. First, the "like terms" must be collected as we attempt to isolate the variable. (<u>Remember</u>, **like terms** are terms we can combine using the distributive property. They actually contain the same variable(s) and the same exponents on those variables.) The easiest way to combine *like terms* is to <u>move</u> terms as required to gather them on the same side of the equal sign. As we have shown, <u>moving</u> can be accomplished by <u>adding the inverse of each term being moved</u>.

For the problem
$$3x - 1 = 2x - 5$$

we'll first collect the "x" terms by moving the $2x$. We'll add the opposite of $2x$ to both sides. (We could have chosen to move the $3x$ instead; either is correct.)

Thus, we have
$$3x - 1 = 2x - 5$$
$$3x - 1 + (-2x) = 2x - 5 + (-2x)$$

$$3x + (-2x) - 1 = 2x + (-2x) - 5 \quad \text{(using the commutative and associative properties)}$$

and, finally,

$$x - 1 = -5.$$

Again using the additive inverse property to isolate x, we add $+1$ to both sides and we have
$$x - 1 + 1 = -5 + 1$$
or
$$x = -4$$

(Check: $3x - 1 = 2x - 5$; $3(-4) - 1 = 2(-4) - 5]$

To summarize, the **formal method** of solving an equation can be thought of as completing three general steps in the order given:

1) On each side of the equation combine like terms, if possible.

2) Move terms using additive inverses; combine like terms.

3) Solve for the variable by dividing by its coefficient (or multiplying by the reciprocal of the coefficient).

This method is advantageous when an equation has two or more terms that contain the variable and when an exact answer is required and the answer appears not to be an integer. In chapter 2 we worked on the equation, 3n + 4 = 5, found the nearest integer solution and then used the property of reciprocal to find the exact answer. We now can find the exact solution by combining the circle and formal methods.

$$\widehat{(3n)} + 4 = 5$$

which implies $3n = 1$

and

Thus, $\left(\frac{1}{3}\right)3n = \left(\frac{1}{3}\right)1$

$$n = \frac{1}{3}$$

Follow along the steps used to solve the next equation and note the justification for each step.

$3(x + 4) = -x + 7$

$3x + 12 = -x + 7 \longrightarrow$ (the distributive property to write out all terms)

$3x + 12 + x = -x + 7 + x \longrightarrow$ (collect the x-terms using the additive inverse of $-x$)

$3x + x + 12 = -x + x + 7$ (the commutative property to gather like terms)
$4x + 12 = 7 \longrightarrow$ (the distributive property to combine like terms)

$4x + 12 + (-12) = 7 + (-12) \longrightarrow$ (collect constant terms using the additive inverse of 12)

$4x = -5 \longrightarrow$ (integer arithmetic)

$\dfrac{4x}{4} = \dfrac{-5}{4} \longrightarrow$ (using 4 as the divisor to isolate the variable)

and $x = -5/4$ or -1.25

[Check: $3(x + 4) = -x + 7$; $3(-5/4 + 4) = -(-5/4 + 7)$]

Sample Problems

1. Solve each equation for x. Use any method you wish.

   a) $5x + 2 = -4x - 3$ b) $2(x - 1) = x + 1/2$ c) $4x - 2 = 9$

   Solutions:

   a)
   $5x + 2 = -4x - 3$
   $5x + 2 + 4x = -4x - 3 + 4x$
   $5x + 4x + 2 = -4x + 4x - 3$
   $9x + 2 = -3$
   $9x + 2 + (-2) = -3 + (-2)$
   $9x = -5$
   $x = -5/9$

   Check:
   $5(-5/9) + 2 = -4(-5/9) - 3$
   and $-.78 = -.78$ (approximately)

b)
$$2(x - 1) = x + \tfrac{1}{2}$$
$$2x - 2 = x + \tfrac{1}{2}$$
$$2x - 2 + (-x) = x + \tfrac{1}{2} + (-x)$$
$$x - 2 = \tfrac{1}{2}$$
$$x - 2 + 2 = \tfrac{1}{2} + 2$$
$$x = 2\tfrac{1}{2} \text{ or } \tfrac{5}{2}$$

Check: $2(\tfrac{5}{2} - 1) = \tfrac{5}{2} + \tfrac{1}{2}$
and $3 = 3$.

c) $\overset{11}{\boxed{4x}} - 2 = 9$

$4x = 11$ Check: $4(\tfrac{11}{4}) - 2 = 9$
$x = \tfrac{11}{4}$ and $11 - 2 = 0$

Additional Comments:
1) When solving any equation, since the <u>addition</u> of the additive inverse is the tool used to move terms, and addition can be done in any order by the commutative property, then we can add those terms in any order we wish on both sides. In other words, in the problem $5x + 2 = -4x - 3$ above, we could have added $4x$ next to the $5x$ right away, resulting in

$$5x + 4x + 2 = -4x + 4x - 3$$

This would have saved one step [the second line in a) above] and is a "legal" move!

2) Finally, the unknown can be isolated on *either* side of the equal sign. We tend to isolate the variable on the left, perhaps because we are taught to read from left to right, but since the statement of equality means that both sides have the same weight, it doesn't matter on which side of the equation the variable is written as long as we gather the constant terms on the opposite side.

<u>Sample Problems</u>

1. Solve for *n* and graph the solution on the number line.

$$-4(n + 3) = -6n - 2$$

Solution:

$$-4(n + 3) = -6n - 2$$
$$-4n - 12 = -6n - 2$$
$$+4n - 4n - 12 = -6n + 4n - 2$$
$$-12 = -2n - 2$$
$$+2 - 12 = -2n - 2 + 2$$
$$-10 = -2n \quad \text{(note here that we divide by } -2 \text{ since we want}$$
$$5 = n \quad \text{the value of } +1n, \text{ not } -1n.)$$

[Check: $-4(5 + 3) = -6(5) - 2$ and $-32 = -32$.]

The graph looks like

$$\leftarrow | \; | \; | \; | \; | \; | \; | \; \bullet \; | \; | \; | \rightarrow$$
$$-2 \; -1 \; 0 \; 1 \; 2 \; 3 \; 4 \; 5 \; 6 \; 7 \; 8$$

which consists of one point, the value 5.

2. Solve the equation, $3 = \dfrac{2x+5}{9}$ for x using both the circle method and the formal method.

Solution:

First, using the circle method, we have

$$3 = \dfrac{\overbrace{\boxed{2x}+5}^{22 \quad 27}}{9} \text{ and } x = 11$$

Using the formal method,

$$(9)3 = \dfrac{2x+5}{9}(9)$$

and

$$27 = 2x+5$$
$$27-5 = 2x+5-5$$
$$22 = 2x$$
so $\quad 11 = x.$

There are some "special case" equations which we encounter on occasion. Let's look at two examples:

1) $3x - 6 = 3(x - 2)$ and   2) $3x - 5 = 3(x - 2)$

In the first problem, removing parentheses leads to

$$3x - 6 = 3x - 6.$$

Using the additive inverse of the variable term, and adding $-3x$ to both sides of the equation, the result is

$$-6 = -6$$

which is certainly a true statement. However, no variable term remains. Because the constant terms are identical, we call this type of equation an *identity* and we say that the solution is **all real numbers**. We can verify this conclusion by choosing *any* value for the variable and substituting that value in the original equation. The result will be a true statement. The graph of the solution to this equation consists of the entire number line!

For the second example, removing parentheses leads to

$$3x - 5 = 3x - 6.$$

Adding $-3x$ to both sides of the equation results in the statement.

$$-5 = -6$$

which is clearly *not* true. In fact, we can describe the statement as a *contradiction* and we can conclude that there is no solution for the equation because any value we may choose for the variable results in the same type of contradiction. (Verify this statement by trying some values for x.)

In summary, in each of these special cases we collected terms containing variables and obtained a statement with no variables remaining. One statement was true, the other false. The true statement led us to conclude that there are an *infinite number of solutions* to the original equation. The false statement led to the conclusion that there are *no* solutions to the original equation.

Sample Problems

Solve each equation for the variable used. If the solution exists, graph it on the number line.

1. $2(n - 5) = 2n + 10$

    Solution:
    $$2(n - 5) = 2n + 10$$
    $$2n - 10 = 2n + 10$$
    $$2n - 2n - 10 = 2n - 2n + 10$$
    $$-10 = 10 \quad \text{a false statement}$$
    Thus, there is no solution.

2. $3x + .45 = 3(x + .15)$

    Solution:
    $$3x + .45 = 3x + .45$$
    $$3x - 3x + .45 = 3x - 3x + .45$$
    $$.45 = .45 \quad \text{an identity}$$
    Thus, the solution is all real numbers.

    The graph is

3. $3x + 4 = 2x + 4$

    Solution:
    $$3x + 4 = 2x + 4$$
    $$3x - 2x + 4 = 2x - 2x + 4$$
    $$x + 4 = 4$$
    $$x + 4 - 4 = 4 - 4$$
    $$x = 0$$
    Check: $3(0) + 4 = 2(0) + 4$
    $$4 = 4$$

    The graph is

Writing Equations from Problems

Previously we have written algebraic expressions by translating words into symbols. Now we'll extend the translation process to include writing equations and then solving them.

For example, if the lengths of the sides of a triangle (in inches) are three consecutive integers, we can represent the sides by n, $n+1$, and $n+2$. If the perimeter of this triangle is 48 inches, we can write an equation for the perimeter in terms of n, the length of the first side as

$$n + n+1 + n+2 = 48$$

and we can find the lengths of each of the sides by solving this equation for n.

$$n + n+1 + n+2 = 48$$
(collecting like terms) $3n + 3 = 48$
(adding -3) $\quad 3n = 45$
(dividing by 3) $\quad n = 15$

If $n = 15$, then $n + 1 = 16$ and $n + 2 = 17$.

Therefore, the sides of the triangle are 15", 16" and 17".

Check: $P = 15" + 16" + 17" = 48"$

Sample Problem

The length of a rectangular garden is 25% greater than the width. The perimeter is 72 yards. Using w to represent the width, write an equation for the perimeter and find the dimensions of the garden.

Solution:
      Let w represent the width.
Then $(1 + .25)w$ or $1.25w$ represents the length.

The equation is: $2w + 2(1.25w) = 72$
$$4.5w = 72$$
$$w = 16$$

Therefore, the width is 16 yards and the length is $1.25(16) = 20$ yards.
Check: $P = 2(16) + 2(20) = 72$ yards.

Problem Set 5.1

1. For each of the following, substitute $x = 2$ to determine whether or not 2 is a solution or root.

   a) $\dfrac{x+2}{2} = 2$ b) $3x - 4 = -1$ c) $5x - 3 = 2x - 3$

2. Solve the equation $6x + 3 = -15$ using the cover-up (circle) method *and* the formal method, step by step, side by side.

3. Draw the pan balance that represents the equation $4x + 2 = x + 8$ and use that method to solve the equation for x. For each step, write the equivalent algebraic step.

4. Write the equation symbolizing each problem statement below. Be sure to define the variable you use and then solve for this variable. Check to make sure your solution is reasonable and then state the answer in a complete sentence.

   a) Three more than twice a number is 9. Find the number.

   b) A rectangle is 6 inches longer than it is wide. The perimeter is 32 inches. Find the length and width of the rectangle in inches. Then find the area of the rectangle.

5. a) Write an expression that represents the perimeter of the region below.

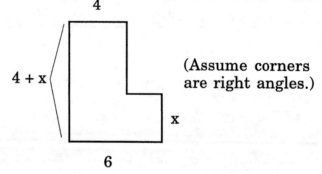

(Assume corners are right angles.)

Given that the perimeter of this figure is 26 units,

   b) Write an equation for the perimeter and find the length represented by x.

   c) Find the area of this region.

6. Solve each of the following equations using either the cover-up or formal method. An exact answer is required.

   a) $\dfrac{x}{5} = 9$ b) $\dfrac{2}{5}n = 10$

   c) $3(x + 1) = 5$ d) $6x - 1 = 2x + 10$

e) $\frac{1}{2}x + 4 = \frac{1}{4}x - 4$ f) $\frac{y}{y+1} = 0$

g) $4(n - 2) = -8$ h) $5(x - 1) = 5x - 5$

i) $x^2 = 25$ (two answers here!)   j) $6x + 4 = 6x - 2$ k) $4(x - 1) = -4 + 4x$

l) $\frac{7a + 3}{3} = 8$ m) $\frac{4}{x + 1} = -2$

7. The population of Slapneck Creek grew by 2% in 1992. If the population of this town at the beginning of 1992 was 825, find the population of Slapneck Creek at the end of 1992. (Round to the nearest whole person.)

8. A sales clerk in a clothing store works for $6.25 an hour plus 7% commission on all sales. If this person worked x hours in one week and sold $d worth of merchandise that week,

   a) write an expression, using x and d, to represent the clerk's salary that week.

   b) If the clerk worked 30 hours that week, write an expression for the clerk's salary that week.

   c) If the clerk worked 30 hours and sold $1225 worth of merchandise that week, what was the clerk's salary that week?

9. A 12-foot board needs to be cut into 2 pieces so that one piece is twice as long as the other. Use s as the length of the shorter piece. Then

   a) write the length of the longer piece in terms of the length of the shorter piece;

   b) write the equation and find the length of each piece.

10. A car on an interstate highway traveled at an average rate of 65 miles per hour. At this rate it took 4 hours to go 260 miles (since d = rt = 65(4)). At the same rate, how long will it take to go 320 miles on this same interstate highway, to the nearest tenth of an hour?

11. The Pizza Palace has a computer that automatically registers the price of the regular pizza menu but the computer is programmed to compute the cost of an individual order (excluding tax) by the equation

    Cost = $6.50 + $.50x

    a) Give the probable meaning of x in this equation.

    b) Write an explanation of the numbers in this equation so that the general public could understand what the equation means.

c) If the Pizza Palace is in a state where the tax rate is 6%, write an expression to represent the amount of tax on an individual order and simplify that expression.

12. Write a word problem, in a real-life context, from which the following equation could be generated. Then solve the problem and write the answer in a complete sentence.
$$3.25x + 40 = 235$$

13. Write an equation in one variable whose solution is represented by the following graph:

14. On four tests in beginning algebra, Pat's scores were: 72, 95, 88, 82. After the fifth test, her average was 85.4. What score did she receive on the fifth test?

Exploration Problems

1. A man spent 18 years of his life at school, living at home. He then spent $1/14$ of his life in college, $3/7$ of his life working at his job, and then spent 24 years in retirement. How old was this man when he died?

2. It took Bill 9 hours to empty all of the sap buckets on the family sugar bush. The next day, Margery helped, and the task took only 5 hours. How long would it take Margery to empty all of the buckets working alone?

Solutions - Problem Set 5.1

1. a) yes   b) no   c) no

2. Cover-up   Formal method

   $\overset{-18}{\boxed{6x}} + 3 = -15$ $6x + 3 = -15$

   $x = -3$ $6x + 3 + (-3) = -15 + (-3)$

                                       $6x = -18$

                                       $\dfrac{6x}{6} = \dfrac{-18}{6}$

                                       $x = -3$

3.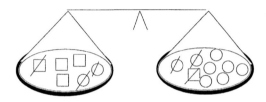

                               $4x + 2 = x + 8$

   so $3 \cdot \square = 6$ $4x - x + 2 = x - x + 8$

   and $\square = 2$ $3x + 2 - 2 = 8 - 2$

                              $3x = 6$

                              $x = 2$

4. a) Let n be the number. Equation: $2n + 3 = 9$. Solution: $n = 3$.
   The number is 3.
   Check: $2(3) + 3 = 9$.

   b) Let x = the width of the rectangle.
   Then the length is x + 6.
   Equation: $x + x + (x + 6) + (x + 6) = 32$.
   Solution: x = 5.
   The width of the rectangle is 5 inches and the length is 11 inches.
   Check: $5 + 5 + 11 + 11 = 32$.
   Area = 55 sq. inches.

5. a) $P = 4 + (4 + x) + 6 + x + 2 + 4$
   $= 2x + 20$

   b) $2x + 20 = 26$ so $x = 3$ units.

   c) Breaking the region into 2 pieces: A = 16 sq. units + 18 sq. units = 34 sq. units

6. a) x = 45   b) n = 25   c) x = $2/3$ d) x = $11/4$

   e) x = –32   f) y = 0   g) n = 0   h) all real numbers

i) $x = +5, -5$ j) no solution   k) all real numbers   l) $a = 3$ m) $x = -3$

7. Approximately 842 people will live in Slapneck Creek at the end of 1992.

8. a) $\$6.25x + .07d$ b) $\$6.25(30) + .07d$ or $\$187.50 + .07d$

   c) $\$6.25(30) + .07(\$1225) = \$273.25$

9. a) The 2nd piece has length 2s.

   b) $s + 2s = 12$ Solution: One piece is 4 feet and the other is 8 feet long.

10. Let t be the time needed. Equation: $65t = 320$, so $t = 4.9$ hours. It will take 4.9 hours to go 320 miles at an average rate of 65 mph on this freeway.

11. a) x is probably the number of extra toppings requested.

    b) The equation, $6.50 + .50x$ gives the price, before tax, of an individual pizza. The 6.50 is the cost of a basic 14" cheese pizza and the .50x term represents the cost of x toppings at .50 per topping.

    c) $(6.50 + .50x).06 = .39 + .03x$

12. Example: In a home business, the cost of producing each large jar of home-made thimbleberry jam is $3.25. One week, if the overhead costs of production, including electricity, etc., totaled $40, and the total cost of producing the jam was $235, how many jars of jam were produced that week?
    For x jars of thimbleberry jam
    $$3.25x + 40 = 235,$$
    and
    $$3.25x = 195$$
    $$\text{so } x = 60$$
    Sixty jars of thimbleberry jam were produced that week.

13. For example: $2x - 4 = -6$.

14. 90 points

Section 5.2 - Development of Algebraic Equations in Two Unknowns

As we continue to work with equations, we need to spend more time on the development and meaning of mathematical relations and on solutions to equations in two unknowns. Consider the table below. The left side and right side are related by a rule. Can you find the rule?

x	y
3	7
1	5
−2	2
0	4

After a little examination, it is not difficult to see that the numbers on the right side are all 4 more than each of the corresponding numbers on the left. We can write this relation between the sides as a mapping:

$$x \longrightarrow x + 4,$$

This illustrates that a number, x, on the left corresponds (**or maps**) to the number, $x + 4$, on the right.

But we can also write this as an equation in two variables, x and y, as

$$y = x + 4$$

[Note that we can show the relationship in the other direction, as $y \longrightarrow y - 4$ or as an equation, $x = y - 4$.]

Let's examine a different kind of chart.

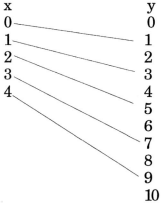

To find the rule relating the x and y values as shown in the chart, observe that the y-values in the "mapping" are all odd numbers. Also note that each y-value is more than double the x-value. In fact, the value on the right is 1 more than double the corresponding x-value, so we can show the mapping as

$$x \longrightarrow 2x + 1.$$

We can again generalize this by writing

$$y = 2x + 1$$

and we can use the equation to extend the table and find the y-value that corresponds to an x-value of 20, for example. This is an **equation** in **two unknowns** since two variables, x and y, are used. Note that, given the rule, we can replace x with any number and generate the corresponding y-value. Again, we could extend the chart indefinitely, and the use of the <u>variable</u> just means that our replacement values for x can <u>vary</u> and the y-values will <u>vary</u> accordingly. This **equation** is a way of expressing the rule given by the **relationship** between the numbers in the chart or table.

Sample Problems

1. Find at least 3 other values of x and y that satisfy the equation, $y = 2x + 1$.

   Examples: $x = 5, y = 11$; $x = \frac{1}{2}, y = 2$; $x = 8, y = 17$

2. The following chart relates the numbers on the left to the corresponding numbers on the right. Describe the rule of this relation using the "arrow" notation and then write the rule as an equation.

L	R
3	8
4	11
6	17
8	23
10	29

   Solution:
   Since the rule has to work for all corresponding pairs of numbers, x and y, we must do some trial and error here. Examining the numbers on the left and corresponding numbers on the right, we notice that doubling the numbers on the left and adding a quantity doesn't consistently generate the numbers on the right. However, if we multiply the left side by 3 and then subtract 1, it does seem to do it. We'll let L represent the number on the left side; then $(3 \cdot L - 1)$ will represent the corresponding number on the right side (*in terms of L*) and the relationship can be shown as

   $$L \longrightarrow 3 \cdot L - 1$$
   (or we can again say that   L **maps to** $3L - 1$)

   The equation is
   $$R = 3 \cdot L - 1.$$

   Check: If $L = 3$, $3 \cdot L - 1 = 8$; if $L = 4$, $3 \cdot L - 1 = 11$, etc.

3. In problem 2 above, if the numbers on the right, R, are given, find the rule that relates these numbers to the corresponding numbers on the left, L, and write the rule as an equation.

   Solution:
   Examining the correspondence from left to right as written above, it would seem reasonable that to go in the opposite direction we should consider doing the

opposite processes, in the *opposite order*. Instead of subtracting 1, we'll *add* 1 (and we'll do this first since subtraction was the *last* process done in problem 1; then we'll divide by 3 instead of multiplying by 3. The resulting equation in terms of R is

$$L = \frac{R + 1}{3}$$

From the correspondences in the original table in problem 1, if $R = 8$, $L = 3$. Substituting $R = 8$ into our equation above

$$L = \frac{8 + 1}{3} = 3 \quad \text{and we have the same result!}$$

(We can easily verify the equality with other pairs of numbers from the table!)

4. If x and y are related by the equation, $y = x + 2$, find four values of x and y that make the statement, $y = x + 2$, true and write the values as ordered pairs.

Examples: (0,2), (–1,1), (5,7), (10,12).

Graphing in Two Variables

Let's again examine the equation $y = 2x + 1$ from our "rule" discussion in this section and write the ordered pairs of x and y values from the original chart. They are:

$$(0,1), (1,3), (2,5), (3,7), \text{ and } (4,9).$$

We'll now extend our original chart by substituting even more values for x into the equation, $y = 2x + 1$, as we did in the sample problem 1 on the previous page.

If $x = \frac{1}{4}$, $y = 2(\frac{1}{4}) + 1 = \frac{3}{2}$ and an ordered pair is $(\frac{1}{4}, \frac{3}{2})$.
If $x = –2$, $y = 2(–2) + 1 = –3$ and the ordered pair is $(–2,–3)$.
If $x = –1$, $y = 2(–1) + 1 = –1$ and we have the ordered pair $(–1,–1)$.

Note that our y-values aren't all integers or positive numbers as we try different values of x, but that all the ordered pairs still make the statement, $y = 2x + 1$, true. Based on our previous discussion, <u>all</u> these ordered pairs are **solutions** to the equation.

We can find even more solutions by drawing a graph of the equation $y = 2x + 1$ on a coordinate system. The ordered pairs we have already generated are labeled.

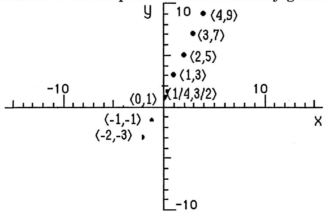

These points appear to lie on a straight line. Substituting other values for x in the equation $y = 2x + 1$, we have the ordered pairs: $(-4,-7)$, $(-\frac{1}{2},0)$, $(-3,-5)$, for example. Plotting these additional points on the graph, we can see that they, too, fall along the same straight line.

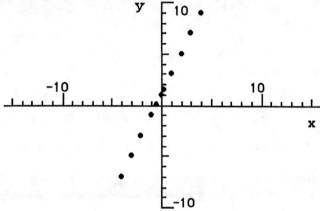

Since x could represent *any* number and we could continue to find ordered pairs, (x,y), that make the statement true, it is appropriate to connect the points on the graph with a straight line. Thus, we have

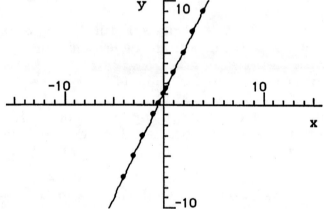

The important conclusion here is that the x and y-coordinates of <u>any</u> point on this line will satisfy the equation $y = 2x + 1$, and that this straight line on the coordinate system is the set of all points, (x,y), that are solutions to $y = 2x + 1$.

Sample Problems

1. Find two other ordered pairs that satisfy the equation, $y = 2x + 1$.

    Examples: $(-4,-7)$, $(5/2, 6)$, etc.

2. From the graph above, what value of x corresponds to a y-value of -5?

    Solution: Find $y = -5$ on the vertical axis and go over to the graph to find the solution, $x = -3$.

3. Is the point $(4,1)$ a solution to the equation, $x + 2y = 5$?

    Solution: No, since substituting $x = 4$ and $y = 1$ into the equation does not result in a true statement, as shown on the top of the next page:

$$4 + 2(1) \neq 5$$

4. Given the following set of ordered pairs:

$$\{(0,1)\ (1,2),\ (2,5),\ (3,10),\ (-1,2),\ (-2,5),(-3,10)\}.$$

a) Find the rule that relates the *x* and *y* coordinates and write the rule in the form of an equation.

b) Graph these points. Does the rule appear to generate a straight line?

Solution:
  a) Studying the values of the *x* and *y* coordinates, it appears the y-values are increasing much faster than the x-values. If we try doubling *x* and adding a number, say, 1, there is still a big gap, especially when x = 4. Let's try squaring the x-value and adding a number. This appears to work and the correspondence that relates these coordinates is

$$x \longrightarrow x^2 + 1 \quad \text{and the equation is} \quad y = x^2 + 1.$$

b) The graph of the ordered pairs given is shown below.

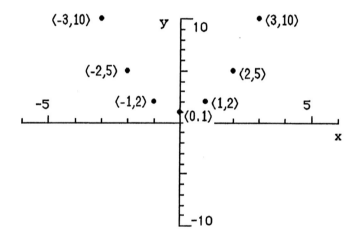

Note that the points do not seem to form a straight line. Since we have a limited number of ordered pairs, we should not immediately connect the points with a curve. However, we could generate more ordered pairs by substituting other values for *x* into the equation, $y = x^2 + 1$, and finding the corresponding *y*-values. We can then plot these additional points on the same coordinate system. All the points do actually fall on the same curve, and after generating many points, we can connect the dots to form the curve as shown on the top of the next page.

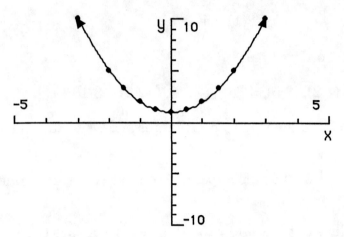

[We have seen a graph of this shape, although in a different position, in chapter 1 when we investigated the relationship between width and area of various rectangles, each with a perimeter of 100 feet!]

5. The diagram below represents a "function machine." The value at the top is put into the machine and the value given at the bottom is the output. Find a rule that would take the value 3 and generate an output of 9. Express the rule in terms of n, the input value.

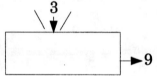

Examples: Output = n^2;   Output = $n + 6$

Graphing Two Linear Equations

Let's extend our discussion of the graphs of linear equations by graphing *two equations*, say

$$y = 2x + 1 \text{ and } y = x - 1$$

on the same coordinate system.

First, we make a chart (sometimes called a **T-table**) of a few ordered pairs, or solutions, for each equation. We choose "easy" values to substitute for x, perform the required computations and we have

```
        y = 2x + 1                          y = x - 1

        x  |  y                             x  |  y
        -------                             -------
        0     1                             0    -1
        1     3                             1     0
        2     5                             2     1
       -1    -1                            -1    -2
```

When we graph each of these sets of ordered pairs on the coordinate system shown,

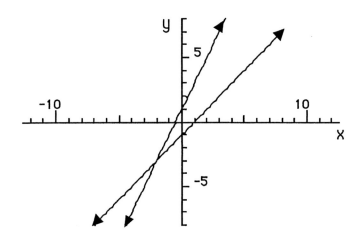

it is apparent that the lines generated intersect. Recall that the graph of a line represents the set of all solutions to that specific equation and, thus, the <u>point of intersection must be the one point that is a **solution** to both equations.</u>

Examining the graph carefully, it appears that the point of intersection is the point $(-2,-3)$. Substituting these coordinates for x and y in each equation above we have

$$y = 2x + 1 \qquad \qquad \text{and} \qquad \qquad y = x - 1$$
$$-3 = 2(-2) + 1 \qquad \qquad \qquad \qquad -3 = -2 - 1$$
$$-3 = -3 \qquad \qquad \qquad \qquad \qquad -3 = -3$$

Since these statements are both true, we have found the **solution** to this **system of equations**.

Sample Problem

Given the equations: $y = 3x + 2$ and $y = 3x$

a) For each equation, list at least 4 ordered pairs that represent solutions to the equation (and, thus, are all points on the graph).

b) Graph both lines on one coordinate system and find the point of intersection, if it exists.

Solution:

For $y = 3x + 2$ and for $y = 3x$

x	y
0	2
1	5
2	8
-1	-1

x	y
0	0
1	3
2	6
-1	-3

and the graphs are

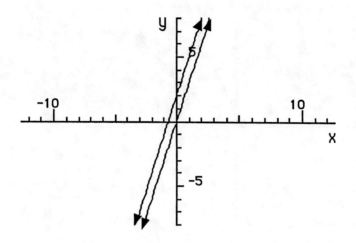

It appears that these lines do not intersect, and, therefore, since no point satisfies both equations, there is *no solution* to this system of equations. The system is said to be **inconsistent** and the lines are called **parallel**.

It should be noted here that one other possibility exists when the graphs of two linear equations are on the same coordinate system. Consider the equations

$$y = 2x - 6 \qquad \text{and} \qquad y = 2(x-3)$$

If the equation on the right is simplified by removing parentheses, we have

$$y = 2x - 6$$

which is exactly the same as the equation on the left. In this case, the line representing the graph of the first equation would exactly coincide with the graph of the second equation. Since the lines have every point in common, this system of equations has an *infinite number of solutions*.

Problem Set 5.2

1. Determine if $1/2$ is a solution to any of the following. Justify your answer.

   a) $4x = 8$ b) $6x - 1 = 2$ c) $\dfrac{2x - 1}{2} = 0$

2. Determine if the ordered pair (1,2) is a solution to any of the following. In each case, justify your answer.

   a) $y - 2x = 0$ b) $y = 3 - 2x$ c) $y = 3x - 1$

3. Discover the rule that relates the two sides of each table or chart below. Let L represent the numbers on the left side and i) write the rule in arrow notation; ii) let R represent the value on the right side; write the rule in the form of an equation.

a) 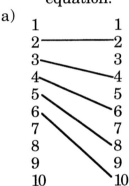   b) 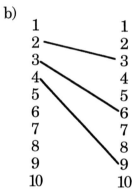   c)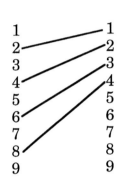

   d) Let L represent the value on the left of part c) above, write the equation relating L in terms of R.

4. a) Give an equation in x and y that describes the relationship between the numbers on the left and the corresponding numbers on the right side of the table below. Write the equation of y in terms of x.

x	y
1	-4
2	-3
3	-2
4	-1
5	0

   b) Write the equation that relates x in terms of y.

5. Write 5 ordered pairs that satisfy the statement $x + y = 6$ and graph these ordered pairs. What kind of graph does this appear to be? Let $x = .25$, 1.5, 2.5, 3.5, and 4.25 and find the corresponding y-values. Plot these points on the same coordinate system. Has the graph now changed its shape? Is it reasonable to connect the points? Justify your answers.

6. For each graph below, write the ordered pairs shown on the coordinate system and use the ordered pairs to generate the equation relating y and x.

   a)

   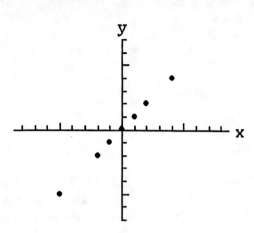

   b)

   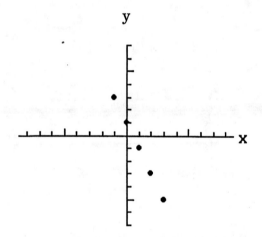

   c) Substitute x = 6 into both equations and find the corresponding value of y in each case.

   d) Find the value of x in both equations for y = 5.

7. Give at least three solutions (as ordered pairs) to each of the following:

   a)
       i) y = 1 − x   ii) 2y − 1 = x

   b) Graph the equation y = 1 − x

8. The formula **d = rt** (which we have used before) gives the relationship between the distance an object travels, the average rate of travel, and the time spent traveling. If Erich bicycles at an average rate of 8 mph, graph the distance he travels with respect to the amount of time he spends traveling at that rate. To accomplish this, do the following:

a) Complete the table below to generate the ordered pairs.

b) Label the distance on the vertical axis, the time on the horizontal axis and construct the intervals on each axis appropriately. [Hint: Make each interval on the vertical axis worth a large value so that the data fits on the first quadrant of the coordinate system.]

time	distance
1 hr.	8 miles
1.5 hrs.	12 miles
2 hrs.	16 miles
2.5 hrs.	_____
3 hrs.	_____
3.5 hrs.	_____
4 hrs.	_____

c) Does it make sense to connect the points on the graph?

d) What kind of graph does this appear to be?

9. John and Kathy together have 15 CD's, but the exact number each has is unknown. The graph below suggests ways the CD's can be distributed between John and Kathy. Answer the following questions with respect to this graph.

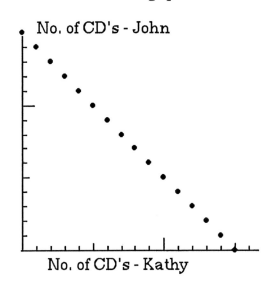

a) Is it reasonable to connect the dots here? Why or why not?

b) From the graph, if Kathy has 7 CD's, how many does John have?

c) From the graph, if John has 15 CD's, how many does Kathy have?

10. Select at least two values for x, the input value, to put into a function machine that uses the rule,
$$\text{Output} = 3x - 2,$$
and find the corresponding output values.

11. Given the following function machine, give two rules that generate the given output for the given input.

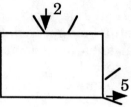

12. Graph each system of linear equations to determine whether or not there is a solution to the system. If there is, give the point of intersection as an ordered pair. In each problem, note any apparent relationship between the lines.

   a) $y = -2x + 5$ and $y = \frac{1}{2}x$

   b) $y = x + 1$ and $y = x - 3$

   c) $y = 3x$ and $y = -3x$

Exploration Problems

1. Use the graphing calculator to graph the line $y = 2x + 1$. Set the range of x between -5 and 5 and the range of y between -10 and 10. Use the trace function to find at least 5 points on this line.

2. Use the graphing calculator to graph $y = x - 2$ on the same coordinate system as the graph of $y = 2x + 1$ in problem 1 above. Use the trace function to find the point(s) at which the two graphs intersect. Substitute the x and y values of this point into both equations and explain what happens and what this means.

3. Use the graphing calculator to graph the equation $y = x^2 + 1$ and then use the trace function to find the point, called the **vertex**, at which the curve changes direction.

4. Use an algebraic method to solve the system of equations

   $$y = -2x + 5 \quad \text{and} \quad y = \frac{1}{2}x$$

   Compare this solution to your results in #12a above.

Solutions - Problem Set 5.2

1. a) no   b) yes   c) yes

2. a) yes   b) no   c) yes

3. a) i) L ➤ 2L – 2; ii) R = 2L – 2   b) i) L ➤ 3L – 3; ii) R = 3L – 3
   c) i) L ➤ $\frac{1}{2}$L; ii) R = $\frac{1}{2}$L   d) L = 2R

4. a) y = x +(–5) (or y = x – 5)   b) x = y + 5

5. Ordered pairs: (6,0), (0,6), (1,5), (8,–2), (3,3), for example.
   The graph appears to be a straight line. Other ordered pairs are
   (.25, 5.75), (1.5, 4.5), (2.5, 3.5), (3.5, 2.5), (4.25, 1.75). All ordered pairs are
   drawn on the system below.

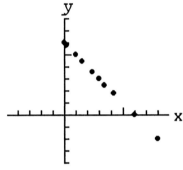

   Connecting the dots is appropriate here because the numbers do not need to
   be integers. Fractional values can be used.

6. a) Ordered pairs: (0,0), (1,1), (2,2), (4,4), (–1,–1), (–2,–2), (–5,–5);
      Rule: x ➤ x   or in equation form, y = x.

   b) Ordered pairs: (0,1), (–1,3), (1,–1), (2,–3), (3,–5);
      Rule: x ➤ –2x + 1   or in equation form, y = –2x + 1.

   c) y = 6 and y = –11
   d) For a), if y = 5, x = 5;
      for b) if y = 5, x = –2

7. a) i) Examples: (0,1), (–1,2), (3,–2)
      ii) Examples: (0,$\frac{1}{2}$), (3,2), (–1,0)
   b) (shown at the top of the next page)

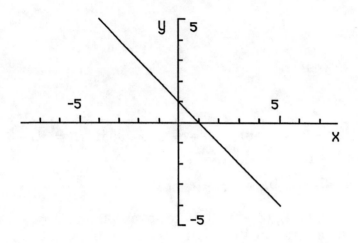

8. a)

time (hrs.)	distance (miles)
1	8
1.5	12
2	16
2.5	20
3	24
3.5	28
4	32

b)

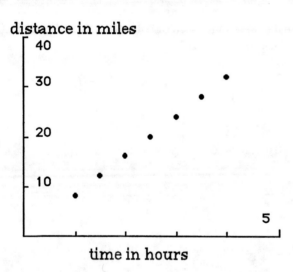

c) Yes, because fractions of an hour are reasonable as are fractions of a mile.

d) A straight line.

9. a) No - neither can own a fraction of a CD.
   b) 8 c) 0

10. Examples: If $x = 1$, output $= 3(1) - 2 = 1$;
    if $x = 0$, output $= 3(0) - 2 = -2$

11. Examples: Output $= x^2 + 1$; Output $= 2x + 1$

12. a) Point of intersection is (2,1). The lines appear to be perpendicular.

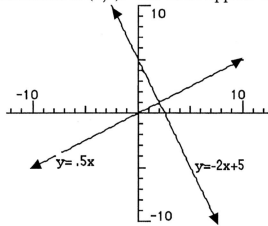

b) Parallel lines, no point of intersection. The system is inconsistent.

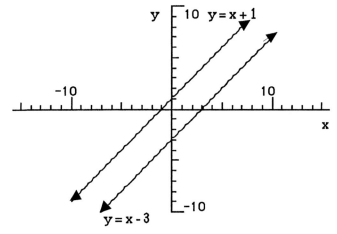

c) The point of intersection is at (0,0). The lines are mirror images in the y-axis.

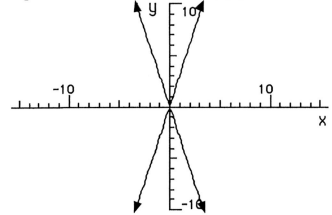

Section 5.3 - Literal Equations

Solving for Other Variables

We can combine the information from sections 1 and 2 of this chapter and apply it to solving what are typically called "literal equations." Sometimes it is useful to take an equation in its given form, rearrange it, and solve for one of the variables in terms of the others. This rearrangement is accomplished through the use of additive and multiplicative inverses.

Consider the equation,
$$2x + y = 3.$$

To make the graphing of this equation a little easier, let's solve the equation for y in terms of x. Note the steps and justifications given.

$$2x + y = 3$$
$$2x + -2x + y = 3 + -2x \quad \text{(adding the inverse of } 2x \text{ to isolate } y\text{)}$$

$$y = 3 + (-2x) \quad \text{or} \quad y = 3 - 2x$$

This process is especially useful when we want to create a **T-table** and graph the equation. By substituting a few values in for x, we can readily generate several ordered pairs. (Note that, by using this form, the arithmetic is all done on one side of the equation!)

x	y
0	3
1	1
2	−1
−1	5

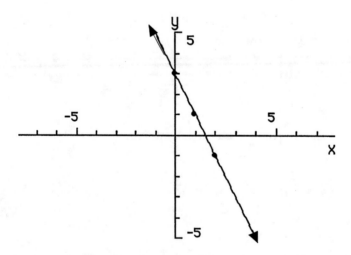

Let's solve for y in terms of x in the equation, $3x + 2y = 8$. This will require a few more steps as outlined below.

$$3x + 2y = 8$$
$$-3x + 3x + 2y = -3x + 8 \quad \text{(additive inverse)}$$
$$2y = -3x + 8$$

$$\frac{2y}{2} = \frac{-3x + 8}{2} \quad \text{(dividing by the coefficient of y)}$$

$$y = \frac{-3x + 8}{2} \text{ or } y = \frac{-3x}{2} + 4 \text{ or } y = \frac{-3}{2}x + 4$$

Again, it will be an easy task to generate a table and graph the equation.

Sample Problems

Solve for y in terms of x and graph each equation.

a) $x - y = 2$

Solution:
$$x - y = 2$$
$$x - x - y = -x + 2$$
$$-y = -x + 2$$
$$(-1)(-y) = (-1)(-x) + (-1)(2)$$
$$y = x + (-2) \quad \text{or} \quad y = x - 2$$

Some ordered pairs are: $(0,-2)$, $(1,-1)$ and $(2,0)$. The graph is

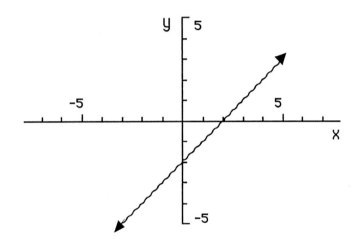

b) $5x + 3y = 6$

Solution:
$$5x + 3y = 6$$
$$3y = -5x + 6$$

$$\frac{3y}{3} = \frac{-5x + 6}{3}$$

$$y = \frac{-5x + 6}{3} \text{ or } y = \frac{-5x}{3} + 2 \text{ or } y = \frac{-5}{3}x + 2$$

A few ordered pairs are: $(0,2)$, $(3,-3)$, $(6,-8)$. The graph is

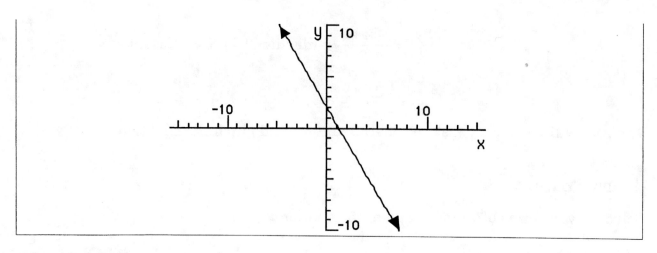

Use of Formulas

Formulas are special equations which apply to a specific situation, and we have used them to solve problems before formally studying algebra. For example, A = lw is used to find the area of a rectangle, given its length and width; I = Prt allows us to calculate how much (simple) interest we might earn (or owe) depending upon the principal amount deposited (or borrowed), the interest rate given in percent, and the time period in years.

Consider the formula for the perimeter of a triangle with sides of a, b, and c.
Then
$$P = a + b + c.$$

To solve for c in terms of the other variables used, we need to isolate the "c" term, and we do so by moving the a and b to the other side of the equation by using the <u>additive</u> inverse of each.

$$P = a + b + c$$
$$P + (-a) + (-b) = a + (-a) + b + (-b) + c$$

$$P - a - b = c$$

A worker received a check from her credit union for the money earned by her investment account. She received $129.61 interest for a three-month period. The amount in her account is $22,744.08 Since that seemed like a small interest payment, she decided to estimate the annual interest rate, based on the amount for three months. We'll use the formula I = Prt, for simple interest, to do the estimation. The equation, I = Prt, can be solved for r in terms of the other variables in order to make the calculation easier. The conversion uses reciprocals and is as follows:

$$I = Prt$$
$$\left(\frac{1}{P}\right)\left(\frac{1}{t}\right) \cdot I = \left(\frac{1}{P}\right)\left(\frac{1}{t}\right) \cdot Prt$$
$$\frac{I}{Pt} = r$$

Now, to discover the interest rate paid on the investment, the worker could substitute the values known as shown below and compute. Note that the principal is the

266

amount in the account, and that, since t represents time in years, 3 months is $3/12$ or $1/4$ or .25 of a year.

$$r = \frac{I}{Pt} = \frac{129.61}{22744.08(.25)}$$
$$r = \frac{129.61}{5686.02}$$
$$r = .0228$$

Since rate is usually expressed as a percent, we can change this decimal value, which is a little greater than 2 hundredths, to its percent form, 2.28%. So the worker was correct; the annual interest rate was very low.

Sample Problem

Solve for r in $d = rt$.

    Solution: Use the multiplicative inverse of t (or divide by t):
$$\frac{1}{t}(d) = \frac{1}{t}(rt)$$
$$\frac{d}{t} = r$$

Rewriting some formulas may require more than a single operation. Before considering such a problem, let's first compare some numerical equations and some literal equations side-by-side, noting the similarities and the differences in the process. Let's solve for x in the following equations:

    $5(x - 3) = 2x + 7$ $a(x - d) = bx + c$

1: Remove parentheses using the distributive property:

    $5x - 15 = 2x + 7$ $ax - ad = bx + c$

2: Move the term containing x from the right side to the left using the additive inverse:

    $5x + (-2x) - 15 = 2x + (-2x) + 7$ $ax + (-bx) - ad = bx + (-bx) + c$
                 or $ax - bx - ad = bx + (-bx) + c$

3: Combine terms using the distributive property. Note the use of factoring in the literal equation to isolate the common factor x.

    $3x - 15 = 7$ $x(a - b) - ad = c$

4: Isolate the term containing the variable x by using the additive inverse; combine like terms:

    $3x - 15 + 15 = 7 + 15$ $x(a - b) - ad + ad = c + ad$
        $3x = 22$ $x(a - b) = c + ad$

5: Divide by the coefficient of x:

$$x = \frac{22}{3} \qquad\qquad x = \frac{(c + ad)}{(a - b)}$$

When you encounter a literal equation in general, ask yourself how you would solve it if the "extra" letters were replaced by numbers as in the example we've just seen. Then follow the same plan of attack to solve the literal equation.

Follow the steps for solving for r in

$$T = P(1 + rt):$$
$$T = P + Prt \quad \text{(distributive property to generate all the terms)}$$
$$T - P = Prt \quad \text{(additive inverse of } P\text{)}$$
$$\frac{1}{Pt} \cdot (T-P) = \frac{1}{Pt} \cdot Prt \quad \text{(reciprocal of } Pt\text{, the coefficient of } r\text{)}$$
$$\frac{T-P}{Pt} = r$$

Sample Problems

1. Solve for w in $P = 2l + 2w$

    Solution: $P + (-2l) = 2l + (-2l) + 2w$

    $$P - 2l = 2w$$
    $$\frac{1}{2}(P - 2l) = \frac{1}{2}(2w)$$
    $$\frac{1}{2}(P - 2l) = \frac{P - 2l}{2} = w$$

2. The formula for the area of the trapezoid pictured is:
    $$A = \frac{1}{2}h(B + b)$$

    Solve for the length of one base, B, in terms of the area, height, and the other base.

    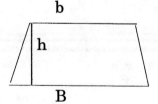

    Possible solution:
    $$A = \frac{1}{2}h(B + b)$$
    $$2 \cdot A = 2 \cdot \frac{1}{2}h(B + b) \quad \text{(multiplying by 2 and dividing}$$
    $$\frac{2A}{h} = B + b \quad\qquad\qquad \text{by } h \text{ to isolate } B \text{ and } b\text{)}$$
    $$\frac{2A}{h} - b = B \quad\qquad\qquad \text{(additive inverse of } b\text{)}$$

Problem Set 5.3

1. Side by side, solve the following pairs of equations for the variable noted.

   a) i) $6 = 7y$ ii) $x = 32y$ both for y

   b) i) $45 = 2 + y$ ii) $x = 32 + y$ both for y

   c) i) $4 - 2x = 6$ ii) $y - .15x = 3$ both for x

   d) i) $25 = 3(2r)$ ii) $C = 2\pi r$ both for r

   e) i) $5 = \frac{1}{2}(3h)$ ii) $A = \frac{1}{2}bh$ both for h

2. The formula for converting degrees Fahrenheit to degrees Celsius is
$$C = \frac{5}{9}(F - 32)$$

   a) Re-write the formula to convert degrees Celsius to degrees Fahrenheit.
   b) If the temperature range on a package of ski wax is given as $-10°$ to $-20°$ Celsius, find the corresponding Fahrenheit temperatures for which the wax is appropriate.

3. Given the following table:

x	y
0	-3
1	-1
2	1
3	3

   a) Write an equation for y in terms of x.
   b) Solve for x in terms of y.

4. Solve the following equations for y in terms of x. Make a T-table and graph each equation. For extra practice, solve each for x in terms of y.

   a) $6x + 2y = 4$

   b) $4x + 3y = -6$

   c) $x - y = 5$

   d) $4x - 2y = 7$

   e) $4x - y = 3$

5. Given the figure shown, find the length x in terms of length y where $P = 40$ in.

6. Over a limited range of temperatures, crickets tend to chirp more rapidly as the temperature rises. In fact, the number of times a cricket chirps appears to depend upon the temperature according to the following formula where t represents temperature in degrees fahrenheit and C represents number of chirps **per minute**.

$$C = \frac{8}{5}t - 4$$

If you hear a cricket chirping and count 19 chirps in 15 seconds, what is the approximate temperature?

7. The sum of two numbers is 45 and their difference is 17. Using x and y to represent the two numbers, write equations for their sum and difference. Then graph the equations and find the values for the two numbers from the graph.

Exploration Problems

1. Given the two figures shown, find the value of x for which the perimeters of the figures will be equal.

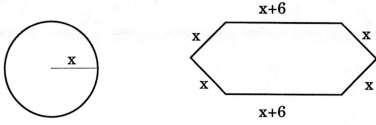

2. a) Find the area of a trapezoid using this geoboard model.

b) On a geoboard, or using dot paper, construct a triangle and a square which have the same area. Compare their perimeters by constructing a "geoboard ruler" to estimate the lengths of nonhorizontal and nonvertical line segments, if necessary.

c) On a geoboard or using dot paper construct trapezoids with areas of 3, 4, and 5 square units.

Solutions - Problem Set 5.3

1. a) i) $y = \dfrac{6}{7}$ ii) $y = \dfrac{x}{32}$

   b) i) $43 = y$ ii) $x - 32 = y$

   c) i) $x = -1$ ii) $x = \dfrac{y - 3}{.15} = \dfrac{3 - y}{-.15}$

   d) i) $\dfrac{25}{6} = r$ ii) $\dfrac{C}{2\pi} = r$

   e) i) $\dfrac{10}{3} = h$ ii) $\dfrac{2A}{b} = h$

2. a) $\dfrac{9}{5}C + 32 = F$

   b) $14°$ to $-4°$ F

3. a) $y = 2x - 3$

   b) $x = \dfrac{y + 3}{2}$

4. a) $y = -3x + 2$

   $x = \dfrac{-2y + 4}{6} = \dfrac{-1}{3}y + \dfrac{2}{3}$

x	y
0	2
1	-1
-1	5

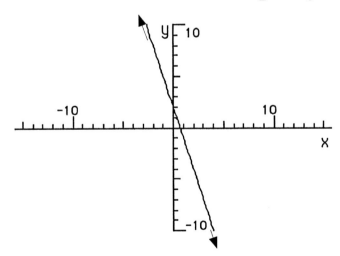

b) $y = \dfrac{-4x-6}{3} = \dfrac{-4}{3}x - 2$

  $x = \dfrac{-3y-6}{4} = \dfrac{-3}{4}y - \dfrac{3}{2}$

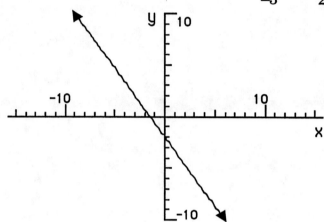

x	y
0	−2
3	−6
−3	2

c) $y = x - 5$

  $x = y + 5$

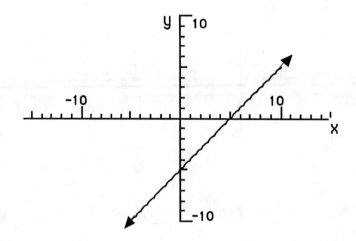

x	y
0	−5
1	−4
5	0

d) $y = \dfrac{-4x+7}{-2} = \dfrac{4x-7}{2} = 2x - \dfrac{7}{2}$

  $x = \dfrac{2y+7}{4} = \dfrac{1}{2}y + \dfrac{7}{4}$

x	y
0	$\frac{-7}{2}$
1	$\frac{-3}{2}$
−1	$\frac{-11}{2}$

e) $y = 4x - 3$
   $x = \dfrac{y+3}{4}$

x	y
0	−3
1	1
−1	−7

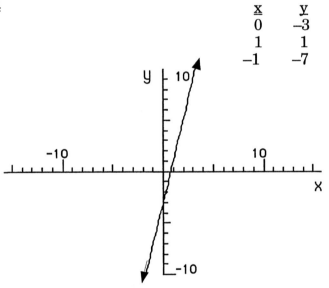

5. If $x + y + 18 = 40$, then $x = 22 - y$

6. $50°$.

7. Let x = the larger no.   and   y = the smaller no.
   $x + y = 45$
   $x - y = 17$

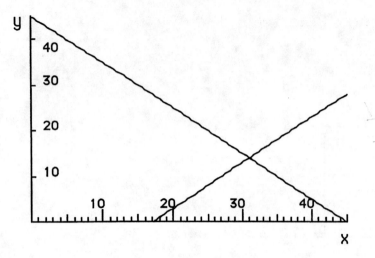

From the graph, the two numbers are 31 and 14.

Section 5.4 - Using Variables in Problem Solving

We employed several problem solving strategies in Chapter One, and we'll now extend the use of these strategies by integrating work with variables and graphs.

The owner of "A Nutty Place" in a large city mall needs to form a mixture of cashews and peanuts to sell at the store. His display bin holds 20 pounds so he wants the mixture to have a total weight of 20 pounds. His regular selling price of peanuts is $2.75 per pound and of cashews is $3.50 per pound. How many pounds of each should be combined so that he can sell the mixture for $3.20 per pound (i.e., $3.20 for one pound)?

A table may guide us to the solution, or help us generate an equation and an algebraic solution. First, we will try different amounts of each kind of nut, realizing that the total weight must always be 20 pounds, and consider the *value* of each of these resulting mixtures. We also need to compare the total value of the mixture to the number of pounds in the mixture in order to find the unit rate, or price for one pound of the mixture.

The following columns and rows will keep track of the information as we try a few different combinations of peanuts and cashews.

No. of lbs. of peanuts	No. of lbs. of cashews	Total weight (lbs.)	Value of Mixture (dollars)	Price per Pound (Value /Total Wt.)
0	20	20	0(2.75) + 20(3.50) = 70	$\frac{70}{20} = 3.50$
5	15	20	5(2.75) + 15(3.50) = 66.25	$\frac{66.25}{20} = 3.31$

Before we continue in this fashion, note the price per pound column. First, it should be apparent that if we use only cashews, the price per pound should still be $3.50 and the table verifies this. Secondly, as we add peanuts in the second row of the table, the price per pound decreases toward our goal of $3.20 per pound. In other words, we are heading in the right direction by increasing the amount of peanuts in the mixture. (It is also reasonable to add more of the cheaper nuts in order to reduce the unit price.) Continuing to add peanuts,

No. of lbs. of peanuts	No. of lbs. of cashews	Total weight	Value of Mixture	Price per Pound
6	14	20	6(2.75) + 14(3.50) = 65.50	$\frac{65.50}{20} = 3.28$
8	12	20	8(2.75) + 12(3.50) = 64	$\frac{64}{20} = 3.20$

By mixing 8 pounds of peanuts and 12 pounds of cashews we have achieved our goal of a mixture worth $3.20 per pound.

Let's use the unknown, n, and the table to help us formulate an equation that may eliminate some of the trial and error.

If n = the number of pounds of peanuts in the mixture,

then the amount of cashews can be represented by

$$20 - n \quad \text{since the "rest" of the mixture must be cashews.}$$
[Note that $n + (20 - n) = 20$.]

As we did in the table, the *value* of the *mixture* is found by first multiplying the weight of each kind of nut by its corresponding price per pound and then adding these quantities together. Thus, the total value is

$$\underbrace{2.75n}_{\text{value of peanuts}} + \underbrace{3.50(20 - n)}_{\text{value of cashews}} = \text{total value of mixture}$$

Since there is a total of 20 pounds of nuts, the price per pound can be found by the **expression**

$$\frac{\text{Total value of mixture}}{\text{Total no. of pounds}} = \frac{2.75n + 3.50(20 - n)}{20}$$

Now this must equal $3.20 so the **equation** representing the price per pound is:

$$\frac{2.75n + 3.50(20 - n)}{20} = 3.20$$

This is an equation in the unknown, n, that we can solve. Follow along the steps below.

$$(20)\frac{2.75n + 3.50(20 - n)}{20} = 3.20(20) \quad \text{(Multiplying both sides by 20 clears the denominators)}$$

$$2.75n + 70 - 3.50n = 64 \quad \text{(using the distributive property)}$$
$$-.75n + 70 = 64 \quad \text{(combining like terms on the left side)}$$

$$-.75n + 70 + (-70) = 64 + (-70) \quad \text{(using the additive inverse)}$$
$$-.75n = -6 \quad \text{(solving for } n \text{ by dividing by } -.75)$$
and $\quad n = 8$

Since we let n represent the number of pounds of peanuts, this means that there are 8 pounds of peanuts in the mixture and $(20 - 8)$ or 12 pounds of cashews. Let's check to see if this is reasonable.

$12 + 8 = 20$ pounds so we have the amount that can fit into the bin.

8 pounds of peanuts are worth $8(2.75) = \$22$,
12 pounds of cashews are worth $12(3.50) = \$42$ and the total value is

$$\$22 + \$42 = \$64.$$

Thus, the value of the mixture is

$$\frac{22 + 42}{20} = \$3.20 \text{ per pound}$$

In this case, we have answers that are integers and the table work was reasonable. It is still important to understand that if the table does not easily generate the answer(s) to a problem, it may at least help in the development of the equation that will generate the algebraic solution.

To highlight the connection among a table, algebraic equation and graph, let's consider making an oil-and-vinegar salad dressing for a group of people. If 1 part of vinegar is added to 9 parts of oil, we have a mixture that is $1/10$ or 10% vinegar. Let's examine how different amounts of vinegar added to oil will affect the mixture. Recalling that the fraction of vinegar is the ratio of the amount of vinegar to the total amount of mixture, we can then find the percent of vinegar by multiplying the decimal fraction by 100, or by setting up a proportion as we did in section 3.2. Our goal will be to achieve a 30% vinegar mixture. Let's set up a table with varying amounts of vinegar added to 16 tablespoons (1 cup) of oil and compute each corresponding percent of vinegar in the mixture.

Amt. of vinegar added (tbs)	Amt. of oil (tbs)	Total amt. (tbs)	Percent vinegar (Amt.of vin./total)•100%
0	16	16	0/16 = 0%
1	16	17	1/17 = 5.88%
2	16	18	2/18 = 11.1%
3	16	19	3/19 = 15.8%

Can you estimate how many tablespoons of vinegar must be added to 16 tablespoons of oil to make the mixture 30% vinegar?

Continuing the table with a few more entries for "amount of vinegar added,"

Amt. of vinegar added (tbs)	Amt. of oil (tbs)	Total amt. (tbs)	Percent vinegar
0	16	16	0/16 = 0%
2	16	18	2/18 = 11.1%
4	16	20	4/20 = 20%
8	16	24	8/24 = 33.3%
15	16	31	15/31 = 48.4%
20	16	36	20/36 = 55.6%

From the table, it appears that we'll need a little less than 8 tbs. of vinegar to make our 30% vinegar mixture.

To generate an exact answer, let's use the pattern in the table and develop an equation. For <u>any</u> percent, n, of vinegar in solution we can use the following proportion

$$\frac{\text{Amount of vinegar}}{\text{Total amount of mixture}} = \frac{n}{100}$$

Let x represent the amount of vinegar added. Since the original problem was to achieve a 30% vinegar mixture, we have

$$\frac{x}{16+x} = \frac{30}{100}$$

(*We must add the amount of vinegar to the amount of oil already there.*)

Using cross-products, we can simplify this equation to

$$100x = 30(16+x)$$

and with the distributive property we have

$$100x = 480 + 30x$$

The remaining steps are described below:

$$100x + (-30x) = 480 + 30x + (-30x) \quad \text{(using the additive inverse)}$$
$$70x = 480 \quad \text{(combining like terms)}$$

$$\frac{70x}{70} = \frac{480}{70} \quad \text{(dividing to isolate } x\text{)}$$

and $\quad x = 6.9$ tbs. $\quad$ which is consistent with our estimate from the table!

At this point, it is important to return to the table and study the last column; it appears that, at first, as the vinegar is added, the percentage increases quite fast, but then as more and more vinegar is added, the percentage increases more slowly.

We can use a graph to explore this pattern in greater detail. Since we'll be using positive quantities, it is reasonable to include only the first quadrant of the coordinate system. The horizontal axis will represent the amount of vinegar added and the vertical axis will represent the percent of vinegar in the mixture.

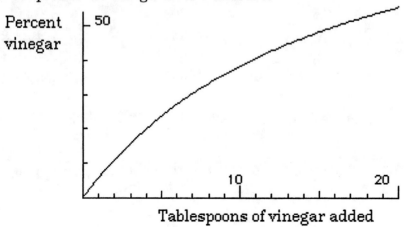

Tablespoons of vinegar added

From the graph, it appears that approximately 7 tablespoons of vinegar must be added to achieve a mixture that is 30% vinegar, confirming our previous work. Note that it will take approximately 11 tablespoons of vinegar to bring the mixture up to 40% vinegar. How many tablespoons will it take to make this mixture 100% vinegar?

Hopefully, you realize that this is a trick question. Since there is oil present in this mixture, it will never be 100% vinegar!!

Sample Problems

1. Antifreeze is ethylene glycol. (We'll use the words interchangeably.) In a car radiator this is mixed with water to keep the coolant from freezing on very cold days and boiling on very hot days. A certain antifreeze solution is 48% ethylene glycol. There are 6 quarts of antifreeze solution in a 10-quart container.

    a) How much ethylene glycol is in this solution?

    b) How much water is in this solution?

    c) How much ethylene glycol should be added to make this a 60% solution?

Solution:

  a) There is 48% of 6, or .48(6) = 2.88 quarts of ethylene glycol in this solution.

  b) There must be 6 – 2.88 or 3.12 quarts of water.

  c) A graph may be used here, but we can again develop an equation with the quantities needed to calculate the percent of antifreeze in solution. The first requirement is the amount of antifreeze in solution, which comes from two factors in this problem:

    1) that already in solution, which is .48(6) = 2.88 qts. *and*
    2) the amount of antifreeze added, in this case, an unknown, or x.

  The second quantity needed is the total amount of solution, which is now $6 + x$.

  Setting up a proportion for calculating the percent we have

  $$\frac{\text{Amount of antifreeze}}{\text{Total solution}} = \frac{2.88 + x}{6 + x} = \frac{60}{100}$$

  Using cross-products and the distributive property results in

  $$(2.88 + x)100 = 60(6 + x)$$
  and
  $$288 + 100x = 360 + 60x$$

  Now follow the next steps to make sure you can justify the reasoning:

$$288 + 100x + (-60x) = 360 + 60x + (-60x)$$
$$288 + 40x = 360$$

$$288 + (-288) + 40x = 360 + (-288)$$
$$40x = 72$$

Finally, $x = \dfrac{72}{40} = 1.8$ quarts of antifreeze must be added.

Let's check the reasonableness of this solution.

If 1.8 qts. of antifreeze are added to 6 quarts of solution, we now have a total of 7.8 quarts of solution, and (2.88 + 1.8) quarts represent the amount of antifreeze. Thus, we have

$$\dfrac{2.88 + 1.8}{7.8} = \dfrac{n}{100}$$

and $\dfrac{4.68}{7.8} \cdot 100 = n$

resulting in

$$60\% = n \quad \text{and our answer is verified.}$$

As a last example, let's consider the following situation. Two cars filled with mathematics instructors going to a conference in Indianapolis, IN from Minneapolis, MN stop at a rest stop on the freeway near Chicago, IL. One car of instructors leaves the rest stop, traveling on the freeway at an average speed of 60 mph, 15 minutes ahead of the other car. If the other car of instructors travels at an average rate of 64 mph, will the second car of instructors catch up to the first car before Car 1 reaches Indianapolis?

Before we make a table, let's draw a diagram to represent the situation described in the problem. The distances driven by the cars 1 hour after leaving the rest stop are

```
        Car 1  ----------------------------------|
                      d = rt = 60(1)
and
        Car 2  --------------------------|
                      d = rt = 64(1 – 1/4)    since 15 minutes = 1/4 hr.
```

Note that after the first hour, the first car has gone 60(1) or 60 miles while the second car has only been traveling for (1 – 1/4) or 3/4 of an hour so it has only covered a distance of 64(3/4) = 48 miles. Varying the amount of time the cars have been on the road after leaving Chicago and comparing the distances each car transversed in that time may help us to answer the question. In table form, and using decimal equivalencies,

Time - Car 1 (hrs.)	Distance - Car 1 (miles)	Time - Car 2 (hrs.)	Distance - Car 2 (miles)
1	60(1) = 60	1 – .25 = .75	64(.75) = 48
1.5	60(1.5) = 90	1.5 – .25 = 1.25	64(1.25) = 80

Before we continue the table in this fashion it is very important to understand how the table may lead to the solution. We must keep in mind that Car 2 has been on the road for a shorter time than the first car, but since it is going faster, it eventually will catch up with Car 1. More importantly, however, when the second car does catch up with the first, they both will have traveled the same distance from Chicago. Again employing the use of a variable, t, to represent the time traveled since leaving the rest stop, and using the table as a guide, we have

Time - Car 1 (hrs.)	Distance - Car 1 (miles)	Time - Car 2 (hrs.)	Distance - Car 2 (miles)
1	60(1) = 60	1 − .25 = .75	64(.75) = 48
1.5	60(1.5) = 90	1.5 − .25 = 1.25	64(1.25) = 80
t	60t	t − .25	64(t − .25)

Setting the distances represented above in columns 2 and 4 equal to each other we have

$$60t = 64(t - .25)$$
and
$$60t = 64t - 16$$

Collecting the variable terms on the left,

$$60t + (-64t) = 64t + (-64t) - 16$$

$$-4t = -16$$

and
$$t = \frac{-16}{-4} = 4 \text{ hours.}$$

(Check: The Car 1-distance of 60(4) miles = Car 2-distance which is 64(4 − .25) or 64(3.75) miles. This means that Car 2 will catch up with Car 1 after the first car has been back on the road for 4 hours (and Car 2 has been on the road for $\frac{1}{4}$ hour less or 3.75 hours).)

We have found that Car 2 will eventually catch up with Car 1. However, we must assess the "reasonableness" of our answer in the context of the distance between Chicago and Indianapolis. Examining a map, Chicago, IL and Indianapolis, IN are about 180 miles apart. How long will it take Car 1 to reach Indianapolis from Chicago? At the average rate of 60 mph, it will take

$$t = \frac{\text{distance}}{\text{rate}} = \frac{180 \text{ miles}}{60 \text{ mph}} = 3 \text{ hrs.}$$

But it will take the second car 3.75 hours to catch up with the first car so Car 1 will reach Indianapolis before Car 2 catches up with it! Therefore, the "real" answer to the problem is that Car 2 will not catch up with Car 1 before it reaches Indianapolis.

_{Have you ever left a site just minutes after another car traveling on the same road and in the same direction and tried to catch up with that car? This problem illustrates this common situation.}

It should also be noted here that when working with a variable to solve a time-rate problem, students sometimes prefer to put the data in a chart as constructed below. This data refers to our "cars of math professors" problem:

	Rate	Time	Distance traveled
Car 1	60	t	60t
Car 2	64	t − .25	64(t − .25)

The advantage of this arrangement is that the data are organized efficiently, and whether the unknown represents rate, time, or distance, the set-up can be adjusted accordingly.

Again, it is important to remember that if you cannot generate a chart right away, using a table with different values for the unknown may be a useful tool for solving the problem directly or for developing an equation.

Finally, let's examine the graphs of the distances as related to hours of travel for the two cars of math professors.

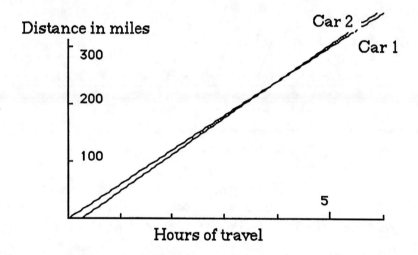

The graphs confirm that Car 2 catches up with Car 1 in 4 hours (that's the point on the graph where the lines intersect). The graph also tells us that, if the cars were traveling farther than Indianapolis, Car 2 would pass Car 1 and the distance between the cars would then increase as the travel time increases.

Problem Set 5.4

1. How much salt is there in 2 gallons of a 15% salt solution? How much water?

2. At the end of a 6-month diet, Jerry had lost 10% of his weight and now weighs 210 pounds. What was his weight before he started the diet?

3. A grocer decided to mix two types of coffee beans: Costa Rican selling for $7.00 per pound and Hawaiian selling for $7.90 per pound.

   a) If he mixes 10 pounds of each, what is the value of the mixture?

   b) If he needs a total of 50 pounds of the mixed beans and he uses n pounds of Costa Rican beans, how many pounds of the Hawaiian beans will he need?

   c) If he intends to sell the 50-pound mixture for $7.27 per pound, how many pounds of each kind of coffee bean should he use?

4. A rectangular pen is to be enclosed for chickens. One side of an existing barn will be used and there is 250 feet of fencing available. The longest side of the enclosed area will be twice the length of a shorter side. The diagram below depicts a top view of this pen. Let x represent the shorter side of this "chicken" pen.

   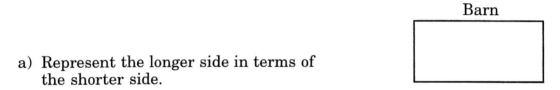

   a) Represent the longer side in terms of the shorter side.

   b) Find the dimensions of this rectangular pen.

   c) Find the amount of area the chickens have in this pen.

5. Ten gallons of a 50% alcohol solution is to be diluted to a 25% alcohol solution by adding water. How much water must be added to the 10 gallons to reduce the amount of alcohol in solution to 25%?

6. If 5 gallons of seawater from the Baltic Sea, which is about 2.4% salt, is mixed with 10 gallons of seawater from the Red Sea, which is about 4.1% salt, what is the percent of salt in the mixture?

7. Your change jar contains only dimes and quarters. If you took it to the bank and the bank's coin counter indicated it contained 82 coins valued at $13, how many of each coin was in the jar? Make a table and try different values for the number of each and/or use an equation.

8. Below is a graph that shows the relationship between the time it takes to travel 200 miles for varying rates. Use the graph below to find the time it takes to travel 200 miles at an average rate of

   a) 45 mph

   b) 60 mph

   c) Use the graph to determine the average rate of speed if the time of travel is 15 hours.

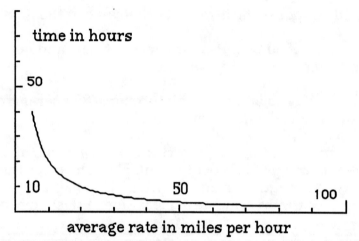

9. Sue bicycles at an average rate of 8 mph and Sam bicycles at an average rate of 6 mph. Sam is in Marquette, MI and plans to travel south on U.S. 41. Sue is in Rapid River, MI and will travel north on the same road. The two locations are approximately 52 miles apart. If they begin riding at the same time, how long will it take for them to meet? You may approximate the time but justify your answer.

10. Two cars leave Denver at the same time and travel in the same direction on an interstate highway. One car is traveling at an average rate of 50 miles an hour and the other at 55 mph. The graph below depicts how the distance traveled by each car depends on the average rate of speed and the time of travel. Approximate the number of hours it would take for the two cars to be 40 miles apart.

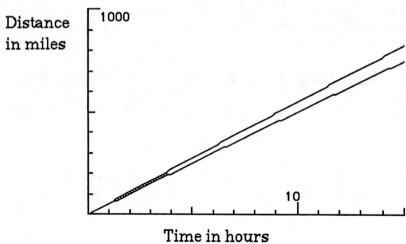

11. Given the roof of a playhouse pictured below with a rise of 3 inches for every horizontal run of 12 inches, commonly called a 3-12 pitch. If a 4 by 8 foot sheet of plywood is to be used to form the triangular piece shown, where must the cut be made on the 4 ft. side so that the triangular piece will be flush with the roof-line?

8 feet

Exploration Problems

1. Refer to the example problem in this section on the two cars of math instructors leaving the rest stop in Chicago. At what rate of speed must the second car travel to catch up with the first car <u>before</u> it gets to Indianapolis? Is this a reasonable rate of speed?

2. The length of this rectangle is shortened in order to make it into a square. Its original perimeter was 114 inches and its area was 810 square inches. What is the dimension of the square?

Solutions - Problem Set 5.4

1. .3 gallons of salt and 1.7 gallons of water.

2. $233 \tfrac{1}{3}$ pounds

3. a) $10(7) + 10(7.90) = \$149$ b) $50 - n$ c) 35 lbs. of Costan Rican beans and 15 lbs. of Hawaiian beans

4. a) $2x$ b) $x + x + 2x = 250$ and $x = 62.5$ feet. The length is 125 feet.

   c) The amount of area for the chickens is 7812.5 sq. feet.

5. 10 gallons of water should be added.

6. 3.5% salt in the mixture.

7. There are 50 dimes and 32 quarters.

8. a) Approx. 4.4 hours   b) Approx. 3.3 hours   c) Approx. 13 mph

9. The time it takes for them to meet is about 3.7 hours (or 3 hrs., 42 min.).

10. Approximately 8 hours

11. Cut the 4-foot side in half, leaving a rectangle that is 2 feet by 8 feet. Then cut on the diagonal as shown below.

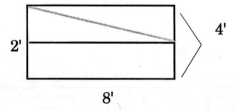

Section 5.5 - Inequalities in One and Two Variables

Meaning of a Solution

Recall that in section 1.1 we looked at graphing an inequality on a number line. For example, we graphed x < 5 and x ≥ −3 as follows:

Some inequalities in one variable need to be written in a different form to make graphing easier. The rules we developed for solving equations will be useful for solving inequalities as well.

Consider the fact that 10 > 6. We'll all agree that 10 is greater than 6. What happens if the same amount, say 5, is added to both sides of this inequality? We'll get a new inequality,
$$10 + 5 > 6 + 5 \text{ or } 15 > 11, \text{ which is also true.}$$

Adding −15 to both sides of the **original** inequality will give us −5 > −9, another obviously true statement.

> **In fact, any quantity can be added to both sides of an inequality and the inequality will still hold.**

What happens when both sides of an inequality are multiplied by the same number? Again use 10 > 6 as our beginning inequality. Multiplying both sides by 2, the result is
$$2(10) > 2(6) \text{ or } 20 > 12.$$

Dividing both sides of the inequality, 10 > 6 by 2 gives

$$5 > 3.$$

Both inequalities are true.

Next try −2 as a multiplier of both sides of the statement 10 > 6. This multiplication leads to the inequality
$$-20 > -12 \text{ which is NOT true.}$$

Dividing both sides of the original inequality by −2 yields

$$-5 > -3 \text{ which is not true, either!}$$

The examples have demonstrated that multiplying or dividing by a positive number maintains the sense or direction of the original inequality. Multiplying or dividing by a negative number, however, reverses the direction of the inequality—that is, ">" becomes "<" and "<" becomes ">". These examples lead to an important rule for solving inequalities:

> **When both sides of an inequality are multiplied or divided by a negative number, the direction of the inequality is reversed.**

Let's try some examples to illustrate these facts.

First we'll solve for x if $2x - 13 > 20$ and graph the solution set on a number line.

$$2x - 13 > 20$$
$$2x - 13 + 13 > 20 + 13 \quad \text{(additive inverse of } -13\text{)}$$
$$2x > 33$$
$$x > 16.5 \quad \text{(divide by 2, the coefficient of } x\text{)}$$

Check: If $x > 16.5$, say if $x = 20$, then $2(20) - 13 = 27$ which is greater than 20, so the value does make the original statement true.

The graph is: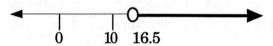

Second, solve for x if $13 - 2x \geq 20$ and graph the solution set on a number line.

$$13 - 2x \geq 20$$
$$13 + (-13) - 2x \geq 20 + (-13) \quad \text{(additive inverse)}$$
$$-2x \geq 7$$
$$x \leq -3.5 \quad \text{(divide by \textbf{negative} 2)}$$

Check: If $x = -10$, $13 - 2(-10) = 33$, and $33 \geq 20$. The value checks.

The graph is:

Note that for these inequalities the solution is not a single value. In the first inequality all values larger than 16.5 make the inequality true, and in the second example all numbers equal to or smaller than negative 3.0 make the inequality true.

Sample Problems

1. Solve for x. Graph the result on a number line.

   a) $3x + 5 < 6$
   b) $3 + \frac{1}{2}x \geq 2$
   c) $3 - \frac{1}{4}x > 5$

   Solutions:

   a) $3x + 5 < 6$
   $3x < 1$
   $x < \frac{1}{3}$

   b) $3 + \frac{1}{2}x \geq 2$
   $\frac{1}{2}x \geq -1$
   $x \geq -2$

   c) $3 - \frac{1}{4}x > 5$
   $-\frac{1}{4}x > 2$
   $x < -8$

In the inequalities considered so far the only variable used was x. What would the graphs of $y < 5$ and $y \geq -3$ look like? Since the y-axis is a vertical number line, the graphs of these new inequalities can be drawn vertically as shown in figures 1 and 2 on the top of the next page

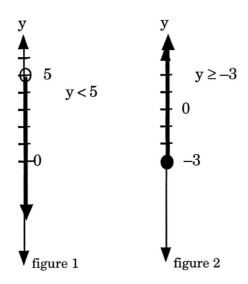

Next we can expand our thinking to include graphing an inequality in one variable on a two-dimensional graph. The graphs of four previously studied inequalities are shown below.

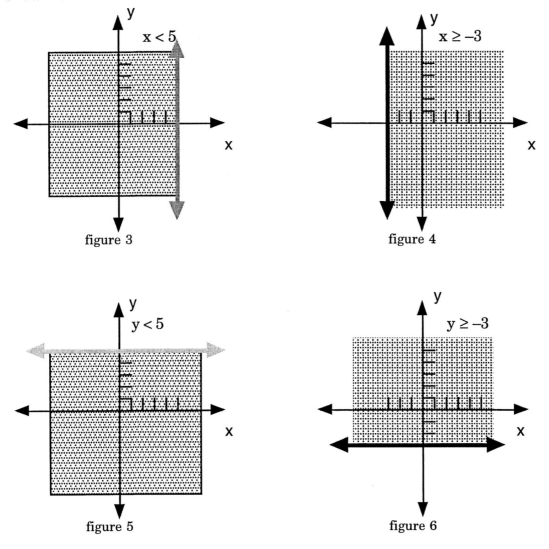

Look at the graph in **figure 3**. All of the points in the shaded region have an x-coordinate less than 5. The vertical boundary line is dotted to indicate that points may be chosen <u>close</u> to it but not <u>on</u> it since the x-coordinate of every point on the line <u>equals</u> 5.

Study the graph shown in **figure 4**. All of the points to the right of the vertical line are shaded because the <u>x-coordinate</u> of each of them is greater than –3. In this case, however, the vertical boundary line is solid because we wish to include points that have an x-coordinate <u>equal</u> to –3 as well as points with x-coordinates that are greater than –3.

All of the points in the shaded region in **figure 5** have a <u>y-coordinate</u> less than 5. All of the points in the shaded region plus the solid line in figure 6 have a <u>y-coordinate</u> greater than or equal to –3. **Note the use of a dotted line when the inequality contains the symbol "<"; the solid line is used when the symbol is "≥".**

<u>Sample Problems</u>

Graph on coordinate axes: a) $x \le 4$ b) $y > 0$

    Solutions: Find the boundary line first by graphing the related equation. As we did in section 5.1, we can make a T-table to generate the boundary.

In part a) note that x is always 4 and we can pick any values for y;
in part b) note that y is always 0 and we can pick any values for x.

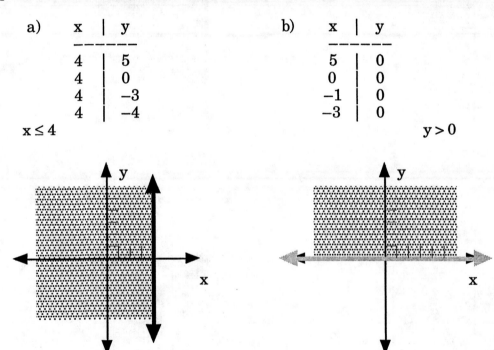

Graphing in Two Variables

Recall that in section 5.2 we looked at graphing <u>equations</u> in two variables. For the equation $x + y = 5$, we found that we could plot several ordered pairs whose x- and y-coordinates have a sum of five—for example, (0,5), (2,3), (6,–1), and (2.5, 2.5)—and then connect the points with a straight line as shown in **figure 7**.

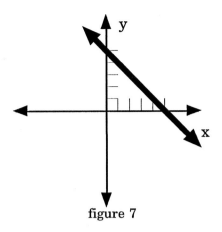

figure 7

Now let's look at the inequality, x + y ≥ 5, which is related to the equation just graphed. The solution to this inequality would consist of all ordered pairs whose x- and y-coordinates have a sum of 5 <u>or more</u>. The straight line graphed in figure 7 would represent part of the solution but not all of it. We must also include ALL of the pairs of real numbers that have a sum *greater* than 5. Some of these pairs are (4,4), (10,–2), (3,7). We could list many more points for which the sum of the coordinates is 5 or more, but we could never list all of them, because there is an infinite number of such pairs. In order to picture all of these points on a graph we need to shade in the region above the line x + y = 5 as shown in figure 8 below.

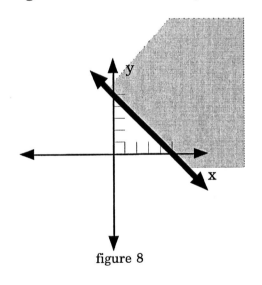

Check: Using (0,0) in x + y ≥ 5, 0 + 0 ≥ 5 is *not* true, so the half-plane containing the origin should not be shaded, and the other side should be. How would you shade x + y ≤ 5?

figure 8

The shaded region represents what is sometimes called a half-plane because "half" of an infinite plane lies on each side of a straight line, and we are choosing the half containing ordered pairs that make the inequality, x + y ≥ 5, true.

Finding points which are solutions to x + y = 5 and x + y ≥ 5 was not difficult, but not all equations and inequalities can be graphed so easily.

Consider, for example, the inequality

$$3y - 2x \leq 12.$$

Ordered pairs which are solutions are not immediately obvious in this case. But, since all points on the y-axis have an x-coordinate of zero, we can find the point where the graph intersects the y-axis, its **y-intercept**, by substituting 0 for x in the related equation 3y − 2x = 12 and finding the corresponding value for y. If x = 0, then

$$3y - 2(0) = 12 \quad \text{and} \quad 3y = 12, \text{ or } y = 4$$

Thus, for the equation 3y − 2x = 12, the y-intercept is the point, (0,4).

Likewise, to find the **x-intercept**, we can substitute 0 for y in the equation, and solve for x. We have

$$3(0) - 2x = 12, \quad -2x = 12 \text{ and } x = -6.$$

and the x-intercept is the point, (−6,0). We can plot these two points and connect them with a solid line because the original inequality contained "≤" as the inequality symbol.

Finally, we need to shade the region that represents all the points whose coordnates make the statement, 3y − 2x ≤ 12, true. Does (0,0) make the statement true? Substitute these values for x and y in the original inequality. The resulting statement, 0 ≤ 12, <u>is</u> true, so the origin is in the half-plane which should be shaded; the final graph is shown in **figure 9**.

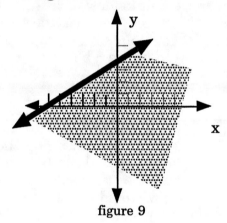

figure 9

Another approach to graphing this equation is first to solve for y in terms of x.

$$3y - 2x \leq 12$$
$$3y \leq 2x + 12 \quad \text{(additive inverse of } -2x\text{)}$$
$$\frac{3y}{3} \leq \frac{2x}{3} + \frac{12}{3} \quad \text{(dividing by the coefficient of y)}$$
$$y \leq \frac{2x}{3} + 4$$

To graph the straight line boundary of the desired region, we now choose values for x (making the task as easy as possible by selecting compatible numbers, i.e., numbers divisible by 3) and substitute them in the related equation in order to find corresponding values for y. Let's try x = 3, x = 6, and x = 0.

If x = 3,
$$y = \frac{2 \cdot 3}{3} + 4$$
$$y = 2 + 4$$
$$y = 6$$

If x = 6,
$$y = \frac{2 \cdot 6}{3} + 4$$
$$y = 2 \cdot 2 + 4$$
$$y = 8$$

If x = 0,
$$y = \frac{2 \cdot 0}{3} + 4$$
$$y = 0 + 4$$
$$y = 4$$

So three ordered pairs are (3,6), (6,8), and (0,4).

The points identified are shown in **figure 10**. Note that the boundary is the same one we obtained using the previous approach.

The "≤" symbol indicates that, since we want to identify points with a y-coordinate less than or equal to the right-hand member of the inequality, the region to be shaded lies <u>under</u> the straight line which is the graph of this inequality's related equation,
$$y = \frac{2x}{3} + 4.$$
Had the symbol been "≥", the region shaded would be <u>above</u> the same straight line.

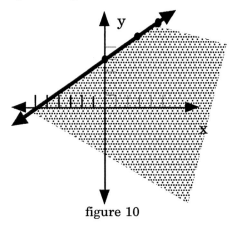

figure 10

The following steps can serve as a guide when graphing any linear inequality in two variables.

1) Unless the inequality is one for which the graph is obvious, find the x-intercept and the y-intercept for the equation which is related to the inequality.

2) Plot the two intercepts.

3) Connect the points with:
   a) a solid line if the original inequality contains "≤" or "≥".
   b) a dotted line if the original inequality contains "<" or ">".

4) Shade the appropriate half-plane, using a test point such as (0,0) for verification, or, if the line passes through the origin, choosing any point <u>not</u> on the line as the test point.

Alternatively:

1) Solve the inequality for y in terms of x.

2) Write the equation related to the inequality.

3) Find at least three ordered pairs which satisfy the equation.

4) Plot the ordered pairs and connect them as noted in 3) above.

5) Shade the appropriate half-plane.

Sample Problem

Graph $2y - 3x > 10$.

Solution, using the second method described: (The intercept method *could* be used, of course.)

1) First solve for y in terms of x.　　$2y - 3x > 10$
$$2y > 3x + 10$$
$$y > 1.5x + 5$$

2) Write the related equation:　　$y = 1.5x + 5$

3) Find points on the boundary line. Let $x = 2$, $x = -4$, and $x = -6$ and solve for y.
　　if $x = 2$,　　　　　if $x = -4$,　　　　　if $x = -6$,
　　$y = 1.5(2) + 5$　　$y = 1.5(-4) + 5$　　$y = 1.5(-6) + 5$
　　$y = 8$　　　　　　$y = -1$　　　　　　$y = -4$

4) Plot the three ordered pairs (2,8), (–4,–1), and (–6,–4) and connect them with a dotted line.

5) Shade above the line. Check: Using (0,0), $2(0) - 3(0) > 10$ but $0 > 10$ is not true, so the region containing the origin should *not* be shaded.

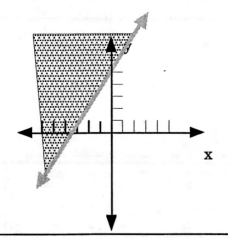

Applications of Inequalities

In problem set 5.2, #10, we investigated a graph which showed the number of CD's John and Kathy each could have if the total of their collections was 15. Let's look at a similar problem involving an inequality.

Let's say that, early in their collection, John and Kathy together have *at most* 10 CD's. We can represent this statement in formal algebraic language as follows:
　　Let J represent the number of CD's John has.
　　Let K represent the number of CD's Kathy has.
　　Then $J \geq 0$
　　　　$K \geq 0$　and　$J + K \leq 10$

The graph of J + K = 10 is much like that for the problem mentioned above. All of the points which satisfy this equation lie along a straight line. What points would be included if the inequality is graphed? The first two inequalities limit the graph region to the first quadrant. Then, since whole numbers whose sum is less than 10 would satisfy the inequality, all of the points with whole number coordinates under the line would be included. It is common to shade a graph to indicate this information, but actually only the whole number values indicated in **figure 11** should be included.

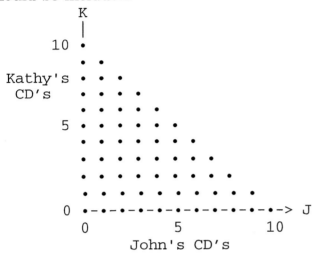

Any of the points plotted has x and y-coordinates that satisfy, $x + y \leq 10$. In other words, all of these points represent possible ways that Kathy and John together have **at most** 10 CD's.

figure 11

As another example, Kris's hockey coach told the team he estimated that they needed to earn at least 12 points to insure a mid-year league championship. A win is worth two points and a tie is worth one. Kris wrote and graphed inequalities that illustrated the possible combinations of wins and losses that would earn the team at least 12 points.

Letting W represent the number of wins and T represent the number of ties,
then $W \geq 0$
$T \geq 0$
$2W + 1T \geq 12$

It makes no difference which variable is placed on which axis, so Kris put W on the x-axis and T on the y-axis. He found the x-intercept and y-intercept for the counterpart equation of the third inequality, $T = -2W + 12$, as follows:

If T = 0, W = 6; if W = 0, T = 12. Kris plotted (0,12) and (6,0) and connected them with a solid line, remembering, however, that only *integer* ordered pairs represented meaningful solutions.

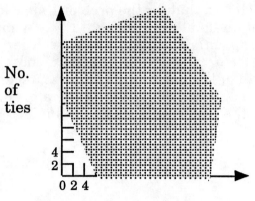

No. of ties

No. of wins

Conclusion: The team needs 12 or more points, so the upper region is shaded.

Check: Using (0,0): $2(0) + 0 \geq 12$ yields $0 \geq 12$ which is not true, so the region containing the origin should *not* be shaded.

We see from the shaded region on the graph that, for example, the team could win 6 or more games with no ties, and win the title. Another possibility is the team could win 7 games, tie 2, and win the title.

Sample Problem

Graph all of the pairs of real numbers whose sum is at least 6 and at most 10.

Solution: Let x represent one number and y the other number.
Then $x + y \geq 6$
and $x + y \leq 10$

The counterpart equations are easily written: $x + y = 6$; $x + y = 10$.

Finding the intercepts of each graph is probably the most efficient method in this case.

$x + y = 6$		$x + y = 10$	
x	y	x	y
0	6	0	10
6	0	10	0

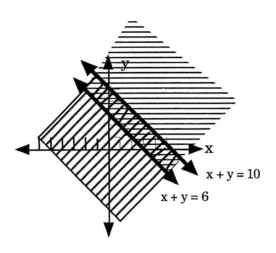

Since $x + y \geq 6$, the area above the line is shaded and since $x + y \leq 10$, the area below that line is shaded. Note the area (in the middle) which is **double-shaded**. All points within this region have x-coordinates and y-coordinates whose sum is greater than or equal to 6 *and* less than or equal to 10.

This problem illustrates how to find the solution to two (or more) inequalities at once, usually described as solving a **system of inequalities**. The points in the double shaded region must satisfy both inequalities.

Problem Set 5.5

1. Solve each of the following inequalities and graph the solution set on a number line.

   a) $x - 5 < -1$

   b) $5 - x \geq 2$

   c) $\frac{3}{4}x > 3$

   d) $0 \leq \frac{1}{3}x + 2$

   e) $2y - 1 < 7$

2. Graph the following inequalities on the coordinate axes.

   a) $x + y \leq 4$

   b) $y < 3x$

   c) $y \geq 2x - 4$

   d) $x \geq 1$

   e) $y \leq 0$

   f) $x - y \leq 13$

3. List 3 ordered pairs in the shaded region.

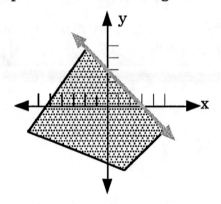

4. List 3 ordered pairs in the **double-shaded** region.

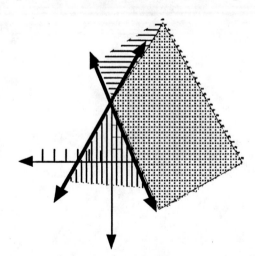

5. On the same coordinate axes graph $x + y > 10$ and $x + y \leq 20$ and find the region which contains all the number pairs that satisfy both inequalities. Name at least two of the solutions to the system of inequalities.

6. Julie is buying supplies for her sister's bridal shower. She wishes to spend no more than $10 on crepe paper and balloons. Crepe paper costs $2 per package, balloons $.75 per bag. Write and graph inequalities to show how many of each item she could buy and stay within her budget.

7. General public theater tickets sell for $10 each and student tickets sell for $5 each.

   a) If the theater holds 300 people, write and graph the inequalities which describe how many tickets of each kind might be sold.

   b) If the overhead for a performance is $600, write and graph, on the same axes, the inequality which describes the amount that must be collected in ticket sales in order for the theater to realize a profit.

8. A Park-and-Ride van holds 15 people who purchase regular tickets for $2 or senior citizen tickets for $1. On a particular run, at least $12 must be collected in order for the van to make the run. Graph the inequalities to determine the possible combinations of types of tickets sold on a permissible run.

Exploration Problems

1. Joel hasn't found a job since graduating, so he's washing cars and mowing lawns until something more lucrative comes along. It takes Joel 15 min to wash a car and one hour to mow an average lawn.

   a) If he works at most 10 hours per day, write the inequalities and graph them to show the possible combinations of cars washed and lawns mowed that he could complete.

   b) If Joel clears $3 per car washed and $10 per lawn mowed, determine the region of the graph which shows how many of each he'll need to do in order to make at least $110.

2. A bakery owner leases his facilities to an organic food co-op whose members work during the owner's "off" hours to produce specialty baked goods. Two products are produced: 7-grain bread and whole-grain healthy cookies.
   A commercial mixer is available 8 hours per shift.
   The large oven is available for 6 hours.
   The mixer will be in use 2 hours during the preparation of each large batch of bread and 1 hour during the preparation of each large batch of cookies.
   The oven will be used 1 hour for each batch of bread and 2 hours for each batch of cookies.

   a) Write and graph inequalities which describe this situation. [Hint: choose variables to represent the two products being made.]

   b) If the profit is $12 per batch of cookies and $15 per batch of bread, how many batches of each should be made to realize the greatest profit?

Solutions - Problem Set 5.5

1. a) $x < 4$

   b) $x \leq 3$

   c) $x > 4$

   d) $x \geq -6$

   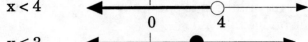

   e) $y < 4$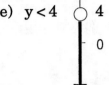

2. a)

x	y
0	4
4	0
1	3

   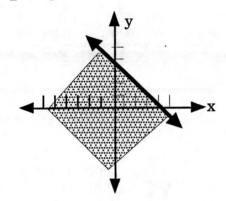

   b)

x	y
0	0
1	3
-1	-3

   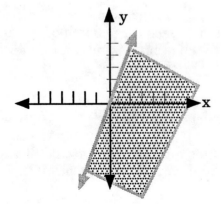

c)

x	y
0	-4
1	-2
2	0

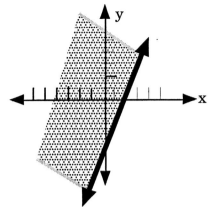

d)

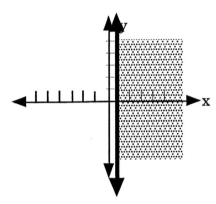

e)

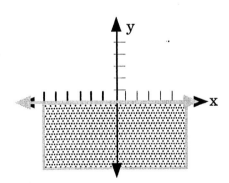

f)

x	y
1	-12
3	-10
5	-8

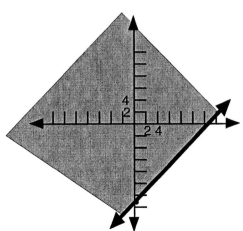

3. Answers will vary;
   Possibilities: (0,0), (1,2), (-1,-1)

4. Answers will vary;
   Possibilities: (3,0), (4,5), (2,6)

5. To graph find the *x* & *y* intercepts for both equations.

   Possible solutions: (15,0), (16,2), (5,14).

   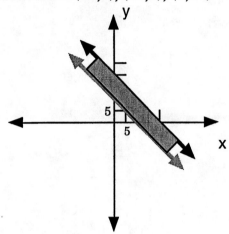

6. Let C rep. no. of pkg of crepe paper
   Let B rep. no. of bags of balloons
   $C \geq 0$
   $B \geq 0$
   $2C + .75B \leq 10$

   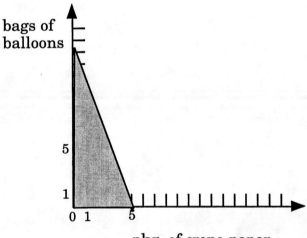

   pkg. of crepe paper

302

7. Let G rep. no. general tickets
Let S rep. no. student tickets

G ≥ 0
S ≥ 0
G + S ≤ 300
10G + 5S > 600

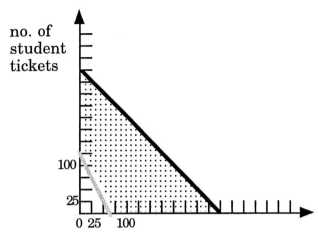

8. Let R rep. no. of regular tickets sold
Let S rep. no. of senior tickets sold

R ≥ 0
S ≥ 0
2R + 1S ≥ 12
R + S ≤ 15

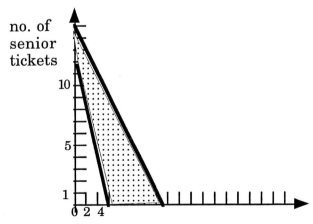

Note: For convenience, regions are shown shaded in the graphs for problems 6-8. However, only points with *whole number coordinates* within the shaded regions represent possible values for the variables.

Chapter Five Review Problems

1. For each of the following, find the rule of correspondence between the two sets of numbers and then write the rule as an equation in two variables.

   a)
   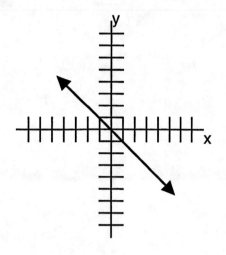

   b)
   ```
   a       b
   0 ————— -1
   1        0
   2        1
   3        2
   4        3
   5        4
   6        5
   7        6
                7
   ```

   c)
t	d
½	4
1	8
3/2	12
2	16
5/2	20

2. Make up a word problem that would generate the equation: $4x + 1.5 = 8.5$.

3. Write the following pan balance configuration in equation form and then solve the equation.

   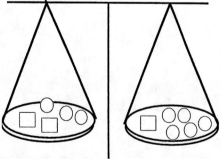

4. Solve the following equation using the pan balance method.

   $$6x + 2 = 4x + 8$$

5. Use an inequality to describe the solution as described below.

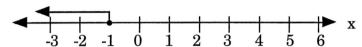

6. Solve the equation, $4n - 5 = 2n + 8$, and graph the solution on the number line.

7. Below is a graph of Kelly's bicycle ride to campus one day with the time plotted on the horizontal axis and her average speed on the vertical axis. Give possible reasons for the fluctuations in her speed at various times.

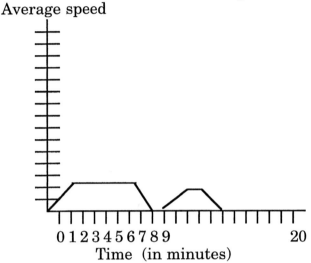

8. Solve each of the following for the variable using any method you wish. An exact answer is required.

   a) $2x + 9 = 3x - 4$

   b) $5(x - 2) = 3x + 1$

   c) $a^2 = 9$

   d) $2(x - 2) = -2(x + 4)$

   e) $5x - 10 = 4 + 5x$

   f) $\dfrac{n+2}{4} = \dfrac{n}{3}$

   g) $\dfrac{1}{3}x - \dfrac{2}{3} = \dfrac{3}{4}x$

   h) $8d = 5d$

9. Graph the equation $2x + y = 6$ on a coordinate system by making a T-table and generating at least 5 ordered pairs. (Hint: To get the ordered pairs, solve for y first.)

10. In a simple interest account paying 4.5%, where $I = Prt$, the interest earned in one year was \$56.25. Find the amount of the principal in that account.

11. A certain plane can travel at a rate of n mph in still air (no appreciable wind blowing).

    a) If there is a tailwind of 74 mph, write an expression to represent the rate of the plane with respect to the ground.

b) If there is a headwind of 80 mph, how fast is the plane moving with respect to the ground?

12. Solve for the variable indicated.

   a) $3x - y = 4$, for x
   b) $d = rt$, for t
   c) $A = \frac{1}{2}bh$, for h
   d) $A = \frac{1}{2}h(B + b)$, for h

13. Write an equation that has a solution of –5.

14. How much water must be added to dilute a 20-gallon, 30% antifreeze solution down to a 10% antifreeze solution?

15. Grandma Rossi made 54 chocolate chip cookies and 48 peanut butter cookies one day. She wanted to keep the two kinds of cookies separated (so the flavors don't mingle) and package the cookies so that each package would have the same number of cookies. What is the greatest number of cookies she could put into each package under these restrictions?

16. Find two consecutive integers whose sum is 43 and whose product is 462.

17. David and Leah bought some new furniture for $982.80, which included a 4% sales tax. They paid $350 down and must repay the remainder in 12 monthly installments of $65 each.

   a) How much will they actually end up paying for the furniture?
   b) How much money would they have saved if they had paid in cash?
   c) What rate of interest is the furniture store charging them for paying on the installment plan (to the nearest tenth of a percent)?

18. Graph the solutions to $x > -2$ on a horizontal number line.

19. Graph the solutions to $y \leq 4$ on a vertical number line.

20. Solve for x and graph the solution:

   a) $6x - 4 \leq 11$
   b) $4x + 1 > 3x + 8$

21. Graph the solution to each of the following on a coordinate system.

   a) $x - y < 2$
   b) $y \leq 2x - 1$

22. If Don and Bob together have fewer than 60 CD's, find a region representing the possible ways the CD's could be divided between the two men.

23. To offer a greater variety of birdseed to its customers, a pet store mixed Niger thistle seed, worth $8.95 per pound with a less expensive seed mix, worth $6.75 per pound. How much of each type of mixture should the owner use so that 25 pounds of new mixture will be worth $7.50 a pound? (Round answers to the nearest tenth of a pound.)

24. The wolf numbers have been slowly increasing in the states bordering Lake Superior. A winter 1978-79 survey done in Minnesota estimated a wolf population of 1235. A survey similar to the 1978-79 survey, utilizing a "large sample of natural resource agencies and personnel..."[1], and geographic computer technology, was conducted in winter 1988-89. The estimate of the wolf population in Minnesota at this time was 1750. Find the percent increase in the population of wolves in Minnesota from 1978 to 1989 (to the nearest tenth of a percent).

25. The graph below depicts the percent of salt in solution as varying amounts of salt are added to 5 kilograms of water. Answer the following questions from the graph.

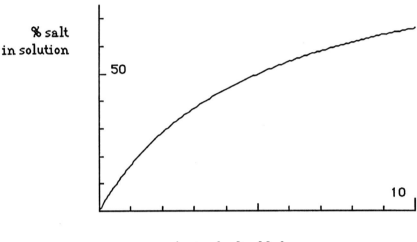

Amt. of salt added

a) What is the approximate percent of salt in solution if 3 gallons of salt are added?

b) Approximately how much salt must be added to produce a 50% salt solution?

Exploration Problem[2]

The following is a sequence of equilateral triangles, where the side of each small triangle is 1 unit in length and all sides of each triangle are the same length.

no. of triangles: 1   4   9

a) Draw the next two figures in the sequence and find the number of triangles in each.

b) Find the perimeter of the first 5 configurations in the sequence.

---

[1] "Wolves in Minnesota: Recovery Beyond Expectations", *International Wolf*, by Wm. E. Berg and Todd K. Fuller, Spring, 1993, pages 15, 16.

[2] From the *Curriculum and Evaluation Standards*, NCTM, page 86

c) Draw a graph to show the pattern growth in perimeter. Put the term number on the horizontal axis and the perimeter of the configuration on the vertical axis.

d) On the same graph, draw the graph of the number of triangles in each configuration as related to the term number.

e) Write all that you can determine from the two graphs.

Solutions - Chapter Five Review

1. a) Ordered pairs on graph: Solution: $x \rightarrow -x$ or $-y = x$ or $y = -x$

   $(1,-1), (4,-4), (5,-5), (-2,2), (-4,4)$

   b) $a \rightarrow 2a - 1$, $b = 2a - 1$

   c) $t \rightarrow 8t$, $d = 8t$

2. If the shipping cost for an order of bulbs is $1.50, and the total cost of 4 packages of tulip bulbs at x dollars each is $8.50, find the cost of each package of tulip bulbs.

3. $2x + 3 = x + 5$; solution: $x = 2$

4.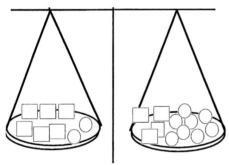

   The solution is: $x = 3$

5. $x \leq -1$

6. $n = 13/2$

7. During the first 2 minutes her speed is increasing, then levels off. After 6 minutes, her speed decreases quickly and she stops (maybe at a busy intersection). After stopping for 1 minute, she starts up again and increases speed for the next 2 minutes. Her speed is constant for a very short time before it declines and she stops (probably at her destination).

8. a) $x = 13$ b) $x = 11/2$ c) $a = 3, -3$ d) $x = -1$

   e) no solution   f) $n = 6$ g) $x = -8/5$ h) $d = 0$

9.
x	y
0	6
1	4
2	2
3	0
-1	8

   $y = -2x + 6$

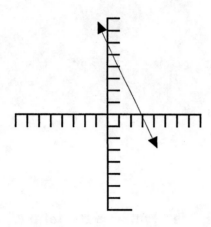

10. P = $1250

11. a) (n + 74) mph   b) (n − 80) mph

12. a) $x = \dfrac{y+4}{3}$ b) $t = \dfrac{d}{r}$ c) $\dfrac{2A}{b} = h$ d) $\dfrac{2A}{B+b} = h$

13. For example, 3x − 2 = −17.

14. 40 gallons

15. 6 cookies

16. The numbers are 21 and 22.

17. a) 1130   b) $147.20   c) 15.0%

18.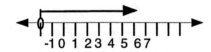

19.

20. a) $x \leq 2.5$

310

b) $x > 7$

21. a)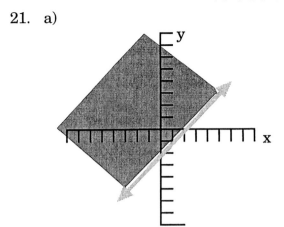

b)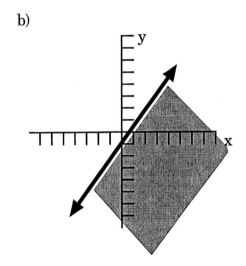

22.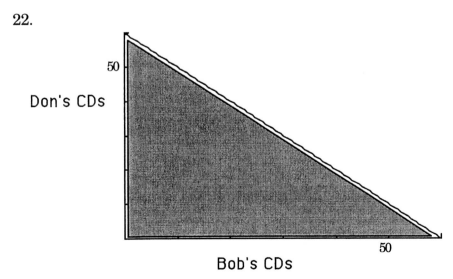

Note that only whole numbers in the shaded region here are possible solutions because fractions of a CD are not reasonable.

23. 8.5 pounds of the thistle seed and 16.5 pounds of the other.

24. 41.7%

25. a) about 38%   b) approx. 5 gals.

Chapter 5 Worksheet 1

1. In each case, write the rule that relates x and y.

   a) | x | y |
       |---|---|
       | 0 | −1 |
       | 1 | −2 |
       | 2 | −3 |
       | 3 | −4 |
       | 4 | −5 |
       | −1 | 0 |

   b) | x | y |
       |---|---|
       | 0 | 1 |
       | 1 | 2 |
       | 2 | 3 |
       | 3 | 4 |
       | 4 | 5 |
       | 5 | 6 |
       | 6 | 7 |

2. For each of the following, use both the cover-up method <u>and</u> the formal method of equation solving.

   a) $2x - 7 = 13$

   b) $8 - 2x = 10$

3. Solve for x, showing work.

   a) $3x - (4x - 5) = 12 - 3(2 - x)$

   b) $-3(x - 5) = 2(x - 10)$

   c) $-5x + 1 = -2x - 5$

   d) $\dfrac{2x + 1}{7} = \dfrac{x - 8}{3}$

   e) $-x + 2 = -2x + 3$

   f) $4x + 1 = 4x - 2$

   g) $3 + 2(6 - 5x) = 5(3 - 2x)$

   h) $2(6 - 4x) = 5(4 - 2x)$

4. The length of a rectangle is 3 feet less than four times its width. Let w represent the width.

   a) Write an expression for the length.

   b) Find the width and length if the perimeter is 64 feet. (Show work.)

   c) Find the area of the rectangle.

5. A ski lift carries a skier up a slope at a rate of 120 feet per minute. The skier skis to the bottom of the hill parallel to the lift at an average speed of 2640 feet per minute. The round trip takes 20 minutes. How long is the ski slope?

6. The Happy Driving car rental company charges $23 a day plus $.12 a mile to rent a compact car. Find the cost of the renting the car for one day if you plan to travel between 250 and 400 miles.

7. Find at least 4 ordered pairs that satisfy each equation below and then graph each equation on coordinate graph paper.

   a) $y = 4x - 2$ b) $y = 2x + 3$

8. What do you know about the solution to the system of equations on each graph below? Justify your reasoning.

   a)

   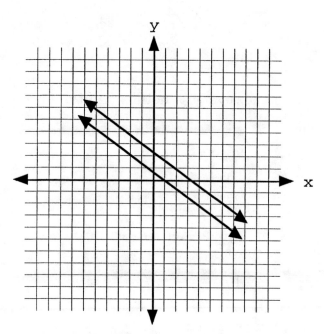

   b)

   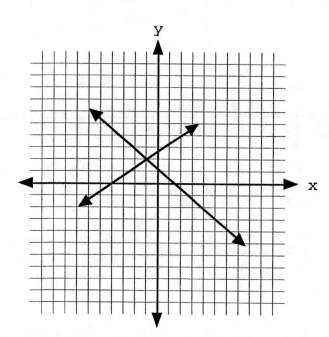

Chapter 5 Worksheet 2

1. Solve the equation $3x + 4 = 4x + 1$ by drawing the pan-balance configuration and using that method. At each step, write the mathematical processes that are illustrated.

2. Solve for the variable indicated.

   a) $7y = -3x - 8$, for y

   b) $\dfrac{x + y}{x} = \dfrac{3}{4}$, for x

   c) $x = 18 - y$, for y

   d) $6a = \dfrac{1}{2}b$, for b

   e) $2a + 3b = c$, for b

   f) $(1/2)bh = A$, for h

   g) $3a - c = d$, for c

3. For the following rectangles, let the width be represented by w. Then write the expression for:

   a) the length, if the length is two and a half times the width.

   b) the length, if the length is five less than twice the width.

   c) the perimeter, if the length is two more than three times the width.

   d) the perimeter, if the length is 4 less than twice the width.

4. If x represents an integer:

   a) represent the next larger integer

   b) represent the next smaller integer

   c) represent three consecutive integers, starting with x

5. If y represents an even integer:

   a) represent the next larger even integer

   b) represent the next smaller even integer

   c) represent the next larger odd integer

6. It takes a passenger train $1\tfrac{3}{4}$ hours less time than it takes a freight train to make a trip from New York City to Boston. The passenger train averages 60 mph, the freight train, 40 mph. How far is it between the cities?

7. Martin has $2.60 in coins, a mixture of pennies, nickels, and dimes. He has the same number of pennies and nickels; he has twice as many dimes as nickels.
   a) Try different values in the chart below to find the number of pennies, nickels and dimes with a value of $2.60.

No. of pennies	No. of nickels	No. of dimes	Total value

   b) Use a variable to write an equation for the total value of the coins and solve for the variable.

8. Jim is three times as old as Tom. Complete the chart.

	Now	4 years ago	5 years from now
Tom's age:			
Jim's age:			

9. An airplane flew on a search mission due east from an airport at 200 mph. It returned at 250 mph. The entire trip took 9 hours flight time. How far did the plane travel?

10. Answer the following questions from the graph below which shows how the cost of carpeting a room depends on the area of the room, in square yds.

    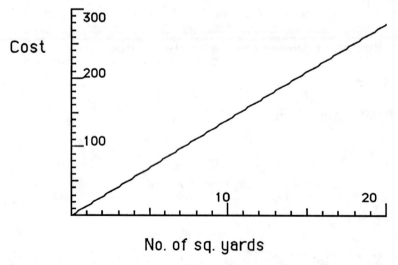

    No. of sq. yards

    a) Find the cost of the carpeting if the area of the room is 18 sq. yards.

    b) Find the area of the room if the cost of the carpeting is approximately $200.

Chapter 5 Worksheet 3

Identify a variable, write an equation, and solve.

1. A large lot will be subdivided into two smaller lots. The larger lot will have 60 feet more along the street than the smaller lot. The current lot has 300 feet of frontage on the street. How much frontage will each of the new lots have?

2. Mrs. Smith has $2346 in a CD that is maturing. She wishes to give the money to her two adult children, Dick and Jane. She wants to give twice as much to Jane, because Jane calls her often while the only time Dick calls is just before his birthday. How much will each child receive?

3. A large meeting room has 4000 square feet of floor space. It is to be subdivided for two groups. The larger group needs four times as much room as the smaller group. How much floor space should be allocated for each group?

4. The owner of a nut store has 23 pounds of cashews which normally sell for $4.80 per pound and 15 pounds of peanuts which normally sell for $3.25 per pound. He wishes to mix the nuts together. What should he charge for the mixture?

5. The owner of the nut store has discovered that the best price for a mixture of cashews and peanuts is $3.99 per pound. He now has 12 pounds of cashews worth $4.80 per pound. How many pounds of peanuts worth $3.25 per pound should he add to the cashews to get a mixture worth $3.99 per pound?

6. a) Ms. Smith has been meeting with an investment advisor who is suggesting how she might invest her money for retirement. The advisor suggested that some money should be invested in a growth mutual fund so that it will maintain its value during times of inflation, while some money should be invested in a bond mutual fund which will pay a higher rate of interest. Ms. Smith has $200,000 which she wishes to split between the two funds. Find her current interest if she puts $30,000 into a bond fund paying 7% and $170,000 into a growth fund paying 3%.

   b) Ms. Smith has decided she needs $10,000 annual income from her $200,000 investment. How much should she put in the growth fund paying 3% and how much should she put in the bond fund paying 7% in order to earn $10,000 per year?

7. For a business course in the spring 2000 semester, Chris and Amy were trying to figure out convenient ways the Post Office could package stamps for letters (33¢) and postcards (20¢). Amy noted that they could sell a booklet for $10 in which the number of 33¢ stamps is three more than the number of 20¢ stamps. How many stamps of each type would be in the booklet?

8. Chris noted that they could sell a larger package containing 16 more letter stamps than post card stamps for $18. How many of each type stamp would be in this package?

9. Kurt needs $2500 in traveler's checks. He decided he wanted to have 20 more $20 checks than $50 checks, How many of each should he get?

10. Joe and Bob met for lunch in Green Bay, WI. When they left, Bob drove 55 mph north toward Marquette while Joe drove 65 mph south toward Chicago on the interstate. How far apart were they after 3 hours, assuming neither stopped?

11. How long did it take until they were 100 miles apart?

12. Suppose Bob left first, driving 55 mph. 15 minutes later, Joe noticed that Bob had left an important package behind, so he left immediately, following at 65 mph. How long (hours and minutes) will it take Joe to catch up with Bob?

13. A tank contains 5 liters of a solution which is 30% antifreeze. How much pure antifreeze must be added to raise the concentration of the solution in the tank to 40%?

14. Solve each of the inequalities below.

    a) $3x + 2 > 8$

    b) $4(x - 1) < 2(x + 2)$

    c) $-3x - 27 \geq -6x - 6$

    d) $7x + 8 \leq 14 + 9x$

15. Graph each of the following on coordinate graph paper.

    a) $y = 2x - 5$ \qquad\qquad b) $y < 2x - 5$

    c) $x + y \leq 2$ \qquad\qquad d) $3x + y < -1$

16. Graph the two inequalities below on the same coordinate system. List at least two ordered pairs in the double-shaded region.

    $$x + y < 5 \quad \text{and} \quad x + 2y < 8$$

CHAPTER SIX - Exponents and Radicals

Section 6.1 - Integral Exponents

Exponents first appeared during our work with the order of operations in Chapter 1. Their use has been concentrated in formulas such as those for finding the area of a square, volume of a cube, area of a circle, etc. However, exponents have a far wider application.

Recall that a positive exponent tells how many times a base number is used as a factor. Thus, 5^4 means $5 \cdot 5 \cdot 5 \cdot 5$.

In general, a^n means a • a • a • a... where a is used as a factor n times.

The exponent form can be a time, energy, and space saver when compared to the multiplication form.

Consider a multiplication problem such as $4^5 \cdot 4^3$. We could write the problem out in the "long form":
$$(4 \cdot 4 \cdot 4 \cdot 4 \cdot 4)(4 \cdot 4 \cdot 4) = 4 \cdot 4 \cdot 4 \cdot 4 \cdot 4 \cdot 4 \cdot 4 \cdot 4 = 65536,$$

which shows "4" being used as a factor a total of 8 times; or we could use the exponent "short-cut" and say that the answer is 4^8 since this expression means "4 used as a factor 8 times."

If an exact final value is needed, a calculator and the y^x (or similar) key, or the ^ key on a graphing calculator, can be used to raise 4 to the eighth power. The same result occurs: 65536. Note that the exponent for the answer is the **sum** of the exponents in the original problem, i.e., $3 + 5 = 8$.

Consider a similar problem, $y^7 \cdot y^4$, for example, in which the base is a variable. Since $y^7 \cdot y^4$ means $(y \cdot y \cdot y \cdot y \cdot y \cdot y \cdot y)(y \cdot y \cdot y \cdot y) = y \cdot y \cdot y \cdot y \cdot y \cdot y \cdot y \cdot y \cdot y \cdot y \cdot y$,
where y is used as a factor 11 times, we can write the answer in exponent form as y^{11}. That's exactly what would result if we took the <u>sum</u> of the original exponents:
$$y^7 \cdot y^4 = y^{7+4} = y^{11}.$$

These two examples provide support for the first of the general principles which apply to exponents.

> **Product Rule: For all real numbers x, m, and n, $x^m \cdot x^n = x^{m+n}$.**

Having previously studied division as the inverse operation of multiplication, let's consider some division problems which involve exponents. In fact, let's look at examples related to the multiplication problems just investigated.
The problem $4^8 \div 4^3$ can be written out in fraction form as follows:

$$\frac{4 \cdot 4 \cdot 4 \cdot 4 \cdot 4 \cdot 4 \cdot 4 \cdot 4}{4 \cdot 4 \cdot 4}$$

Since common factors appear in numerator and denominator, we can simplify as follows:

$$\frac{\cancel{4}\cdot\cancel{4}\cdot\cancel{4}\cdot 4\cdot 4\cdot 4\cdot 4\cdot 4}{\cancel{4}\cdot\cancel{4}\cdot\cancel{4}}$$

and the answer is 4^5 or 1024.

Likewise, for $y^7 \div y^{11}$, we could write:

$$\frac{\cancel{y}\cdot\cancel{y}\cdot\cancel{y}\cdot\cancel{y}\cdot\cancel{y}\cdot\cancel{y}\cdot\cancel{y}}{\cancel{y}\cdot\cancel{y}\cdot\cancel{y}\cdot\cancel{y}\cdot\cancel{y}\cdot\cancel{y}\cdot\cancel{y}\cdot y\cdot y\cdot y\cdot y} = \frac{1}{y^4}$$

If we note what has happened, it should be clear that a lot of time and energy can be saved on division problems involving exponents by taking the <u>difference</u> between the exponents. If the larger exponent appears in the numerator, the base and the resulting exponent are written in the numerator; if the larger exponent appears in the denominator, the base and its resulting exponent are written in the denominator. Thus,

$$4^8 \div 4^3 = 4^{8-3} = 4^5 \quad \text{and} \quad y^7 \div y^{11} = \frac{1}{y^{11-7}} = \frac{1}{y^4}$$

We can now summarize the second principle for using exponents.

Quotient Rule: For all real numbers x, m, and n, $x \neq 0$,

$$x^m \div x^n = x^{m-n}, \text{ if } m > n,$$

$$\text{or} = \frac{1}{x^{n-m}}, \text{ if } n > m.$$

Sample Problems

1. Find the product: $2x^2 \cdot 3x^3$
   Solution: $2 \cdot x \cdot x \cdot 3 \cdot x \cdot x \cdot x = 2 \cdot 3 \cdot x^5 = 6x^5$

2. Find the quotient: a) $10y^{10} \div 5y^5$ b) $14a^4 \div 2a^8$

   Solutions: a) $\dfrac{10y^{10}}{5y^5} = 2y^5$ b) $\dfrac{14a^4}{2a^8} = \dfrac{7}{a^4}$

Zero and Negative Exponents

There are two special types of division problems which can lead to other discoveries regarding exponents. The first type requires us to consider a number divided by itself. From the time division was introduced in arithmetic, it was made clear that a nonzero number divided by itself equals one, or in more formal mathematical language, when the dividend and the divisor are identical, the quotient equals one. Consider a problem involving exponents: $3^4 \div 3^4$. If we write the problem in fraction form we have:

$$\frac{3^4}{3^4} \text{ or } \frac{3 \cdot 3 \cdot 3 \cdot 3}{3 \cdot 3 \cdot 3 \cdot 3} = 1.$$

Using the quotient rule,

$$3^4 \div 3^4 = 1 = 3^{4-4} = 3^0.$$

Since there are two answers for the same problem, this means that $3^0 = 1$. To check this calculation, key in 3^0 on a calculator. In fact, key in any value to the zero power (except zero itself) and the result will be 1. We can state another exponent rule as follows:

> **For all non-zero real numbers, x, $x^0 = 1$.**

Now consider another special division problem—one in which the exponent in denominator is greater than that in the numerator.

For example, $4^3 \div 4^8$, could be written $\dfrac{4^3}{4^8}$ or $\dfrac{4 \cdot 4 \cdot 4}{4 \cdot 4 \cdot 4 \cdot 4 \cdot 4 \cdot 4 \cdot 4 \cdot 4}$

If the problem is simplified as before, the result is $\dfrac{1}{4^5}$. But a division problem containing exponents can also be simplified by subtracting the exponent in the denominator *from* the exponent in the numerator—in this case by writing $4^{3-8} = 4^{-5}$.

Thus, $\dfrac{4^3}{4^8} = \dfrac{1}{4^5} = 4^{-5}$

The calculator can be used to verify that $\dfrac{1}{4^5}$ is equal to 4^{-5}.

[Key in 4^{-5} and $1 \div 4^5$. Both key sequences should yield .0009765625 or a rounded off version of this value. Also, most calculators have a reciprocal key, labeled $1/x$, which can be used to evaluate $\dfrac{1}{4^5}$. Key in 4^5, press =, and then press the $1/x$ key. Your result should be the decimal value given above.]

Let's try another example such as $10^3 \div 10^9$. If we write

$$\frac{10 \cdot 10 \cdot 10}{10 \cdot 10 \cdot 10 \cdot 10 \cdot 10 \cdot 10 \cdot 10 \cdot 10 \cdot 10}$$ and simplify, we'll arrive at $\dfrac{1}{10^6}$.

On the other hand, if the exponent in the denominator is subtracted from the exponent in the numerator, the result is 10^{3-9} or 10^{-6}. Again, a calculator will show us that $1 \div 10^6$ is equal to 10^{-6} (or one one-millionth.)

Although it is not possible to verify this principle if the base number is a variable instead of a constant, enough examples should convince us that

$$y^2 \div y^5 = \frac{y^2}{y^5} = \frac{1}{y^3} = y^{-3}.$$

There is another way to look at powers of a number which may help make negative exponent notation easier to understand. Study the following pattern.

$2^5 = 32$
$2^4 = 16$
$2^3 = 8$
$2^2 = 4$
$2^1 = 2$
$2^0 = 1$

Examining the sequence of exponents in the left column and the values in the right column, note that each exponent on the left is one less than the exponent above it in the chart; each right-hand value is half the value above it.

$2^{-1} = \dfrac{1}{2^1} = \dfrac{1}{2}$

$2^{-2} = \dfrac{1}{2^2} = \dfrac{1}{4}$

$2^{-3} = \dfrac{1}{2^3} = \dfrac{1}{8}$

$2^{-4} = \dfrac{1}{2^4} = \dfrac{1}{16}$

$2^{-5} = \dfrac{1}{2^5} = \dfrac{1}{32}$

Thus, the foregoing can be summarized:

$$\text{For all real numbers } x \text{ and } m, x \neq 0, \quad \dfrac{1}{x^m} = x^{-m}.$$

In other words, x^{-m} is equivalent to the reciprocal of x^m.

Sample Problems

Simplify the following expressions using the exponent rules. Only positive exponents should be used in the answers.

1. $(-5)^{10} \div (-5)^{10}$

   Solution: $\dfrac{(-5)^{10}}{(-5)^{10}} = (-5)^{10-10} = (-5)^0 = 1$

2. $(x+y)^2 \div (x+y)^8$

   Solution:

   $\dfrac{1}{(x+y)^{8-2}} = \dfrac{1}{(x+y)^6}$

Other principles which apply to exponential expressions and their simplification are expansions of the Product and Quotient Rules. Consider, for example, $(2x)^3$. In this example the base, 2x, is used as a factor 3 times. Therefore,

$(2x)^3$ means $2x \cdot 2x \cdot 2x$, which can be written

$2 \cdot 2 \cdot 2 \cdot x \cdot x \cdot x$ (using the commutative and associative properties)

or $2^3 x^3$ or $8x^3$.

Thus, to simplify $(2x)^3 \cdot (2x)^7$, we can write $(2x)^{3+7} = (2x)^{10}$ and then remove the parentheses and write $2^{10} x^{10}$ or $1024 x^{10}$.

Similarly, for $(2x)^{10} \div (2x)^7$ we can write $(2x)^{10-7} = (2x)^3$ or $8x^3$.

The form in which the final answer is given depends upon the situation in which such an expression appears.

[Note here the difference between the expressions $2x^3$ and $(2x)^3$. $2x^3$ means $2 \cdot x \cdot x \cdot x$, while $(2x)^3$ means $2x \cdot 2x \cdot 2x$.]

The expression $(2x)^3$ can also be written by including the exponent *1* for each of the factors *2* and *x*. Thus 2x equals $2^1 x^1$, and therefore $(2x)^3$ would be $(2^1 x^1)^3$. We saw above that $(2x)^3$ equaled $2^3 x^3$. Therefore, $(2^1 x^1)^3$ must also equal $2^3 x^3$. We can arrive at the exponents for this expression by multiplying each of the exponents within the parentheses by the exponent outside the parentheses.

Another example will help to verify this principle. Let's simplify $(3^2)^4$.

$(3^2)^4 = 3^2 \cdot 3^2 \cdot 3^2 \cdot 3^2$ or $3 \cdot 3 \cdot 3 \cdot 3 \cdot 3 \cdot 3 \cdot 3 \cdot 3$

where three is used as a factor a total of 8 times.

3 times itself 8 times equals 6561. If we multiply the exponents in the original problem we have $(3^2)^4 = 3^8 = 6561$.

Consider also the simplification of $(5x^3)^2$. If the *5* is written with its invisible exponent as 5^1, then the expression is $(5^1 x^3)^2$, and multiplying each exponent inside parentheses by the exponent outside will yield $5^2 x^6$ or $25 x^6$. That's exactly the same result as the one obtained by writing $(5x^3)^2 = 5x^3 \cdot 5x^3 = 5 \cdot 5 \cdot x^3 \cdot x^3 = 25 x^6$.

Finally, since an exponent outside parentheses applies to all factors in parentheses:

$$\left(\frac{x^2}{2y}\right)^3 = \frac{(x^2)^3}{2^3 y^3} = \frac{x^6}{8y^3}$$

This principle can be summarized as follows:

> **Power Rule:** For all real numbers x, a, m and n,
>
> $$(ax^m)^n = a^n x^{mn}$$
>
> and for $x \neq 0$, $\left(\dfrac{a}{x^m}\right)^n = \dfrac{a^n}{x^{mn}}$

Note that the Power Rule applies only when a product or quotient is used as a base, but not when the base contains a sum or difference. For example, $(a + b)^3 \neq a^3 + b^3$. We'll look at this idea more in sample problem 6 below.

<u>Sample Problems</u>

Simplify each problem by removing parentheses. Multiply out numerical factors.

1. $(3a^3b^2)^3$

   Solution: Re-write $(3a^3b^2)^3 = 3^3 a^{3 \cdot 3} b^{2 \cdot 3} = 27 a^9 b^6$

2. a) $(-2x)^3$ b) $(-2x)^4$

   Solution: a) $(-2x)^3 = (-2)^3 x^3 = -8x^3$ b) $(-2x)^4 = (-2)^4 x^4 = 16x^4$

3. $(a^2 b)^3 \div (a^2 b)^6$

   Solution: $\dfrac{(a^2 b)^3}{(a^2 b)^6} = \dfrac{1}{(a^2 b)^{6-3}} = \dfrac{1}{(a^2 b)^3} = \dfrac{1}{a^6 b^3}$

   or $\dfrac{a^6 b^3}{a^{12} b^6} = \dfrac{1}{a^6 b^3}$

4. $\dfrac{a^5 b^2}{a^2 b^5}$

   Solution: $\dfrac{a^{5-2}}{b^{5-2}} = \dfrac{a^3}{b^3}$ or $a^{5-2} \cdot b^{2-5} = a^3 b^{-3} = \dfrac{a^3}{b^3}$

5. $\left(\dfrac{10a^2}{20b^3}\right)^2$

   Solution: $\left(\dfrac{10a^2}{20b^3}\right)^2 = \dfrac{10^2 a^4}{20^2 b^6} = \dfrac{100 a^4}{400 b^6} = \dfrac{a^4}{4 b^6}$

6. Use the distributive property to show that $(a+b)^2 \neq a^2 + b^2$.

   Solution: Using our model from chapter 1,

	a	b
a	a^2	ab
b	ab	b^2

   $(a+b)^2 = (a+b)(a+b) = a^2 + 2ab + b^2$, not $a^2 + b^2$

Solving Equations Containing Exponents as Variables

The exponent principles (rules) can be used to solve exponential equations, that is, equations in which the variable is an exponent. We utilize the following fact:

$$\text{If } a^m = a^n, \text{ then } m = n.$$

Therefore, if an equation includes the same base in each term, the exponents can be compared and the value of the variable determined. For example, let's find the value of x in the equation $2^x = 2^3$. Since the bases are identical, the exponents must have the same value. Therefore, $x = 3$.

What if we have $2^x = 8$? If we can re-write 8 as *2 to a power*, we can solve as before. Since $8 = 2^3$, then $2^x = 2^3$, and we know that x must equal 3.

Sample Problems

Find the value of x in each equation. Try making the bases match and then compare the exponents.

1. $5^x = 125$

   Solution: $125 = 5^3$ so
   $5^x = 5^3$
   $x = 3$

2. $3^{-x} = 9$

   Solution: $3^{-x} = 3^2$
   $-x = 2$
   $x = -2$

3. $4^x = 8^2$

   Solution: Re-write both terms with a base of 2; compare exponents.
   $(2^2)^x = (2^3)^2$
   $2^{2x} = 2^6$
   $2x = 6$ and $x = 3$

Exponents and Polynomials

We can now extend our work with the addition, subtraction, multiplication, and division of polynomials.

Consider the example:
$$2x(4x^2 + 5x - 6).$$

Using the distributive property and the exponent rules, this expression simplifies to:
$$8x^3 + 10x^2 - 12x$$

To simplify the product of two binomials such as $(3x - 1)(2x + 5)$, distribute both terms of the first binomial as follows and then combine like terms:

$$3x(2x + 5) + (-1)(2x + 5)$$
$$= 6x^2 + 15x - 2x - 5$$
$$= 6x^2 + 13x - 5$$

For the problem $(6x^3 + 9x^2 - 27x) \div 6x$, we isolate the common factor $3x$ from each term of the trinomial and then simplify the result as follows:

$$\frac{3x(2x^2 + 3x - 9)}{6x} = \frac{(2x^2 + 3x - 9)}{2}$$

Sample Problems

Simplify by removing parentheses and combining like terms.

1. $6x(x - 2) + 5x(x + 4)$

    Solution: $6x(x) + 6x(-2) + 5x(x) + 5x(4)$
    $= 6x^2 - 12x + 5x^2 + 20x$
    $= 11x^2 + 8x$

2. $(6x - 1)(3x - 2) - (4x + 3)(x + 1)$

    Solution: $6x(3x - 2) + (-1)(3x - 2) - [4x(x + 1) + 3(x + 1)]$
    $= 18x^2 - 12x - 3x + 2 - [4x^2 + 4x + 3x + 3]$
    $= 18x^2 - 15x + 2 - [4x^2 + 7x + 3]$
    $= 18x^2 - 15x + 2 - 4x^2 - 7x - 3$
    $= 14x^2 - 22x - 1$

Problem Set 6.1

1. Find the value of $(-2)^3$, $(-2)^2$, $(-2)^1$, $(-2)^0$, $(-2)^{-1}$, $(-2)^{-2}$.
   What conclusion(s) can you draw from your results?

2. Find the <u>greatest</u> integer value for x such that:

   $$5^x < 1000$$

3. Find the <u>least</u> integer value for x such that:

   a) $5^x > 1000$ b) $(-5)^x > 1000$ c) $(-5)^x < -1000$

4. Explain 3 ways to arrive at the answer to $7^6 \div 7^6$.

5. Compare each pair of expressions using <, >, or =.

   a) 12^0 -12^0
   b) $(-15)^6$ $(15)^6$
   c) $(.5)^{10}$ $(.5)^9$
   d) $(.5)^{10}$ $(-.5)^9$

6. Simplify as much as possible, using only positive exponents in the final answer.

   a) $a^3 b^{-2} c^5$ b) $x^5 \cdot x^{-7}$ c) $5y^{-2}$ d) $(5y)^{-2}$

   e) $(x^2)^3$ f) $x^2 \cdot x^3$ g) $(b^6 c^5)^2$ h) $(a^2 b)^3 (a^3 b^2)$

   i) $\dfrac{8a^8 b^2}{2a^2 b^5}$ j) $\dfrac{(5x-2y)^2}{(5x-2y)^5}$ k) $\dfrac{(3a)^2}{(-4ab)^3}$

   l) $2x(x^2+4)$ m) $-3x(2x^2 - 3x + 1)$ n) $\dfrac{8x^4 - 16x^3 - 20x}{4x^2}$

   o) $(3a-1)(2a+3)$ p) $(y-2)(y+3) - 2(y-1)$

7. Find the value for x using the exponent rules.

   a) $5^6 = 5^2 \cdot 5^x$ b) $5^6 = (5^2)^x$ c) $2^x = 64$ d) $2^6 = 4^x$ e) $2^8 = 4^x$

   f) $(3y)^4 = \dfrac{(3y)^x}{(3y)^3}$ g) $3^x = \dfrac{1}{3^3}$ h) $2^x = \dfrac{1}{16}$

8. Explain the difference in meaning between $(-3)^2$ and -3^2.
   [Hint: what is the base?]

9. Give an example that illustrates the fact that, for $n \neq 1$, $(x+y)^n \neq x^n + y^n$.
   Justify your answer.

10. How many times larger is 3^{20} than 3^5?

Exploration Problems

1. Which is greater, 2^{3000} or 3^{2000}? Explain how you arrived at your answer.

2. What is the one's place digit in 3^{99}? Justify your answer.

Solutions – Problem Set 6.1

1. $(-2)^3 = -8$; $(-2)^2 = +4$; $(-2)^1 = -2$; $(-2)^0 = +1$; $(-2)^{-1} = -.5$; $(-2)^{-2} = +.25$
   Raising a negative expression to an even power gives a positive result; raising a negative expression to an odd power gives a negative result. The absolute value of the result continues to decrease as the exponent decreases.

2. $x = 4$

3. a) $x = 5$ b) $x = 6$ c) $x = 5$

4. A number divided by itself equals 1; using the quotient rule, $7^{6-6} = 7^0 = 1$; write out expressions and simplify like factors.

5. a) $12^0 > -12^0$
   b) $(-15)^6 = (15)^6$
   c) $(.5)^{10} < (.5)^9$
   d) $(.5)^{10} > (-.5)^9$

6. a) $\dfrac{a^3 c^5}{b^2}$ b) $\dfrac{1}{x^2}$ c) $\dfrac{5}{y^2}$ d) $\dfrac{1}{(5y)^2} = \dfrac{1}{25y^2}$

   e) x^6 f) x^5 g) $b^{12}c^{10}$ h) $a^6 b^3 \cdot a^3 b^2 = a^9 b^5$

   i) $\dfrac{4a^6}{b^3}$ j) $\dfrac{1}{(5x-2y)^3}$ k) $\dfrac{9}{-64ab^3}$ l) $2x^3 + 8x$

   m) $-6x^3 + 9x^2 - 3x$ n) $\dfrac{2x^3 - 4x^2 - 5}{x}$ o) $6a^2 + 7a - 3$ p) $y^2 - y - 4$

7. a) $5^6 = 5^2 \cdot 5^x$
      $6 = 2 + x$
      $4 = x$

   b) $5^6 = (5^2)^x$
      $6 = 2x$
      $3 = x$

   c) $2^x = 64$
      $2^x = 2^6$
      $x = 6$

   d) $2^6 = 4^x$
      $2^6 = (2^2)^x$
      $2^6 = 2^{2x}$
      $6 = 2x$
      $3 = x$

e) $2^8 = 4^x$
$2^8 = (2^2)^x$
$2^8 = 2^{2x}$
$8 = 2x$
$4 = x$

f) $x = 7$

g) $x = -3$

h) $x = -4$

8. $(-3)^2$ means raising -3 to the second power so the answer is positive 9.
-3^2 means raising 3 to the second power and taking the opposite of the answer, that is the opposite of 9, so the answer is negative 9.

9. Answers will vary. Verification should be through the geometric (area) model or the distributive property.

10. 3^{15} times larger.

Section 6.2 – Exponents in Applications

Scientific Notation

Place value in our decimal number system can be represented in more than one way. The following chart is similar to the one used for powers of 2 in the previous section.

$$10^5 = 10 \cdot 10 \cdot 10 \cdot 10 \cdot 10 = 100000$$
$$10^4 = 10000$$
$$10^3 = 1000$$
$$10^2 = 100$$
$$10^1 = 10$$
$$10 = 1$$

As the exponent decreases by 1, the right-hand value is one-tenth of that above.

$$10^{-1} = \frac{1}{10^1} = \frac{1}{10} = .1$$

$$10^{-2} = \frac{1}{10^2} = \frac{1}{100} = .01$$

$$10^{-3} = \frac{1}{10^3} = \frac{1}{1000} = .001$$

$$10^{-4} = \frac{1}{10^4} = \frac{1}{10000} = .0001$$

$$10^{-5} = \frac{1}{10^5} = \frac{1}{100000} = .00001$$

Note, for example, that ten thousand can be written as 10,000 or 10^4; one ten-thousandth can be written as .0001, $\frac{1}{10000}$, $\frac{1}{10^4}$, or 10^{-4}.

Based on this chart, consider the different ways to write each of the following.
  70,000 = 7000 • 10 or 700 • 100 or 70 • 1000 or 7 • 10000 or **7 • 10^4**.
  26,000 = 2600 • 10 or 260 • 100 or 26 • 1000 or even 2.6 • 10000 or **2.6 • 10^4**.
  375,000 = 37500•10 = 3750•100 = 375•1000 = 37.5•10000 = 3.75•100000 = **3.75 • 10^5**.
  6230 = 623 • 10 = 62.3 • 100 = 6.23 • 1000 = **6.23 • 10^3**.

Note that the expressions in bold type are similar in appearance. They consist of two parts: the first, a value greater than or equal to one but less than ten; the second, a value expressed as a power of 10.

Consider some additional examples as follows:

.5 equals $\frac{5}{10}$ or $5 \div 10$ or $5 \cdot \frac{1}{10}$ or **$5 \cdot 10^{-1}$**

.09 equals $\frac{9}{100}$ or $9 \div 100$ or $9 \cdot \frac{1}{100}$ or **$9 \cdot 10^{-2}$**

.0043 equals $\frac{43}{10000}$ but it is also equivalent to $\frac{4.3}{1000}$.
  (Verify this fact with a calculator.)

325

Then: $\qquad \dfrac{4.3}{1000} = 4.3 \div 1000 = 4.3 \cdot \dfrac{1}{1000} = 4.3 \cdot \dfrac{1}{10^3} = \mathbf{4.3 \cdot 10^{-3}}$

and $\quad .0207 = \dfrac{207}{10000} = \dfrac{20.7}{1000} = \dfrac{2.07}{100} = 2.07 \div 100 = 2.07 \cdot \dfrac{1}{100} = 2.07 \cdot \dfrac{1}{10^2} = \mathbf{2.07 \cdot 10^{-2}}$

Once again, the values in bold type are similar in appearance. They are of the form

$$\boxed{x \cdot 10^n \text{ where } 1 \leq x < 10 \text{ and } n \text{ is an integer.}}$$

This is the formal symbolic definition for what is termed **scientific notation**. Scientists who deal with very large or very small numbers use scientific notation a great deal, but non-scientists as well are likely to encounter numbers in this form. For example, the sun is ninety-three million miles from the earth. We could write that value as 93,000,000 or, in scientific notation, as $9.3 \cdot 10^7$. The thickness of a soap bubble is approximately .00001 cm or, in scientific notation, $1 \cdot 10^{-5}$ cm.

When using a scientific calculator to compute 2,144,000 x 12,452, the result usually appears as 2.6697 ^{10}. Most calculators display a number in scientific notation without the "times 10" portion of the expression. This means that $9.3 \cdot 10^7$ is likely to appear as 9.3 $\quad$ 7, or 9.3 $\quad$ E7, or 9.3 $\quad$ E 07, or in a similar form. Likewise, $1 \cdot 10^{-5}$ may be displayed as 1^{-05} or 1 $\quad$ E–05. However, we MUST include the " $\cdot 10$" when we write numbers in scientific notation. The expressions are incomplete otherwise.

Keys labeled EE or EXP are used on many calculators to enter a number in scientific notation. Consult the handbook that came with your calculator.

Sample Problems

1. Write in scientific notation:
   a) 30,000,000,000,000,000
   b) .000 000 000 000 000 006
   c) 723 billion
   d) 45 thousandths

   Solutions: a) $3 \cdot 10^{16}$ b) $6 \cdot 10^{-18}$ c) $723{,}000{,}000{,}000 = 7.23 \cdot 10^{11}$
   d) $^{45}/_{1000}$ or $.045 = 4.5 \cdot 10^{-2}$

2. Write in decimal notation:
   a) $1.6 \cdot 10^4$
   b) $1.6 \cdot 10^{-4}$
   c) $-5.76 \cdot 10^0$
   d) $-4.1 \cdot 10^{-1}$

   Solutions: a) 16000   b) .00016   c) -5.76 d) $-.41$

3. Using a calculator, find the value in scientific notation, of $(3 \cdot 10^7)(6 \cdot 10^8)$ and of $(3 \cdot 10^7) \div (6 \cdot 10^8)$.
   Solution: $1.8 \cdot 10^{16}$; $5 \cdot 10^{-2}$ or .05 which is equivalent to $5 \cdot 10^{-2}$

Scientific notation provides a quick way to compare numbers or put them in order.

a) If two *positive* expressions in scientific notation have different exponents, for example $1.9 \cdot 10^{11}$ and $9.8 \cdot 10^{10}$, or $2.5 \cdot 10^{-3}$ and $8.7 \cdot 10^{-5}$, the expression with the exponent larger in value is the greater.

b) If the value of both expressions is negative, for example, $-3 \cdot 10^4$ and $-4 \cdot 10^2$, the one with the larger exponent would be found further to the left from zero on the number line and would therefore be smaller.

c) If the exponents are identical, then we compare the decimal values to determine their order. For example, $6.37 \cdot 10^{-15} > 5.91 \cdot 10^{-15}$, and $-6.37 \cdot 10^7 < -5.91 \cdot 10^7$.

Inflation and Depreciation

The increasing price of a new car or boat reflects *inflation*. The value of the car or boat after purchase exemplifies depreciation. Each of these changes can be described using expressions which include exponents. For example, assume that a boat manufacturer has produced a new model which sells for $10,000 when first put on the market. Then, if for the next several years the basic price of this model increases at the same rate each year, say 4%, let's see what happens to the cost of the boat.

year	cost add 4% of price new price
0	$10,000
1	$10,000 + 10,000(.04) = 10,000(1 + .04)$ or $10,000(1.04) = \$10,400$
2	$10,000(1.04) + 10,000(1.04)(.04) = 10,000(1.04)(1 + .04) = 10,000(1.04)^2$
	$= \$10,816$
3	$10,000(1.04)^2 + 10,000(1.04)^2(.04) = 10,000(1.04)^2(1 + .04)$
	$= 10,000(1.04)^3 = \$11,248.64$

A pattern seems to be developing. Note that 10,000, the original cost, is multiplied by the quantity 1.04 raised to a power corresponding to the number of years since the boat was introduced. A calculator can be used to verify that the "formula" works for any number of years. Therefore, after 10 years, if the basic cost of this boat continues to increase 4% per year, the expression $10,000(1.04)^{10}$ should reveal what the boat will cost at that time: $14,802.44.

Consider the reverse process—*depreciation*. If you purchased one of the above boats when it was first introduced, and it depreciates at a constant rate, say 5% per year, what will it be worth after 1 year? 2 years? 3 years? 5 years? Let's examine another chart to help analyze the problem.

year	value
0	$10,000
1	$10,000 - 10,000(.05) = 10,000(1 - .05)$ or $10,000(.95) = \$9500$
2	$10,000(.95) - 10,000(.95)(.05) = 10,000(.95)(1 - .05) = 10,000(.95)(.95)$ or
	$10,000(.95)^2 = \$9025$
3	$10,000(.95)^2 - 10,000(.95)^2(.05) = 10,000(.95)^2(1 - .05) = 10,000(.95)^2(.95)$ or
	$10,000(.95)^3 = \$8573.75$

Again a pattern has developed. Therefore, to find the value of the boat after 5 years, we could repeatedly subtract 5% of each previous value, or, since 1–.05 = .95, we have 10,000(.95)^5 = $7737.81.

The patterns in these two examples can be summarized in a general formula:

New Value = Original Value(1 ± rate of change)$^{\text{time in years}}$ or

$$\boxed{\text{New} = \text{Original}(1 \pm r)^t}$$

Note: If rate of change is given in percent per *month*, then the exponent in the formula, t, should reflect the elapsed time in *months*. If rate of change is given per *day*, the exponent should represent the elapsed time in *days*.

Sample Problems

1. Last year tuition for a nursery school was $1800. An increase for the coming year is predicted to be 3%. If that annual rate of increase continues over the next five years, find the cost of tuition for each of those years.

   Solution: Using a calculator:
   first year: $1800 + 3% of $1800 = $1854;
   second year: $1854 + 3% of $1854 = $1909.62;
   third year: $1909.62 + 3% of $1909.62 = $1966.91;
   Using the general formula for years 4 and 5:
   1800(1.03)^4 = $2025.92 for the year
   1800(1.03)^5 = $2086.70 for the year

2. A stereo system was purchased for $2000. If its value depreciates 10% per year, what will it be worth in 8 years? (Use the general formula.)

   Solution: New value = 2000(1 – .10)^8 = $860.93

While inflation and depreciation seldom continue at constant rates, we can use a non-changing rate to *estimate* answers to many problems, and we will see some situations in which the rate of increase or decrease will remain constant.

Population Growth and Decline

Population growth and decline follow the same general formula already given for inflation and depreciation. Changing the terms appropriately, we have:

$$\boxed{\text{New Population} = \text{Original Population}(1 \pm \text{rate of change})^{\text{time in years}}}$$

If a metropolitan area reported a population of 95,000 during the 1990 census and it lost an average of 2% per year, we can predict the approximate population of the area in the year 2000. Although it is easy to subtract 2% of each previous year's population using a calculator, if intermediate values are not needed, it is quicker to evaluate the expression:

New = $95{,}000(1-.02)^{10} \approx 77{,}622$ predicted as the population in the year 2000.

A bacterial colony numbering 1000 is expected to grow at an average rate of 20% *per hour* until it reaches the maximum density which can survive in a particular environment. If this maximum population occurs after *16 hours* of growth, about how many bacteria will be present at that time?

Again using the general formula relating new and old populations:

New = $1000(1+.20)^{16} \approx 18{,}488$ bacteria.

Population growth can be pictured on a graph. The growth of the colony just described is pictured below. This is an example of an **exponential graph**. It shows the population increasing slowly at first and then more and more rapidly.

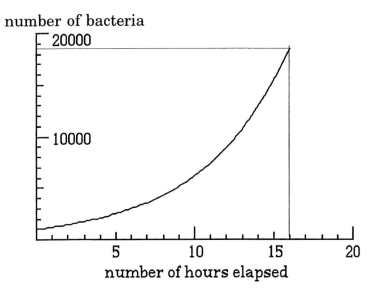

Sample Problem

a) If the bacteria colony considered above begins to decline in population at a rate of 6% per hour after maximum growth is reached, how many bacteria will remain after 8 hours? (Make sure to round off to a whole number of bacteria!)

b) From the graph (on the next page) of this problem, how many will remain after 12 hours?

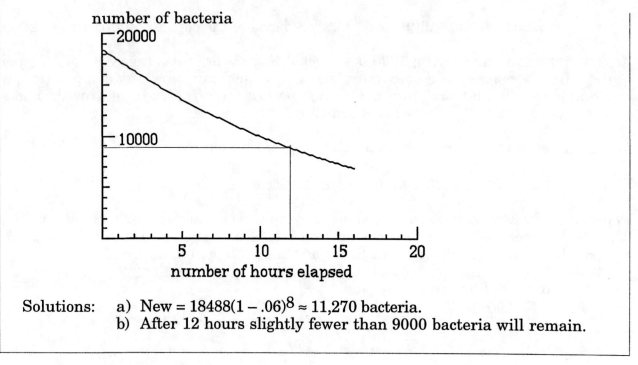

Solutions: a) New = $18488(1-.06)^8 \approx 11{,}270$ bacteria.
b) After 12 hours slightly fewer than 9000 bacteria will remain.

Let's look again at the graph of the bacteria colony which began with a population of 1000 and grew at 20% per hour.

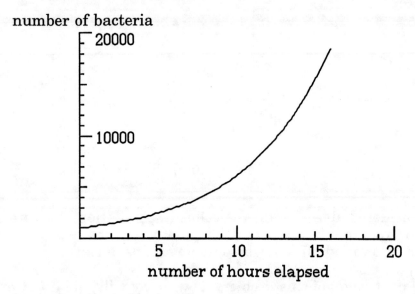

From the graph, can we predict about how long it will take for the population to increase to 5000? We can draw a horizontal line from 5000 on the y-axis to the graph line, and from this point draw a vertical line to the x-axis as shown on the next page.

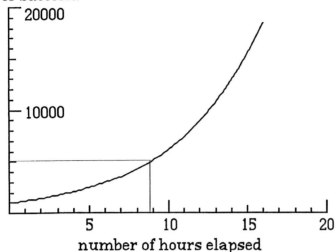

number of hours elapsed

It appears that the population will reach 5000 shortly before 9 hours have elapsed.

Stating the question another way, what value of t in the formula

$$\text{New} = 1000(1 + .20)^t$$

would result in a new population of 5000?

A problem of this type can be explored with a calculator. Since we want to know how long it will take for the population to reach 5 times the original size, we are solving the equation

$$5(1000) = 1000(1 + .20)^t \quad \text{for } t.$$

We can simplify the equation by dividing both sides by 1000, yielding

$$5 = 1(1 + .20)^t.$$

Now we'll try various values for t until we find the one which makes the value of the expression as close to 5 as possible.

If $t = 5$, $(1 + .20)^t \approx 2.49$ which is too small
If $t = 10$, $(1.2)^t \approx 6.19$ which is too large
If $t = 8$, $(1.2)^t \approx 4.30$
If $t = 9$, $(1.2)^t \approx 5.16$
If $t = 8.5$, $(1.2)^t \approx 4.71$
If $t = 8.7$, $(1.2)^t \approx 4.89$
If $t = 8.9$, $(1.2)^t \approx 5.07$
If $t = 8.8$, $(1.2)^t \approx 4.98$

Therefore, by trial and error, we have found that, to the nearest tenth of an hour, it will take about 8.8 hours for the population to reach 5000. That corresponds to our previous conclusion, reached by examining the graph, that it would take slightly less than 9 hours for the population to reach 5000.

Consider a similar example. If a small town in Alaska has a population of 829 in 1992, how long will it take the population of that small town to double, assuming that the population growth rate is 2% annually? Using our population growth formula with r = .02, and t, the number of years it will take to double the population, we have

$$2(829) = 829(1 + .02)^t$$

This equation simplified is

$$2 = 1(1.02)^t$$

and it is now a matter of trying various values of t until the answer is as close as possible to 2. Here are some values for t substituted into the formula and the corresponding results:

If t = 10, $(1.02)^{10} \approx 1.22$
If t = 20, $(1.02)^{20} \approx 1.49$
If t = 30, $(1.02)^{30} \approx 1.81$
If t = 40, $(1.02)^{40} \approx 2.20$

We are now over 2 which means the population has more than doubled after 40 yrs. This implies that t must be <u>between</u> 30 and 40 years.

To approximate the answer more closely,
If t = 34, $(1.02)^{34} \approx 1.96$
If t = 35, $(1.02)^{35} \approx 1.999889...$
and   if t = 36, $(1.02)^{36} \approx 2.04$

It appears that the closest to doubling that we can generate is 1.999889... and thus, it will take approximately 35 years for the population of this small town to double. (Whether it is a reasonable assumption for the population growth rate to stay the same for this length of time is questionable, however.)

Sample Problems

1. Find n, to the nearest tenth.

   a) $3^n = 10$
   b) $\left(\dfrac{1}{2}\right)^n = 9$

   Solutions: a) If n = 2, $3^2 = 9$ which is less than 10 so using 2.1 we get $3^{2.1} = 10.04$ and this is close. Thus, n = 2.1.

   b) n must be a negative exponent since the left side is a fraction less than one and the right side is an integer.

   Letting n = –3, we have $\left(\dfrac{1}{2}\right)^{-3} = 8$.

   If n = –3.1, $\left(\dfrac{1}{2}\right)^{-3.1} = 8.57$ which is less than 9, and

   for n = –3.2, $\left(\dfrac{1}{2}\right)^{-3.2} = 9.19$.

   Since 9.19 is closer to 9 than is 8.57, the value of the exponent, to the nearest tenth, is –3.2.

Simple and Compound Interest

If you own a certificate of deposit (CD) which pays a fixed amount of interest every year, the amount of interest is calculated using the simple interest formula which we looked at in section 5.3. That is, interest = principal • rate • time or

$$\boxed{I = Prt.}$$

Interest and principal are given in dollars and cents, rate is given in percent but utilized in its decimal form, and time is given in years. So, if the CD is worth $500, and the rate of interest is 6%, each year you should receive a check for $500•.06•1 or $30. As long as the rate of interest remains unchanged, the amount of the interest payment will also remain unchanged, because simple interest depends upon the original investment.

Now let's see what would happen if you cashed in that CD and invested the money in a different type of account that paid interest, i, at the rate of 6%, but instead of being paid out, this interest is added to the account, increasing the principal upon which the next year's interest will be calculated. This is no longer a simple interest account. It is called a **compound interest** account, and the amount in the account can be calculated as follows.

year	amount in account
0	$500
1	$500 + 500(.06) = 500(1 + .06) = $530
2	$500(1.06) + 500(1.06)(.06) = 500(1.06)(1 + 06) = 500(1.06)(1.06) = 500(1.06)^2 = $561.80
3	$500(1.06)^2 + 500(1.06)^2(.06) = 500(1.06)2(1 + .06) = 500(1.06)^2(1.06) = 500(1.06)^3 = $595.51

After 3 years, instead of earning 3($30) = $90 total interest you've earned $595.51 – $500 or $95.51. That's not a big difference, but it would be more impressive if your original deposit had been much greater, say $500,000.

The pattern shown in the chart is the same as we've seen in other examples in this section. The compound interest formula for accounts compounded annually (once a year) is

Total = Principal (1 + rate of interest)$^{\text{time in years}}$ or

$$\boxed{T = P(1 + r)^t.}$$

However, not all accounts pay interest just once a year. Interest may be compounded semiannually, quarterly, monthly, or daily. (Some banks advertise interest compounded continuously, but we will not be considering this type of account at this time.) Since interest rates are usually given in terms of the annual rate, an adjustment must be made in the formula we use when calculating the amount of interest paid in each interest period. In the general formula, another variable, n, is included to represent the number of interest payments, called <u>compounding periods</u>, per year.

The general compound interest formula becomes:

Remember:
T = Total; P = Principal,

$$T = P\left(1 + \frac{r}{n}\right)^{n \cdot t}$$

r = annual rate; t = time;
n = no. of comp. periods/yr.

Note also that the exponent, n•t, can also be thought of as representing the total number of interest payments made.

The following chart summarizes some useful information about different kinds of compound interest accounts.

compounding period	number of interest payments per year (n)	rate paid *each* time
annually	1	$\frac{r}{1}$ or r
semiannually	2	$\frac{1}{2}r$ or $\frac{r}{2}$ (In other words, ½ the annual rate is earned each time.)
quarterly	4	$\frac{r}{4}$
monthly	12	$\frac{r}{12}$
daily	365	$\frac{r}{365}$

Let's look at that same $500 which was originally invested in the CD paying simple interest, and let's calculate the total in the account after 3 years if the compounding period is changed as shown. (Find the exponent value first as shown.)

semiannually:
$$T = 500\left(1 + \frac{.06}{2}\right)^{2 \cdot 3} = \$597.03 \quad \rightarrow 6$$

quarterly:
$$T = 500\left(1 + \frac{.06}{4}\right)^{4 \cdot 3} = \$597.81 \quad \rightarrow 12$$

monthly:
$$T = 500\left(1 + \frac{.06}{12}\right)^{12 \cdot 3} = \$598.34 \quad \rightarrow 36$$

daily:
$$T = 500\left(1 + \frac{.06}{365}\right)^{365 \cdot 3} = \$598.60 \quad \rightarrow 1095$$

Clearly, the more frequently interest is compounded, the greater the amount in the account for the same length of time.

Sample Problems

Calculate the interest earned in 10 years if $1000 is invested in an account which pays

a) a simple interest of 4.25%

b) 4.25% compounded daily.

Solutions:

a) $I = 1000(.0425)(10)$
 $ = \$425$ interest earned.

b) $T = 1000\left(1 + \dfrac{.0425}{365}\right)^{365 \cdot 10} = \$1529.55$

$\$1529.55 - \$1000$ originally invested $= \$529.55$ interest earned.

Problem Set 6.2

1. Write the following in scientific notation.
   a) 3276
   b) $3276 \cdot 10^3$
   c) $3276 \cdot 10^{-2}$

2. Write the following in standard decimal notation.
   a) $1.23 \cdot 10^{12}$
   b) $1.23 \cdot 10^{-12}$
   c) $5 \cdot 10^7$
   d) $5 \cdot 10^{-7}$

3. Calculate, giving the answer in scientific notation.
   a) $2.5 \cdot 10^{10} \div 5 \cdot 10^{10}$
   b) $\dfrac{(3.2 \cdot 10^3)(4.1 \cdot 10^{-2})}{5 \cdot 10^5}$
   c) $\dfrac{6 \cdot 10^{-8}}{(2 \cdot 10^4)(3 \cdot 10^5)}$

4. Compare each pair of expressions using >, <, or =.
   a) $1.1 \cdot 10^{10}$ $9.9 \cdot 10^9$
   b) $45.2 \cdot 10^5$ $452 \cdot 10^4$
   c) $7.3 \cdot 10^{-6}$ $7.3 \cdot 10^{-7}$
   d) $2.15 \cdot 10^2$ $-2.15 \cdot 10^2$

5. Find the value of each missing exponent:
   a) $800000000 = 8 \cdot 10^a$
   b) $.00000312 = 3.12 \cdot 10^a$
   c) $-725 = -7.25 \cdot 10^a$
   d) $-.00604 = -6.04 \cdot 10^a$

6. If one molecule of water weighs $\approx 5.4 \cdot 10^{-23}$ grams, find the weight of thirty million molecules, expressing the answer in scientific notation.

7. The mass of the earth is approximately $5.98 \cdot 10^{24}$ kg. The mass of the sun is approximately $1.99 \cdot 10^{30}$ kg. How many times heavier is the sun than the earth?

8. The mass of an electron is approximately $9.1 \cdot 10^{-28}$ grams. The mass of a hydrogen atom is approximately $1.7 \cdot 10^{-24}$ grams. Which is heavier? How many times heavier?

9. If a can of tomato soup cost 11¢ in January, 1960, predict its cost in January, 2000 if the average annual inflation rate for the period is 4%.

10. During a week-long end-of-season clearance sale, a department store advertised that prices would be reduced each morning by 10% from the previous day's price. A jacket you really liked was marked $75 before the sale started. You decided to take a chance and wait until the last (7th) morning of the sale to buy it. If it's still available, how much will it cost, including 6% sales tax?

11. A high school built to house 750 students is losing 6% of its students per year. If the student body drops below 425, the building will cost too much to be kept open and the remaining students will be divided among other schools. Your nephew is in third grade. Will the school still be open during his senior year?

12. The population of rabbits in a wooded area is estimated at 400. The numbers are growing at a rate of 2.5% every six months. There is not enough food to maintain more than 700 rabbits, and the State Department of Natural Resources is contemplating issuing permits outside the regular hunting season to control the population.

   a) From the graph below, determine approximately how long the department can wait before implementing such a control measure.

   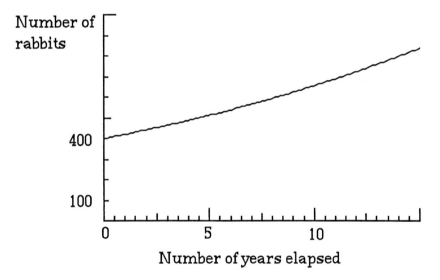

   b) Use a calculator and trial and error to determine when the population will reach 700.

13. In 1900, if one of your ancestors had invested $100 in an account paying 3% compounded annually, how much money would be in that account in the year 2000?

14. Determine how much new parents would need to invest in an account paying 4.5% annual interest compounded monthly in order for the account to contain a minimum of $10,000 in 18 years when their child is ready for college. (Make use of the memory feature of your calculator for this one.)

15. Using trial and error, determine how long it will take to double an investment placed in an account paying 6% compounded annually. How long will it take if interest is compounded daily?

16. The graph below shows how the total amount in an account grows as the interest is compounded.

   a) How much was the original deposit?
   b) What is the total in the account after 10 years? after 30 years?
   c) How long would it take for the original deposit to triple?

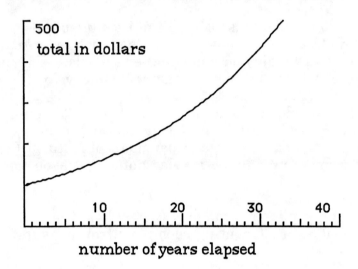

number of years elapsed

17. As treasurer of an investment club, you are responsible for finding the best **8-month** investment for $1000. You may choose between one account paying 5.25% annual interest compounded monthly or another paying 5.3% annual interest compounded quarterly. Which should you choose? Show calculations to justify your answer.

Exploration Problems

1. $500 was deposited in an account 10 years ago. Interest has been compounded quarterly at a constant rate since that time. The account now contains $801.76. At what rate has interest been calculated?

2. In 1980, two friends graduated and were fortunate to find jobs which utilized their newly acquired skills. Beginning in December, 1980, one of the two deposited $1200 per year in a retirement account which paid 5% tax-exempt interest, compounded annually. She continued to make deposits regularly through December, 1989. The other friend decided to wait to begin retirement savings until December of the year 2000. At that time he also planned to deposit $1200 per year in a similar account, and to continue making deposits through December, 2019. How much money will each of these two friends have in their respective accounts when they both retire January 1, 2021?

Solutions - Problem Set 6.2

1. a) $3.276 \cdot 10^3$
    b) $3.276 \cdot 10^6$
    c) $3.276 \cdot 10^1$

2. a) 1 230 000 000 000
    b) 0.000 000 000 001 23
    c) 50 000 000
    d) 0.000 000 5

3. a) $5 \cdot 10^{-1}$
    b) $2.624 \cdot 10^{-4}$
    c) $1 \cdot 10^{-17}$

4. a) > b) = c) > d) >

5. a) $a = 8$ b) $a = -6$ c) $a = 2$ d) $a = -3$

6. $1.62 \cdot 10^{-15}$

7. ≈ 332,776 times heavier

8. Hydrogen atom is heavier. ≈ 1868 times heavier.

9. about 53¢

10. $35.87 + tax = $38.02

11. Yes with about 430 students.

12. a) About 11-11.5 years.
    b) $400(1.025)^n = 700$ where n is the number of 6 month time periods.
        $n \approx 23$ 6 month time periods, or 11.5 years.

13. $1921.86

14. $4455.33

15. 12 years; 4217 days.

16. a) $100
    b) about $160; about $430
    c) about 23 years

17. $1000\left(1 + \dfrac{.0525}{12}\right)^8$ vs. $1000\left(1 + \dfrac{.053}{4}\right)^2$

    $1035.54 vs. $1026.67. Choose the 5.25% account.

Section 6.3 - Radicals and Rational Exponents

The Pythagorean Theorem

According to one legend, in the 500's B.C., a Greek mathematician named Pythagoras noticed an arrangement of mosaic tiles on a floor similar to the one shown below.

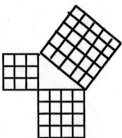

He apparently saw this not as a mere design but as a mathematical statement. Specifically, for the right triangle (a triangle that contains one right angle) on the inside of the design, if a and b are the two sides of the triangle forming the right angle, and c is the side opposite the right angle,

the relationship between the sides can be written as

$$\boxed{a^2 + b^2 = c^2}$$

It is easy to see from the design above that if the tiles from a 3 by 3 square and a 4 by 4 square are combined, they form exactly one 5 by 5 square. In other words,

$$3^2 + 4^2 = 5^2 \quad \text{or} \quad 9 + 16 = 25$$

This relationship, between sides a and b, called <u>legs</u>, and c, the <u>hypotenuse</u>, is referred to as the **Pythagorean Theorem**. The numbers 3, 4, and 5, are called Pythagorean triples since they make the statement, $a^2 + b^2 = c^2$, true. The second sample problem on the following page refers more to Pythagorean triples.

It is important to remember that this theorem applies *only* to right triangles and not to triangles in general. If a triangle is right, the theorem holds, and conversely, if the statement holds, the triangle is a right triangle.

The theorem is extremely useful in a variety of contexts. For example, in construction, "square" corners are necessary.

To check if the corner is "square", or 90°, we measure out 3 feet on one side and 4 feet on another, mark both distances, and then check whether the diagonal, straight-line distance between these two marks is 5 feet. This is shown below.

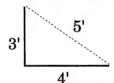

If the distance is less than five feet, the corner angle is less than 90° and if it is greater than 5 feet, the angle is greater than 90°. Corners that are not 90° can cause big problems when walls and siding must be put up!

(It should be noted that instead of measuring each corner, a rope can be marked, and knots tied every foot so that there are 4 knots, or 3 one-foot intervals on one side and 5 knots, or 4 one-foot intervals, on the other side. The configuration may look like:

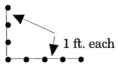
1 ft. each

With the rope laid along the walls forming the corner, the only measurement needed is the length of the diagonal which can be made with another similarly knotted rope.)

Sample Problems

1. Find the length of the missing side in this right triangle.

   Solution: $8^2 + n^2 = 10^2$
   or $64 + n^2 = 100$ so $n^2 = 100 + (-64)$.
   Thus, $n^2 = 36$ and $n = 6$.
   (n also $= -6$, but in distance, this is unreasonable.)

2. Can you find other Pythagorean triples? Is there a pattern in these numbers? What if $a = 20$ and $b = 21$? Is c then an integer?

   Possible Solutions: 6, 8, and 10; 12, 16, and 20, etc. It appears we can generate triples by using whole number multiples of 3, 4, and 5 each time.
   However, some other triples are 5, 12, and 13 (and multiples of these numbers). For $a = 20$ and $b = 21$, $c = 29$, an integer. Can you find more?

Consider the suitcase with measurements as indicated below. What is the longest umbrella that will fit into the bottom of the suitcase?

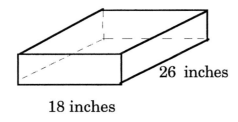
26 inches
18 inches

Looking at the suitcase from the top, we have

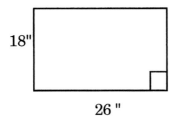
18"
26 "

341

We could put an 18" or a 26" umbrella in, but let's check another possibility. The diagonal from one corner to the opposite corner could represent the length of the umbrella, x, and it also is the hypotenuse of a right triangle.

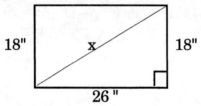

Using the Pythagorean Theorem,

$$18^2 + 26^2 = x^2$$

and $\quad 324 + 676 = x^2$

so $\qquad 1000 = x^2$

Recalling the meaning of x^2, we must find a number that, when multiplied by itself, gives a product of 1000. Since

$$30 \cdot 30 = 900 \quad \text{and} \quad 40 \cdot 40 = 1600$$

the number (in other words, the length of the umbrella), must be somewhere between 30 and 40 inches. Since 30^2 is closer to 1000 than 40^2, it is reasonable to assume the length is closer to 30. Trying 32, $32^2 = 1024$, and we have an umbrella that is still too long. Using 31, we have $31^2 = 961$ which means that a 31-inch umbrella would fit. To find the *longest possible umbrella* that would fit in this suitcase, let's try a decimal such as 31.5.

Squaring 31.5, we have 992.25. For 31.6^2, we get 998.56 and we're even closer to 1000. How about 31.7? If we square 31.7, we have 1004.89, which is greater than 1000. An umbrella of this length would not fit. Thus, to the nearest tenth, we can use an umbrella that is 31.6" long.

It is possible to find the length of an umbrella that would fit into this suitcase to any accuracy desired by continuing to "zero-in" on possible values. Trying numbers such as 31.62, 31.65, etc., squaring each, and finding the largest such decimal number that is less than 1000 will work. However, instead of this approximation method, there is a notation, $\sqrt{}$, called a **radical sign**, that symbolizes the quantity we are looking for, namely, a number that when multiplied by itself exactly equals 1000. We call this the **principal square root** of 1000 and write it as $\sqrt{1000}$.

Calculators are programmed to find square roots, as well as cube roots, fourth roots, etc. Experiment with your calculator to determine the key sequence needed to find $\sqrt{1000}$. The display should read $\qquad 31.6227766...$

Depending on the length of display on a calculator, a different number of digits to the right of the decimal point may appear. It should be apparent, however, that, in this case, the digits do not end.

For our umbrella problem, the accuracy of an actual measurement limits the number of decimal places we need to the right of the decimal point.

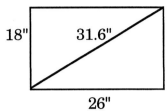

Thus, it is reasonable to say the longest possible umbrella that would fit into this 18" by 26" suitcase is 31.6 inches, or 31 inches, but definitely not 32"! Note that this agrees with our previous result.

Exponents and Radicals

There is another way to use the calculator to find the value of $\sqrt{1000}$, and it involves the relationship between exponents and radicals.

Exponents that are rational numbers can be keyed into the calculator—check your manual, although the y^x or x^y key is typically used and the exponent is keyed in as *(a÷b)* or its decimal equivalent. We can find the value of $25^{1/2}$, for example, by using the "square root" key <u>or</u> by keying in 25 and an exponent of .5:

$$\sqrt{25} = 5 \text{ and } 25^{1/2} = 25^{.5} = 5$$
$$\text{Also, } \sqrt{100} = 10 \text{ and } 100^{.5} = 10.$$

It is not coincidence that using $1/2$ or .5 as the exponent *or* the radical sign indicating the square root generates the same answer. In fact,

> **The principal square root of a nonnegative number, *b*, can be written as $\sqrt{b}$ or as $b^{1/2}$, and it represents a number, *a*, such that $a^2 = b$.**

This opens up opportunities for finding the cube root, or third root, of a number. If a calculator does not have a cube root key, we now have a way to find that root by using a rational number exponent. Let's try finding the third root of 27. Since $\sqrt{25} = 25^{1/2}$, it would seem reasonable that $\sqrt[3]{27} = 27^{1/3}$.

Try using the cube root key to find this root of 27. Then key in $27^{1/3}$ using the y^x key. Your result in each case should be *3*. In general,

> **The nth root of a number, *b*, can be written as $\sqrt[n]{b}$ or as $b^{1/n}$.**
>
> (Note: for *n* even, *b* > 0)
>
> and for $\sqrt[n]{b} = a$, $a^n = b$.

Let's try $27^{2/3}$. Since there is no $2/3$ root key on most calculators, we must use the rational exponent equivalent. Keying this in we have $27^{2/3} = 9$
[Note here that more accuracy is achieved if $(2 \div 3)$ is keyed in as the exponent instead of .6 or .66 or even .666.]

Returning to the rules for exponents, since $2 \cdot 1/3 = 2/3$, we'll write $27^{2/3}$ as $\left(27^{1/3}\right)^2$ or as $\left(27^2\right)^{1/3}$ and perform the inner operation first in each case.

Since $27^{1/3} = 3$ then $\left(27^{1/3}\right)^2 = 3^2$ or 9

or

$27^2 = 729$ and $(729)^{1/3} = 9$

In other words, on most calculators we can use the exponent, $2/3$, directly, **or** rewrite the exponent as a product in two different ways. We've really given an illustration of a rule regarding rational number exponents:

$$b^{m/n} = \left(b^{1/n}\right)^m = \left(b^m\right)^{1/n} \text{ for } n \neq 0$$
(and b nonnegative when n is even)

We must clarify why we have written that b must be nonnegative in $\sqrt{b}$, and in $\sqrt[n]{b}$ when n is even. Perhaps the best way is through an example. To find $\sqrt{-4}$, we must be able to find, by definition of the radical sign, a number multiplied by itself that equals -4. But a number times itself is always positive, and so $\sqrt{-4}$ does not exist as a real number. (Later mathematics courses consider this type of expression when complex numbers are studied.)

It should now be apparent why $b^{m/n}$ will not be defined, again in our context, when n is 2, or <u>any</u> even number, and b is negative. For example, we cannot find $(-4)^{1/4}$ since there is no number that can be multiplied by itself *four* times and generates a product of -4. In general, $\left(\sqrt[n]{b}\right)^m$, or $b^{m/n}$, will not exist in the real number system when b is negative and n is even, because multiplying any number by itself an even number of times always generates a positive product.

<u>Sample Problems</u>

1. Write $27^{2/3}$ as a radical.

    Solution: $\left(27^{1/3}\right)^2 = \left(\sqrt[3]{27}\right)^2$ or $\left(27^2\right)^{1/3} = \sqrt[3]{27^2}$.

2. Write $(-8)^{1/3}$ using a radical sign and find its value.

    Solution: $\sqrt[3]{-8} = -2$, *since* $(-2)(-2)(-2) = -8$.

    (Note here that we can find an *odd*-numbered root of a negative radicand.)

3. Using your scientific calculator, find the value (to the nearest hundredth) of

   a) $4^{2/3}$ b) $(-8)^{1/4}$ c) $125^{1/3}$ d) $-\sqrt{81}$

   Solutions: a) 2.52
   b) Calculator yields an error. There is no answer because there is no real number multiplied by itself 4 times that will give a negative 8.
   c) 5.00
   d) −9.00 (Note here that the negative refers to the "inverse" of $\sqrt{81}$.)

4. Find the value of x in
   $$x^4 = 81$$

   Solution: $(x^4)^{1/4} = (81)^{1/4}$ (taking the fourth root of each side)
   and $x = 3$

The Real Numbers

Let's consider the unit interval on the number line below.

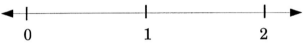

Drawing a 1 x 1 square on this interval and constructing one of the diagonals, of length n, we have

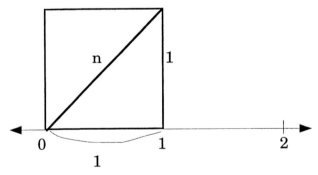

Finding the length of this diagonal requires the Pythagorean Theorem.

$$a^2 + b^2 = c^2 \quad \text{and} \quad 1^2 + 1^2 = n^2.$$

Thus, $\quad n^2 = 1 + 1 = 2 \quad \text{and} \quad n = \sqrt{2}$

If we extend the length of this diagonal down to the number line using a compass, for instance, we have a point on the number line whose distance from 0 represents the length of the diagonal, which in this case is $\sqrt{2}$ units.

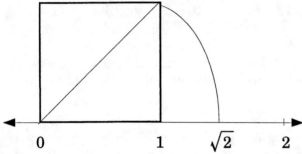

Note that $\sqrt{2}$ ≈ 1.414 and the construction verifies that this point is a little less than $\frac{1}{2}$ way between 1 and 2 on the number line.

Using the calculator, the decimal approximation of $\sqrt{2}$ is 1.414213562.... The decimal component will vary according to the amount of display available on the calculator, but it should be apparent that this decimal *does not terminate and does not repeat*. (A computer can easily verify this by carrying the decimal out to many more places.) It is interesting that even though we cannot find an <u>exact</u> decimal to represent this number, we can find its <u>definite position</u> on the number line. This type of number whose decimal equivalent does not terminate and does not repeat, is called an **irrational number**. Pythagoras and the Pythagorean Society discovered this number line representation of irrational numbers but kept the matter secret because they believed whole numbers governed everything and they were worried about the consequences of their discovery. Obviously, the secret eventually became public knowledge.

Sample Problems

Use your calculator and determine whether each of the following is rational or irrational.

    a) $\sqrt{5}$ b) $2/7$ c) $1 + \sqrt{3}$

    Solutions: a) irrational because the decimal equivalent is 2.236067978...
                       b) rational because it's in the form of a over b, but also because it has a repeating decimal, .285714285714...
                       c) irrational because the decimal equivalent is 2.732050808...

We can make several very important statements at this point.

> 1) **Rational numbers have decimal equivalents that either terminate or repeat in a block of one or more digits;**
>
> 2) **Irrational numbers have decimal approximations that do not terminate or repeat;**
>
> 3) **The rational numbers together with the irrational numbers form the system of *real numbers*.**

Finally, there are some interesting properties of radicals, actually based on the work we have done with exponents. For example, to add $2\sqrt{5}$ and $\sqrt{5}$, we can write

$$2\left(5^{1/2}\right) + \left(5^{1/2}\right) = (2+1) \cdot \left(5^{1/2}\right) = 3\left(5^{1/2}\right) \text{ or } 3\sqrt{5}.$$

In other words, if the radicands, the quantities under the radical sign, are the same, the *coefficients* can be added or subtracted just as they can be when we are adding or subtracting like terms such as *2x* and *x*.

To multiply radicals, consider the following equivalencies, justified by the rules of exponents.

$$2\sqrt{3} \cdot \sqrt{8} = 2 \cdot 3^{1/2} \cdot 1 \cdot 8^{1/2} = (2 \cdot 1)(3 \cdot 8)^{1/2}$$

But $2(3 \cdot 8)^{1/2}$ is the same as $2(24)^{1/2}$ or $2\sqrt{24}$ and so

$$2\sqrt{3} \cdot \sqrt{8} = 2\sqrt{24}$$

Thus, when multiplying radicals, multiply the radicands together and write the product as a radicand, then if there are any coefficients, multiply them together, and leave this value as the coefficient.

It should be noted that sometimes the radicand can be simplified. That is, any factor that is a perfect square (has an integer root) can be isolated and written as an integer coefficient. For example,

$\sqrt{24}$ can be simplified by writing

$$\sqrt{24} = \sqrt{4 \cdot 6} = \sqrt{4} \cdot \sqrt{6} = 2\sqrt{6}$$

Also, $\quad 2\sqrt{3} \cdot 3\sqrt{3} = 6\sqrt{9} = 6 \cdot 3 = 18$

and $\quad \sqrt[3]{16} = \sqrt[3]{8} \cdot \sqrt[3]{2} = 2\sqrt[3]{2}$

(Use your calculator to verify that both sides of each equation are equivalent.)

Let's do another example where simplifying the radicand is required as a first step. To simplify $\sqrt{50} + 2\sqrt{2}$, we isolate a perfect square factor in the radicand, 50, and we have

$$\sqrt{50} + 2\sqrt{2} = \sqrt{25}\sqrt{2} + 2\sqrt{2}$$
$$= 5\sqrt{2} + 2\sqrt{2} \quad \text{(Once the radicands are}$$
$$= 7\sqrt{2} \qquad \quad \text{the same, the coefficients}$$
$$\qquad \qquad \qquad \text{can be combined.)}$$

Sample Problems

1. Use like terms to subtract $2\sqrt{5}$ from $5\sqrt{5}$.

   Solution: $\quad 5\sqrt{5} - 2\sqrt{5} = 3\sqrt{5}$

2. Use the properties of exponents to multiply $\sqrt{2}$ and $\sqrt{6}$.

   Solution: $\quad 2^{1/2} \cdot 6^{1/2} = (2 \cdot 6)^{1/2}$
   $$= 12^{1/2}$$
   which can be simplfied as $\quad = 4^{1/2} \cdot 3^{1/2}$
   $$= 2 \cdot 3^{1/2} = 2\sqrt{3}$$

3. Use the previous examples of radical addition, subtraction, and multiplication, and simplify:

   a) $\sqrt{2}\left(2\sqrt{2} + \sqrt{6}\right)$ b) $\sqrt{3x}\sqrt{8xy^2}$ c) $3\sqrt[3]{2} - \sqrt[3]{16}$

   Solutions:

   a) $\sqrt{2}\left(2\sqrt{2} + \sqrt{6}\right) = 2\sqrt{4} + \sqrt{12}$
   $= 2 \cdot 2 + \sqrt{4} \cdot \sqrt{3}$
   $= 4 + 2\sqrt{3}$

   b) $\sqrt{3x}\sqrt{8xy^2} = \sqrt{24x^2y^2}$
   $= \sqrt{4x^2y^2} \cdot \sqrt{6}$
   $= 2xy\sqrt{6}$

   $\left(\sqrt{4x^2y^2} = 2xy \text{ since } (2xy)^2 = 4x^2y^2\right)$

   c) $3\sqrt[3]{2} - \sqrt[3]{16} = 3\sqrt[3]{2} - \sqrt[3]{8}\sqrt[3]{2}$
   $= 3\sqrt[3]{2} - 2\sqrt[3]{2}$
   $= \sqrt[3]{2}$

4. Find a fraction with a rational denominator that is equivalent to:

   $$\frac{2\sqrt{8}}{\sqrt{3}}$$

   Solution: $\frac{2\sqrt{8} \cdot \sqrt{3}}{\sqrt{3} \cdot \sqrt{3}} = \frac{2\sqrt{24}}{\sqrt{9}} = \frac{2\sqrt{4}\sqrt{6}}{3} = \frac{2 \cdot 2\sqrt{6}}{3} = \frac{4\sqrt{6}}{3}$

   (Note: This is called **rationalizing the denominator**.)

Problem Set 6.3

1. Write a word problem that would require the use of the Pythagorean Theorem in its solution. Solve your problem.

2. In each part below, find the missing side (to the nearest tenth of a cm.).

   a)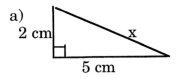
   b)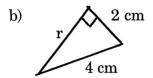

3. For the suitcase with dimensions as shown, find the longest umbrella that will fit on the diagonal from the lower left (front) corner to the top right (back) corner.

   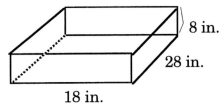

4. Write without the radical sign

   a) $\sqrt{3}$ b) $3\sqrt[3]{2}$ c) $\left(\sqrt[3]{6}\right)^2$ d) $\sqrt{x^5}$

5. Write as a radical expression and put in simplest radical form, if possible.

   a) $27^{1/3}$ b) $6^{1/2} \cdot 3^{1/2}$ c) $5^{1/2} + 2(5^{1/2})$ d) $8^{2/3}$ e) $(200a^2)^{1/2}$

6. Use your calculator to approximate the value of each of the following to the nearest hundredth.

   a) $(10)^{1/3}$ b) $\sqrt{125}$ c) $2\sqrt{6}$ d) $\sqrt[4]{-9}$ e) $-\sqrt{24}$

7. Determine whether or not it is possible to have a square with area of 12 sq. inches. Justify your answer.

8. Give two examples of an irrational number.

9. Explain how the decimal equivalent of a rational number differs from that of an irrational number.

10. Write an expression equivalent to $\dfrac{3}{\sqrt{2}}$ without a radical sign or fractional exponent in the denominator.

11. Evaluate $\sqrt{a^2 + b^2}$, to the nearest tenth, when $a = 41$ and $b = 50$ units.

12. The area of a square is 90 square centimeters. Find:
    a) the length of one side (to the nearest hundredth of a centimeter)
    b) the length of the diagonal (to the nearest hundredth of a centimeter).

13. Solve $a^2 + b^2 = c^2$ for b, if b is positive.

14. Solve for the variable in each of the following.

    a) $a = \sqrt{144}$ b) $x^2 = 196$ c) $6^2 + x^2 = 10^2$ d) $x^5 = 10$

15. One day, to get some exercise, Vince left his home on the corner of Lincoln and Bluff streets and walked 1 mile east. He then turned and walked 1.5 miles north before stopping for a rest.

    a) What is the straight line distance from his resting point to his home?

    b) If he could walk this straight line distance back to his home, how much shorter would his walk be than if he returned along his first route?

16. Simplify each of the following.

    a) $\sqrt{25ab^2}$ b) $\sqrt{75}$ c) $\sqrt{24x^3y}$

    d) $\sqrt[3]{3^3}$ e) $\sqrt{5^3}$

17. Perform the operations and simplify, if possible.

    a) $\sqrt{6xy^2}\sqrt{2xy}$ b) $\sqrt{10}\sqrt{8}$ c) $\sqrt{27} + \sqrt{12}$ d) $\sqrt[3]{24} - \sqrt[3]{3}$

Exploration Problems

1. Problem 2b) of this section refers to a triangle that is pictured as

    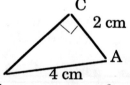

    a) Use the inverse trig function keys on your calculator to find the measures of angles A and B.

    (Hint: cosine of an angle = $\dfrac{\text{length of adjacent side}}{\text{length of hypotenuse}}$

    and

    sine of an angle = $\dfrac{\text{length of opposite side}}{\text{length of hypotenuse}}$)

    b) Now draw another triangle where the hypotenuse is twice the length of the shorter side. Find the measures of the non-right angles. Explain your results.

2. Use the Pythagorean Theorem and the method shown in this section to draw the exact point on the number line representing the distance $\sqrt{5}$ units from 0.

3. If the side of a square is doubled, what happens to the length of the diagonal? Justify your answer.

Solutions - Problem Set 6.3

1. Possible answer: The end of a 2-inch thick board is to be cut on a 45° angle as shown. Find the length of the cut.
   Solution: $2^2 + 2^2 = c^2$
   $8 = c^2$
   $c = \sqrt{8} \approx 2.8$ in.

2. a) 5.4 cm   b) 3.5 cm

3. An umbrella of 34.2 inches is the longest that would fit into this suitcase.

4. a) $3^{1/2}$ b) $3(2^{1/3})$ c) $(6^{1/3})^2 = 6^{2/3}$ d) $(x^5)^{1/2} = x^{5/2}$

5. a) $\sqrt[3]{27} = 3$ b) $\sqrt{6}\sqrt{3} = 3\sqrt{2}$ c) $\sqrt{5} + 2\sqrt{5} = 3\sqrt{5}$
   d) $(\sqrt[3]{8})^2 = 4$ e) $\sqrt{200a^2} = 10a\sqrt{2}$

6. a) 2.15 b) 11.18   c) 4.90   d) no real solution   e) –4.90

7. Yes; each side has length $\sqrt{12}$ and $\sqrt{12} \cdot \sqrt{12} = 12$.

8. Examples: .121121112... , $\sqrt{5}$.

9. The decimal equivalent of a rational number terminates or the digits repeat in a block. The decimal equivalent of an irrational number neither terminates nor repeats.

10. $\dfrac{3\sqrt{2}}{2}$

11. 64.7 units.

12. a) 9.49 cm   b) 13.42 cm.
    Since $\sqrt{90} \approx 9.49$ gives the length of a side, to find the diagonal we need to square that number and $(\sqrt{90})^2 = 90$. To find the diagonal, just add 90 and 90 and take the square root of the sum. $[(\sqrt{90})^2 + (\sqrt{90})^2 = 90 + 90 = c^2]$.

13. $b = \sqrt{c^2 - a^2}$

14. a) $a = 12$ b) $x = 14, -14$ c) $x = 8, -8$ d) $(10)^{1/5} = 1.5848...$

15. a) 1.8 miles   b) Shortest distance is 4.3 miles, the longest is 5 miles, so Vince saves .7 mile by not retracing his route.

16. a) $5b\sqrt{a}$ b) $5\sqrt{3}$ c) $2x\sqrt{6xy}$ d) 3   e) $5\sqrt{5}$

17. a) $\sqrt{12x^2y^3} = 2xy\sqrt{3y}$ b) $\sqrt{80} = 4\sqrt{5}$
    c) $3\sqrt{3} + 2\sqrt{3} = 5\sqrt{3}$ d) $2\sqrt[3]{3} - \sqrt[3]{3} = \sqrt[3]{3}$

Section 6.4 - Applications of Rational Exponents and Radicals

Length of Line Segments

In section 1.1 we found the length of both horizontal and vertical line segments using a number line and a coordinate system and in section 1.5, we did more work with line segments on the geoboard. We'll now extend that work and use the Pythagorean Theorem to find the length of *any* line segment.

To find the perimeter of the square drawn on the coordinate system below,

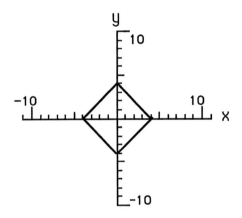

we first must find the length of one side. Since the sides of the square are not horizontal or vertical lines, we must use a different tactic here. Restricting our discussion to one quarter of the square, we have the right triangle shown.

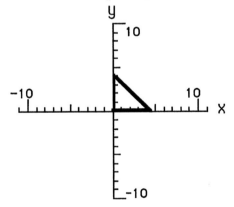

The legs of the triangle are horizontal and vertical lines and are both 4 units in length. We can use the Pythagorean Theorem to find the length of the hypotenuse, c, which represents one side of the square.

Thus,
$$4^2 + 4^2 = c^2$$
$$16 + 16 = c^2$$
$$32 = c^2$$

and
$$\sqrt{32} = c \text{ or } c = \sqrt{16} \cdot \sqrt{2} = 4\sqrt{2} \text{ or } 5.66 \text{ graph units.}$$

Since a square has four equal sides, the perimeter of the square can be written in radical form as $4(4\sqrt{2})$ or $16\sqrt{2}$ or approximately 22.64 graph units.

(Note that this answer is equal to 4(5.66) graph units.)

What if we want to find the length of the line segment below?

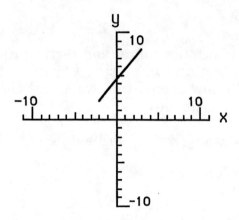

The endpoints of the line segment have coordinates (3,8) and (−2,2). We can sketch a right triangle using this segment as the hypotenuse by dropping a vertical line from point B and a drawing horizontal line through point A as shown below. The point of intersection, C, is the vertex of the right angle.

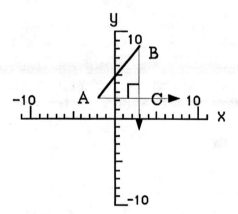

and the coordinates of point C are (3,2). (Can you see why?) We can find the length of the horizontal line segment by finding the distance between the coordinates (−2,2) and (3,2) that form the line segment. Thus, the length of the horizontal leg, segment **AC**, is

$$|-2-3| = |-5| = 5 \text{ graph units.}$$

Similarly, the length of the vertical line segment, **BC**, is the distance between the coordinates (3,2) and (3,8). This length is

$$|8-2| = |6| = 6 \text{ graph units.}$$

Finally, to find the length of the original line segment **AB**, our "hypotenuse," we'll use these lengths in the Pythagorean Theorem and we have

$$c^2 = 5^2 + 6^2 = 25 + 36 = 61$$
and so $c = \sqrt{61}$ or about 7.8 graph units.

(Rounding to the nearest tenth is appropriate with integer data.)

Thus, our line segment with endpoints at (3,8) and (–2,2), has a length of approximately 7.8 graph units.

This seems like a fairly long process for finding the length of a line segment. Since we formed a right triangle in both examples, we should be able to consolidate some of the steps and shorten the process.

In the last example, to find the length of the horizontal line segment we actually found the difference, $|-2-3|$, in the x-coordinates. For the length of the vertical line segment, we found the difference, $|8-2|$, in the y-coordinates. Since these values represent the length of the legs, we then used the Pythagorean Theorem to find the length of the original line segment by

$$(\text{difference in } x\text{-coordinates})^2 + (\text{difference in } y\text{-coordinates})^2 = c^2$$
$$|-2-3|^2 + |8-2|^2 = (-2-3)^2 + (8-2)^2 = c^2$$

[It is important to realize that parentheses can replace the absolute value signs since the squaring process will eliminate the negative values anyway. (This also implies that the order of subtraction does not matter.)]

Finally, taking the square root of both sides we have the value of c:

$$\sqrt{(-2-3)^2 + (8-2)^2} = \sqrt{(-5)^2 + 6^2}$$
$$= \sqrt{61} = \sqrt{c^2} = c$$

Since we described a possible short-cut process here, let's generalize to find the length of *any* line segment (i.e., find the distance between any two points) on a coordinate system. We'll use the points (x_1, y_1) and (x_2, y_2) as shown below.

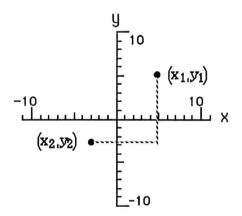

First, the length of the horizontal line segment is $|x_1 - x_2|$.
The length of the vertical line segment is $|y_1 - y_2|$.

Again using the Pythagorean Theorem and replacing the absolute value signs with parentheses, the length, d, of the original line segment is

$$d = \sqrt{(x_1 - x_2)^2 + (y_1 - y_2)^2}$$

Sample Problems

1. Find the perimeter of the figure on the coordinate system below.

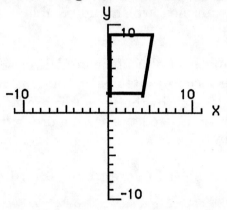

Solution:
The vertices of the figure are, starting at the lower left and going counterclockwise: (0,0), (4,0), (5,7) and (0,7).

The horizontal line segments are 4 graph units and 5 graph units in length. The length of the vertical line segment is 7 graph units.

To find the length of the line segment formed by the points (4,0) and (5,7), we'll use our "distance" formula:

$$d = \sqrt{(4-5)^2 + (0-7)^2} = \sqrt{1+49} = \sqrt{50} \text{ or approximately 7.1 graph units.}$$

Thus the perimeter of the figure is about
4 + 5 + 7 + 7.1 = 23.1 graph units.

Finding Midpoints

Up to now, we have restricted our discussion of midpoints to horizontal and vertical lines. In section 1.4, we found the midpoint of a line segment in several ways, but "averaging" the two endpoints was the most efficient.
Given this line segment,

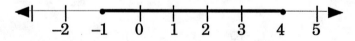

the midpoint is at

$$\frac{-1+4}{2} = 1.5$$

Consider the following line segment defined by the endpoints E = (–5,4) and F = (2,–2).

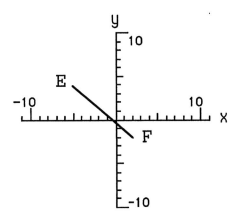

We again draw a right triangle so that this line segment is the hypotenuse.

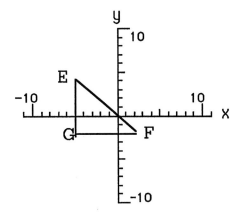

The coordinates of the vertices of the triangle are now G = (–5,–2), F = (2,–2), and E = (–5,4). The midpoint of the horizontal line segment formed by the points (2,–2) and (–5,–2) is

$$\frac{2+(-5)}{2} = -1.5 \quad \text{[In coordinate form, } (-1.5, -2)\text{]}$$

The midpoint of the vertical line segment formed by points (–5,4) and (–5,–2) is

$$\frac{4+(-2)}{2} = 1 \quad \text{[In coordinate form, } (-5,1)\text{]}$$

We'll illustrate the relationship between these calculated midpoints and the original line by drawing the vertical line, x = –1.5, and the horizontal line, y = 1, as dashed lines, on the diagram following.

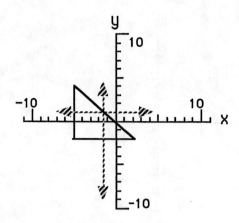

It should be apparent that the point of intersection of these two dashed lines, the point (−1.5, 1), is actually the midpoint of the original line. In fact, the midpoint of *any* line can be found by taking the average of the x-coordinates and the average of the y-coordinates and writing these values as an ordered pair.

In formula form, the **midpoint** of a line segment formed by the points (x_1, y_1) and (x_2, y_2) can be written as

$$\left(\frac{x_1 + x_2}{2}, \frac{y_1 + y_2}{2} \right)$$

Sample Problems

1. Find the midpoint of the line segment formed by the points (5,4) and (−3,−4).

    Solution: The x-coordinate of the midpoint is

    $$\frac{5 + (-3)}{2} = 1$$

    and the y-coordinate is

    $$\frac{-4 + 4}{2} = 0$$

    Thus, the midpoint is (1,0). (Verify this on the coordinate system.)

2. If a line segment has a midpoint at (1,2) and one endpoint has coordinates (4,−1), find the coordinates of the other endpoint.

    Solution: $\frac{x_2 + 4}{2} = 1$ implies $x_2 + 4 = 2$, so $x_2 = -2$

    and $\frac{y_2 + (-1)}{2} = 2$ implies $y_2 + (-1) = 4$, so $y_2 = 5$.

    Therefore, the other endpoint is at (−2,5).

Perimeter and Area Revisited

When we found the perimeter and area of geometric figures on the geoboard, we dealt first with horizontal and vertical segments as sides. However, recall that we also found the perimeter of figures such as the one shown below,

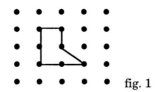
fig. 1

but the length of the diagonal line segment was given because we did not have enough information at that time to calculate its length. However, we can now apply our work with the Pythagorean Theorem to find the perimeter of this figure.

Focusing our discussion on this line segment,

we can think of it as the "hypotenuse" of a right triangle formed by connecting the dots as shown below.

Since each leg of the triangle is 1 geoboard unit, the hypotenuse is

$$c = \sqrt{1^2 + 1^2} = \sqrt{2} \approx 1.4 \text{ geoboard units}$$

(Recall that in our geoboard problems, that was the value given for the length of this specific line segment.)

Finally, the perimeter of **fig. 1** above is then

$$2 + 2 + 1 + 1 + 1.4 \approx 7.4 \text{ geoboard units}$$

Let's find the perimeter of triangle ABC drawn on the following geoboard model.

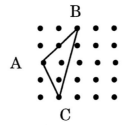

fig.2

Since none of the sides is a vertical or horizontal line, we must find the length of each line segment through our generalized method, each time visualizing the line segment as a "hypotenuse". (We'll round the answers to the nearest tenth unit.)

For segment **AB**, the length is $\sqrt{2^2 + 2^2} = \sqrt{8} \approx 2.8$ geoboard units, based on this triangle

For segment **AC**, the length is $\sqrt{2^2 + 1^2} = \sqrt{5} \approx 2.2$ geoboard units, based on this triangle

For segment **BC**, the length is $\sqrt{4^2 + 1^2} = \sqrt{17} \approx 4.1$ geoboard units, based on this triangle

Therefore, the perimeter of triangle ABC in **fig. 2** on the previous page is

$$2.8 + 2.2 + 4.1 \approx 9.1 \text{ geoboard units.}$$

How can we find the area of triangle ABC? Examining it again on the geoboard model, let's draw a rectangle around this triangle.

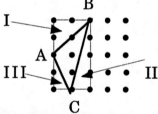

The area of the outside rectangle is 2•4 or 8 square geoboard units. To find the area of triangle ABC, we will have to <u>subtract</u> the areas of the three outer triangles, I, II, and III, within the rectangle. Visualize each of these triangles as half a rectangle.

The area of *triangle I* is $(1/2)(2 \cdot 2) = 2$ square geoboard units;

The area of *triangle II* is $(1/2)(4 \cdot 1) = 2$ square geoboard units;

The area of *triangle III* is $(1/2)(2 \cdot 1) = 1$ square geoboard unit.

So the area of triange ABC is
$$8 - (2 + 2 + 1) = 8 - 5 = 3 \text{ square geoboard units.}$$

Sample Problems

Find the perimeter and area of triangle EFG shown below on the geoboard model.

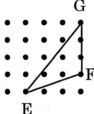

Solution: The length of segment $GF = 3$ geoboard units;

the length of segment $EG = \sqrt{4^2 + 3^2} = 5$ geoboard units; and

the length of segment $EF = \sqrt{3^2 + 1^2} \approx 3.2$ geoboard units.

Thus, the perimeter of triangle EFG is $3 + 5 + 3.2 \approx 11.2$ geoboard units.

To find the area, draw a rectangle as shown below and subtract the areas of triangles I and II.

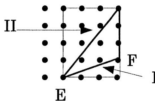

Area of *triangle I* $= \left(\frac{1}{2}\right)(3) = 1.5$ sq. geoboard units and

Area of *triangle II* $= \left(\frac{1}{2}\right)(3 \cdot 4) = 6$ sq. geoboard units.

Therefore, the area of triangle EFG is

$(3 \cdot 4) - (1.5 + 6) = 4.5$ sq. geoboard units.

2. Find the area of the figure given on the coordinate system below.

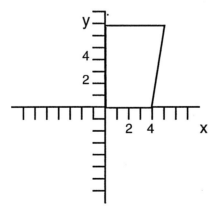

Solution:
Separating the figure into two regions,
$A = 4 \cdot 7 + \left(\frac{1}{2}\right)(1 \cdot 7)$
$A = 28 + 3.5 = 31.5$ sq. units

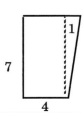

Problem Set 6.4

1. Find the length of each line segment below.

    a)

    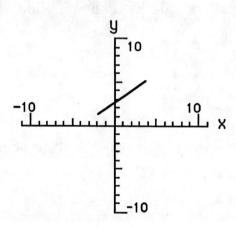

    b)

    c) Defined by the points (6,4) and (–5,2).

2. Given the triangle below,

    a) Find its perimeter, to the nearest tenth of a graph unit.

    b) Find the area of the triangle.

    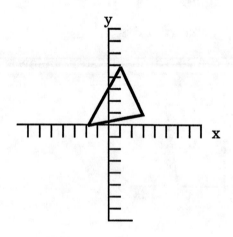

3. Sketch and estimate the coordinates of the midpoint of each line segment below. Then check each estimate by using the formula to find the coordinates of the midpoint.

   a)
   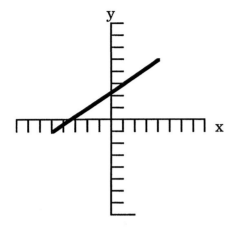

   b)
   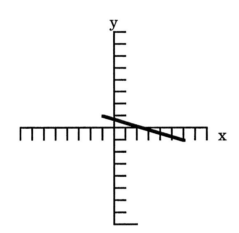

4. A median of a triangle is a line segment that connects a vertex to the midpoint of the opposite side. Find the length of median **AD**, to the nearest tenth of an inch.

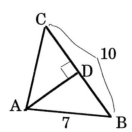

5. a) Find the coordinates of the point F which lies on segment TR and is an endpoint of the median **SF**.

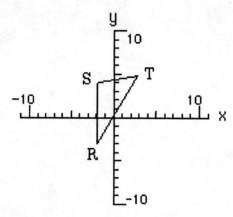

   b) Find the length of median **SF**.

6. Find the perimeter and area of the following figures on the geoboard model.
   a)    b)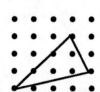

7. Draw two triangles, each with an area of 3 square units, on a geoboard model. Then find the perimeter of each.

8. The bottom of a ladder is placed 5 feet from the base of a wall. The ladder is 16 feet long. Approximately how far above the ground does the ladder touch the wall? (Round to the nearest tenth of a foot.)

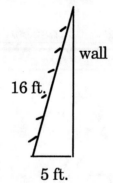

9. The midpoint of a line segment is at (3,0) and one endpoint is given by the coordinates (4,5). Find the coordinates of the other endpoint of the line segment.

Exploration Problems

1. Given the triangle below, calculate the midpoint of each side and plot those points on the graph. Draw the three medians of the triangle. What do you find?

   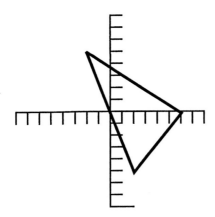

   Now draw a different triangle on a coordinate system, find the midpoint of each side, and draw the medians. Explain your results. If you have access to the program *Geometer's Sketchpad*, verify your results.

2. A square tile, □ , can be used to represent a table. Chairs can be placed on each side of the table for seating guests. Sketch different arrangements for placing 4 tables in a room where each table touches a side or corner of another table.

   a) Which arrangement gives the maximum seating?

   b) Which arrangement seats the smallest number of people?

Solutions - Problem Set 6.4

1. a) $d = \sqrt{52} \approx 7.2$ graph units;   b) $\sqrt{8} \approx 2.8$ geoboard units   c) 11.2 graph units

2. a) The lengths of the sides are 5.1, 4.5, and 5.8 graph units, so the perimeter = 15.4 graph units.
   b) The area is approximately, 25 − (2.5 + 7.5 + 4) = 11 sq. graph units

3. a) (−.5, 2)   b) (2.5, 0)

4. Segment **AD** is 4.9 inches.

5. a) ($\frac{1}{2}$, 1)   b) 3.9 units

6. a) P = 13.4 geoboard units;   A = 8.5 sq. geoboard units
   b) P = 4.1 + 4.2 + 2.2 = 10.5 geoboard units;
      A = 12 − (1 + 2 + 4.5) = 4.5 sq. geoboard units

7. One solution: The perimeter of this triangle is 8.6 geoboard units.

   Another possibility: The perimeter here is 8.3 geoboard units.

8. The ladder touches the wall approximately 15.2 feet above the ground.

9. The other endpoint is at (2, −5).

Chapter Six Review

1. Write each as a mathematical expression and then simplify, if possible.

    a) The product of 6x and $5x^2$

    b) The reciprocal of 4^2

    c) The quotient of $8y^5$ and $4y^2$

    d) The term $3a^2$ taken to the fourth power.

2. Write b^0 as two different expressions.

3. Simplify. Make sure your answers contain no zero or negative exponents.

    a) $(-4)^2$ b) -4^2 c) $(x+y)^5(x+y)$

    d) $\dfrac{(5x)^2 y^4}{(xy)^3}$ e) $\dfrac{(a-b)}{(a-b)^3}$ f) $\left(\dfrac{8x}{2y}\right)^{-1}$

    g) $(6ab^2)^2(2a^2b)^{-2}$ h) $(3x)^{1/2}(4x)^{1/2}$ i) $5x^2(x^2 - x + 2)$

    j) $(x+y)(x+y)$ k) $(x+y)(x-y)$ l) $(x^2+2)(x^3-3)-(x^6-6)$

    m) $\dfrac{6xy + 2x^2y - 4xy^2}{2x^2y}$

4. Use a calculator to evaluate the following expressions.

    a) $(-5)^6$ b) -5^6 c) $(-2187)^{1/7}$ d) $(81)^{2/3}$

5. Simplify the expression below and write your answer in scientific notation.

$$\dfrac{(6 \cdot 10^{-4})(2 \cdot 10^{12})}{3 \cdot 10^4}$$

6. The sun is 9.3×10^7 miles from the earth. If light travels at approximately 186,000 miles per second, how long (in minutes) does it take the light from the sun to reach the earth?

7. The planet Pluto is approximately 4.644×10^9 miles from the earth. The planet Mercury is approximately 1.36×10^8 miles from the earth. How many times farther from the earth is the planet Pluto compared to the planet Mercury?

8. Dan deposited $2000 he saved from his summer job into a savings account paying 5% annual interest. If Dan makes no deposits or withdrawals for the next two years, find the amount in the account after that time if the interest is compounded:

   a) quarterly;
   b) monthly.
   c) Compare your results in parts a) and b) above and explain why the amount in part b) is larger.

9. a) Complete the matrix below and calculate the value of each entry. A few numbers have been put in as examples.

•	2^0	2^1	2^2
3^0	$3^0 \cdot 2^0$	$3^0 \cdot 2^1$	$3^0 \cdot 2^2$
3^1	$3^1 \cdot 2^0$		

   b) Note that $3^0 \cdot 2^0 = 1$ and $3^0 \cdot 2^1 = 2$, etc. and that $2^2 \cdot 3^1 = 12$. Study the results in the table. How do they relate to 12?

10. The population of Pike Lake in 1990 was 1492.

    a) If the population is projected to grow at an annual rate of 1.5% for the next few years, what will be the population of Pike Lake in the year 2000?

    b) If the population of Pike Lake in 1991 was 1511, what was the actual rate of population growth from 1990 to 1991? (to the nearest tenth of a percent)

11. The population of a colony of bacteria reached 1000; then the bacteria began to die at a rate of 10% per hour. Draw a graph which shows the population of the colony over a 12 hour period. Show on the graph how you can estimate the population of the colony after 10 hours.

12. A set of living room furniture originally sold for $1129.95 in June, 1994. If it depreciated 10% each month for the first 6 months after purchase, what was the value of the furniture in December, 1994?

13. Label each of the following numbers as rational or irrational.

    a) $\frac{.1}{2}$ b) $.\overline{63}$ c) $\sqrt{3}$ d) π e) $.3252252225...$

14. Simplify:

    a) $\sqrt{2x} \sqrt{2x}$ b) $\sqrt{24x^2y^3}$ c) $(\sqrt{5x})(\sqrt{10x^2})$

    d) $6\sqrt{3} - 2\sqrt{3}$ e) $5\sqrt{a}(2\sqrt{a} + 3\sqrt{b})$

15. Rewrite $27^{2/3}$ using a radical and then simplify.

16. Rewrite $\sqrt{4x^2}$ using a rational exponent and simplify.

17. Find the length and the midpoint of each line segment given below.

   a)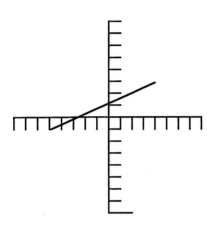

   b)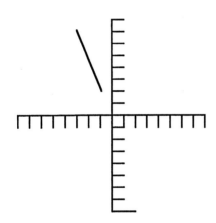

   c) The line segment defined by the points $(2,4)$ and $(-1,1)$.

18. Find the perimeter and area of triangle ABC on the coordinate system below. Round your answers to the nearest tenth.

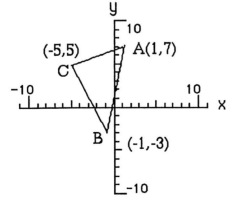

19. Find the perimeter and area of the figure on the geoboard model below.

20. Find the perimeter and area of the square below.

21. On the geoboard model, draw a square with a perimeter of 8.94 units and an area of 5 square units.

22. Simplify the expression $\dfrac{x}{\sqrt{3}}$ so there is no radical or fractional exponent in the denominator.

23. Solve for x: a) $3^x = 9^4$ b) $(3 \cdot 2^x)^2 = 144$ c) $.00000612 = 6.12 \cdot 10^x$

24. Naseef wishes to send his nephew a kite as a birthday gift. The length of the kite is 28 inches. Naseef found a carton which measures 3" by 25" by 12". Will the kite fit in the carton? Justify your answer.

Exploration Problems

1. Measure the sides of triangle DEF below to the nearest tenth of a centimeter and check to see if the Pythagorean Theorem holds. Then explain your results.

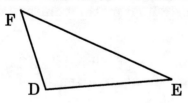

2. Which is larger, $\sqrt[8]{17}$ or $\sqrt[10]{20}$? Justify your answer.

Solutions - Chapter 6 Review

1. a) $6x(5x^2) = 30x^3$
   b) $\dfrac{1}{4^2} = \dfrac{1}{16}$
   c) $\dfrac{8y^5}{4y^2} = 2y^3$
   d) $(3a^2)^4 = 81a^8$

2. 1 and $\dfrac{b}{b}$

3. a) 16 b) –16 c) $(x+y)^6$
   d) $\dfrac{25y}{x}$ e) $\dfrac{1}{(a-b)^2}$ f) $\dfrac{2y}{8x} = \dfrac{y}{4x}$
   g) $\dfrac{9b^2}{a^2}$ h) $12^{1/2}x$ or $2x\sqrt{3}$

   i) $5x^4 - 5x^3 + 10x^2$
   j) $x^2 + 2xy + y^2$
   k) $x^2 - y^2$
   l) $x^5 - 3x^2 + 2x^3 - 6 - x^6 + 6$
      $= -x^6 + x^5 + 2x^3 - 3x^2$
   m) $\dfrac{3 + x - 2y}{x}$

4. a) 15625 b) –15625 c) –3 d) ≈18.72

5. $40000 = 4 \cdot 10^4$

6. approximately 8.3 minutes

7. approximately 34 times as far

8. a) $2208.97 b) $2209.88
   c) When interest is compounded monthly as compared to quarterly, the interest is calculated and added to the principal more often, resulting in a larger amount of interest for the consumer.

9. a) Second row entries, from left to right are: $3^1 \cdot 2^1 = 6$ and $3^1 \cdot 2^2 = 12$.
   b) The table entries are all factors, or divisors, of 12.

10. a) approximately 1732 b) 1.3%

11.

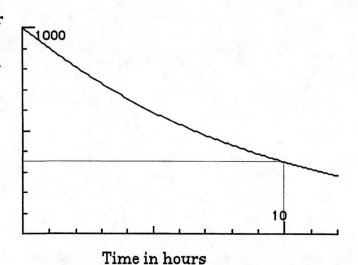

Time in hours

It appears there would be approximately 350 bacteria after 10 hours.

12. $600.50

13. a) rational   b) rational   c) irrational   d) irrational
    e) irrational

14. a) $2x$ b) $2xy\sqrt{6y}$ c) $5x\sqrt{2x}$ d) $4\sqrt{3}$ e) $10a + 15\sqrt{ab}$

15. $\left(\sqrt[3]{27}\right)^2 = 9$

16. $\left(4x^2\right)^{1/2} = 2x$

17. a) Length = $\sqrt{97}$ or about 9.8 graph units; midpoint: $(-1/2, 1)$

    b) Length = $\sqrt{29}$ or about 5.4 graph units; midpoint: $(-2, 4.5)$

    c) Length = $\sqrt{18}$ or about 4.2 graph units; midpoint: $(.5, 2.5)$

18. P = $2\sqrt{10} + 2\sqrt{26} + 4\sqrt{5}$ or about 25.5 graph units

    A = 60 − (16 + 6 + 10) = 28 sq. units

19. P = $\sqrt{2} + 2 + 2 + \sqrt{8} + \sqrt{5} + \sqrt{5}$ which is about 12.7 geoboard units.
    Area = 5.5 square units.

20. Perimeter = 4(2.83) = 11.3 geoboard units; Area = 8 sq. units.

21.

22. $\dfrac{x\sqrt{3}}{3}$

23. a) x = 8   b) x = 2   c) x = −6

24. The kite will not fit. The longest diagonal is about 27.89 inches.

Chapter 6 Worksheet 1

1. Show the meaning of $(3x^2)^3$ and then simplify the expression.

2. Simplify each of the following. Make sure your answers contain no negative or zero exponents. (no calculator needed)

   a) $2x^3y^2 \cdot 3xy^3$

   b) $(a+b)^2(a+b)^3$

   c) $3x^{-1}$

   d) $2(x-y)^{-2}$

   e) $\dfrac{(3x-y)^4}{(3x-y)}$

   f) $\dfrac{6a^3 \cdot 4ab^2}{2ab^6}$

   g) $\dfrac{2x^3y^5}{(3x)^3(x^2y)}$

   h) $\dfrac{15x^2 - 10x^4 + 5x}{5x}$

3. Write in scientific notation:

   a) 32,300,000

   b) .00000000045

4. Write in decimal form:

   a) $3.4 \cdot 10^4$

   b) $4.07 \cdot 10^{-6}$

5. The population of Vulcan is projected to grow at a rate of 1.5% for the next few years. If the population is currently 782, how long will it take for it to grow to at least 850?

6. A savings account pays an annual interest rate of 4.75%. If the interest is compounded quarterly, find the amount that must be put into the account now so that it will total at least $5,000 in 5 years.

7. One tire of a certain brand cost $83 in 1948. Assuming a 2.5% yearly rate of inflation for the next few years, find the cost of the same tire in each of the next 6 years. Draw a graph to show the relationship between the cost of the tire and the number of years since 1998.

8. Sam put $1500 in a savings account paying 4.8% interest, compounded quarterly. If he makes no deposits or withdrawals, find the amount in the account after

   a) 3 years

   b) 5 years

9. If $1200 was originally put into an account, paying interest that was compounded daily, and the amount in the account after 3 years is $1394.19, find the rate of interest for the account.

Chapter 6 Worksheet 2

1. Explain how the decimal equivalents of rational numbers and irrational numbers differ.

2. Give a rational number between $.\overline{24}$ and $.\overline{25}$.

3. Rewrite each of the following in radical form and simplify.

   a) $20^{1/2}$ b) $8^{2/3}$

4. Simplify using simplest radical form where appropriate.

   a) $\sqrt{75}$ b) $\sqrt{200}$ c) $\sqrt{5} \cdot \sqrt{10}$ d) $2\sqrt{3} + 3\sqrt{3}$

   e) $\sqrt[3]{-27}$

5. Find n, to the nearest tenth.

   a) $2^n = 9$ b) $(1/3)^n = 12$

6. Find the length of a diagonal of a 4" by 4" square.

7. Find the perimeter and area of the triangle on the geoboard model below.

8. Find the perimeter of the triangle below and then find the coordinates of the midpoints of the sides.

   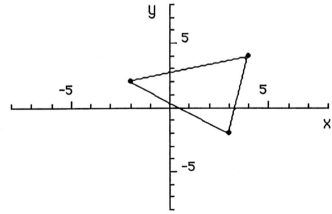

CHAPTER SEVEN - Functions

Section 7.1 - The Function Concept

In earlier chapters we examined the relationship between different sets of numbers, sometimes in table form, such as

$$\begin{array}{c|c} x & y \\ \hline 3 & 5 \\ 4 & 7 \\ 5 & 9 \end{array}$$

where we described the relationship between x and y using a rule or equation. We also discussed extending a table indefinitely and projecting the value of y, for a given value of x. Recall that for the table above we generated the equation

$$y = 2x - 1$$

and if $x = 7$, we know that, based on this rule, $y = 13$, and if $x = 20$, $y = 39$. In other words, we are saying that the *y-value* <u>depends on</u> *the rule and on the x-value* chosen.

As another example, the amount of money an hourly employee earns depends on the number of hours that employee works. Letting x = the number of hours worked, if the hourly wage for this employee is $7.25, then the gross pay, y (before taxes, insurance, etc. and excluding overtime), for this person can be written as

$$y = \$7.25x$$

We say that the employee's wage "depends on," or is a "function of," the number of hours worked. The common mathematical notation is

$$W(x) = 7.25x$$

where $W(x)$ represents the wage, or y-value. The $W(x)$ notation offers a reasonable way of writing the employee's wages as a **function** of x, the number of hours worked. [Note: It does <u>not</u> mean W times x!]

To calculate the employee's gross pay for 40 hours of work, we write

$$W(x) = W(40) = 7.25(40) = \$290 \quad \text{and}$$

for 20 hours of work, the gross pay is

$$W(x) = W(20) = 7.25(20) = \$145$$

Note that if we set up a table by substituting a few values in for x, we can calculate the pay each time. Using the equation $W(x) = 7.25x$, and ignoring the possibility of overtime pay for work beyond 40 hours, we have

x	W(x)
0	0
10	72.5
15	108.75
20	145
30	217.50
40	290
50	362.50

The graph of this equation is a straight line as shown. Since it is possible for an employee to work fractions of an hour, and the pay to be in fractions of a dollar, it is appropriate to have a solid line here.

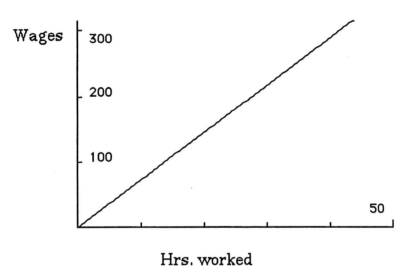

It should be noted at this time that it is by convention that the function value, in this case $W(x)$, the gross pay, is placed on the y-axis since that value <u>depends on</u> the number of hours worked. In fact, the y-variable, or, more generally, the variable plotted on the vertical axis, is referred to as the **dependent variable** and the variable plotted on the x or horizontal axis is called the **independent variable**.

Let's now consider the *function machine* pictured below.

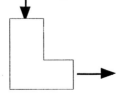

As we have shown in a previous example in chapter 5, the first arrow represents an *input* value and the last arrow represents an *output* value. What <u>this particular machine</u> does to numbers must be determined by a few examples of input and corresponding output values. Assume the function machine has generated this t-table:

Input	Output
−2	−7
−1	−5
0	−3
2	1

As we did in chapter 5, w're really looking for a rule that relates these input and output values. After a little trial-and-error, it appears the rule can be expressed as

$$\text{Output value} = (2 \cdot \text{Input value}) - 3$$

If n represents any input value from the set of numbers on the left, called the **domain**, then $2n-3$ generates the corresponding set of output values, called the **range**. Mathematically, in function notation, we write

$$f(n) = 2n - 3$$

where $f(n)$ is the name of this particular rule, $2n - 3$, given in terms of n. (We could have called this rule g, or h, or F, or G, or H, which are all common function names.)

Again, each output value depends on, or is a "function of", a specific input value and every input value produces only one output value. The graph of all possible ordered pairs, $(n, f(n))$, that satisfy $f(n) = 2n-3$. is shown below. Since we have not restricted the values we may choose for n in this case, it is appropriate to connect the points with a solid straight line.

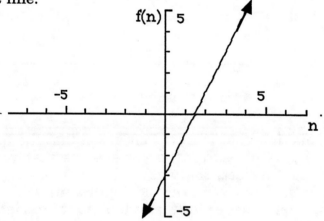

Sample Problems

1. Use x as the input and $g(x)$ as the output and write a rule relating the ordered pairs: (1,4), (2,6), (3,8), (4,10).

   Solution: After some trial-and-error, we have: $g(x) = 2x + 2$

2. Let $x = -1, 5, 20$ and find the corresponding values of $g(x)$ in sample problem 1 above.
   Solution: If $x = -1$, then $g(-1) = 2(-1) + 2 = 0$
   If $x = 5$, then $g(5) = 2(5) + 2 = 12$ and
   if $x = 20$, then $g(20) = 2(20) + 2 = 42$.

3. For *x* a number in the domain, {1, 3, –4}, find the corresponding numbers in the range of the function, f(x) = 3x – 1.

Solution: If x = 1, f(1) = 3(1) – 1 = 2;
If x = 3, f(3) = 3(3) – 1 = 8;
If x = –4, f(–4) = 3(–4) – 1 = –13

Consider the relationship between sets A and B, shown as a "mapping" of elements in set A to elements in set B.

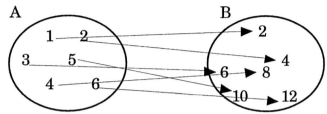

From the diagram it is apparent that 1 "maps" to 2, 2 "maps" to 4, 3 "maps" to 6, etc. Note that every element in set A maps to exactly one element in set B.

Specifically, for *x* = a number in set A and *F(x)* the corresponding number in set B, the rule can be written as

$$F(x) = 2x$$

We can now define

a mathematical function as a relationship between sets of numbers such that every element of the first set, the domain, corresponds to exactly one element of the second set, the range.

It can be easily verified that our previous examples of f(n) = 2n – 3 and W(x) = 7.25x are, in fact, functions in the formal sense. Since both of these equations produced graphs that were straight lines, we call them **linear functions**. (Can you see why g(x) = 2x + 2, F(x) = 2x and f(x) = 3x – 1 are also functions? Are they linear functions?)

Another example of a function is given by the equation

$$f(x) = x^2$$

Note that in the table of input and output values below some of the numbers in the range, *f(x)*, are duplicated (both 1 and –1 map to 1, for example).

x	$f(x) = x^2$
0	0
1	1
2	4
3	9
–1	1
–2	4
–3	9

However, **since each x-value maps to exactly one f(x)-value, the equation describes a function**.

In contrast, consider the relationship between the numbers in the domain on the left and the values in the range on the right of the next table. Here *a* is the input and *b* is the output value.

a	b
9	3
9	-3
4	2
4	-2
0	0

The rule relating *a* and *b* here is

$$b = \pm\sqrt{a}$$

(Verify this by substituting values from the table above.
Also recall that $a \geq 0$)

We do have an equation, or rule, relating *a* and *b*, but note that for an input value of 4, for example, we have two output values, namely 2 and –2.
The graph of this equation is

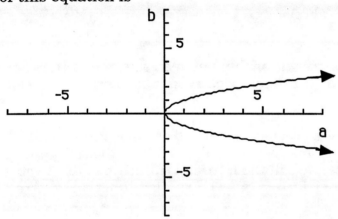

which is not a straight line. More important, however, since we found at least *one* input value corresponding to *two* output values, this equation does *not* represent a function. The function notation was purposely not used here.

The comparison of the graphs of $f(n) = 2n - 3$ and $b = \pm\sqrt{a}$ offers some insight into identifying equations, or rules, that exhibit the special functional relationship. In the first graph, for any *n*-value that can be chosen, the $f(n)$-value is unique. In other words, a vertical line drawn through any *n* on that graph hits the graph in only one place (representing exactly one $f(n)$-value corresponding to that particular *n*). This is shown on the coordinate system below.

For $f(n) = 2n - 3$,

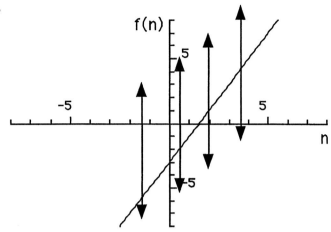

Examining the graph of $b = \pm\sqrt{a}$ reproduced below, and drawing a vertical line through several a-values, note that each vertical line hits the graph in two places for all values of a except 0.

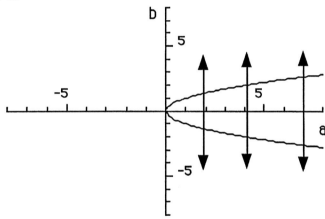

Since a function is formally defined as a relationship between sets of numbers such that every element of the first set maps to exactly one element of the second set, the **"vertical line test"** is sufficient for determining whether or not the graph, and, thus, the equation, indeed represents a function.

It is important to mention that the word "function" is often used in an informal mathematics sense. We often say that increased memory loss is a "function" of age, that the price of an airline ticket is a "function" of the purchase date, that the monthly phone bill is a "function" of the number of long distance calls made that month. For some of these relationships, a mathematical function can be defined, as the problem below illustrates.

Sample Problems

1. Give a real-life example of the concept of function, that is, how one thing depends on another. Then write the function using function notation.

    One example: The cost of apples, at $.99 per pound, depends on the number of pounds purchased. Letting n = the number of pounds of apples purchased, the cost function can be given by :
    $$C(n) = .99n$$

2. Determine whether or not each of the following represents a function. For the functions, identify the domains and ranges.

   a) The set of ordered pairs: (1,3), (4,3), (6,1), (−4,2).

   b)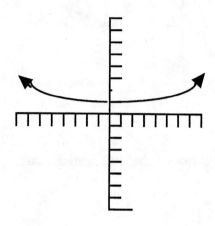

   c) 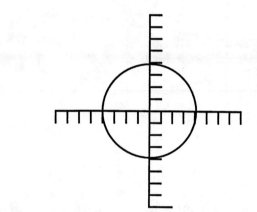   d)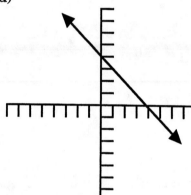

   e)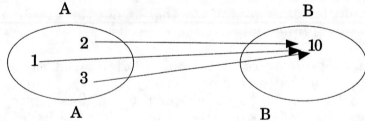

   f)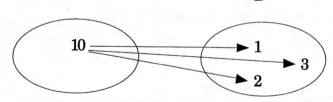

   Solutions:
   a) Yes, because each first coordinate is paired with a unique second coordinate and it does not matter if two different x-values map to the same y-value.
   The domain = {1,4, 6,−4} and the range = {3, 1, 2}.

b) Yes, the graph passes the vertical line test. Since the graph could be extended, the domain could be any real number. However, the range consists of only positive real numbers ≥ 1.

c) No, some input values such as x = 2, have two output values. The domain is all values between −4 and 4; the range values are between −4 and 4.

d) Yes, the graph passes the vertical line test. Again, the domain as well as the range could be any real number.

e) Yes, since the definition of function is satisfied. The domain = {1, 2, 3} and the range is = {10}.

f) No, since one input value maps to 3 different output values.

Problem Set 7.1

1. Using function notation, write the distance traveled as a function of time for a constant rate of 50 mph.

2. Given the chart and graph below, for each, write the rule relating the two variables using function notation. Identify the domain and range of each function.

   a)
   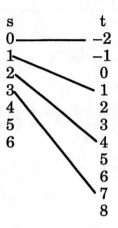

   b)
   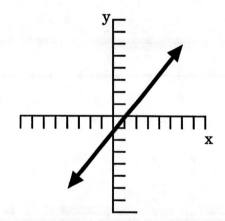

3. Write a function that might result in the graph below. Justify your answer.

   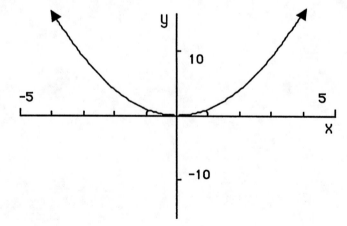

4. If $f(x) = 3x - 2$, find the value of x such that $f(x) = 3$.

5. Find the value of f(x) = 4x + 2 for

   a) x = 0   b) x = –2   c) x = $\frac{1}{2}$ d) x = a

6. For g(x) = –4x + 1, find x such that g(x) =

   a) –5   b) 5

7. Given the mapping below between sets U and V, list the domain and range and then write the mapping as a rule using function notation.

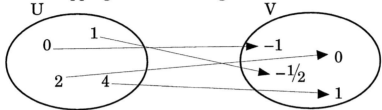

8. Determine whether or not each of the following is a function. Explain your reasoning in each case.

   a) {(2,3), (4,3), (5,3), (0,3), (–1,3)}

   b)
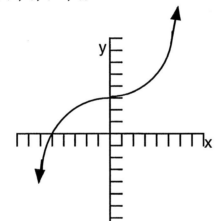

   c)
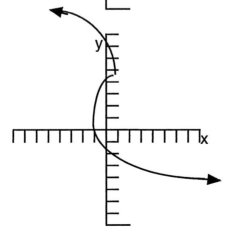

9. Draw the graph of the equation, $y = x^2 - 1$ and determine if it is a function. Justify your answer.

383

10. The Gitchee-Gumee telephone company has an interesting rate structure for its customers. The basic monthly charge is $15.95. Additionally, each long distance call costs $2.50 for up to three minutes, and $.45 per minute or partial minute after that.

   a) Find the cost of a 5-minute long distance call.

   b) Find the cost of an n-minute long distance call for $n \geq 3$.

   c) Write the cost function for making x long distance calls, each of which is 5 minutes in length.

   d) Write the cost function for Jan's August telephone bill if x long distance phone calls are made, each 5 minutes long.

11. Give an example of a set of ordered pairs that does not describe a function.

12. Consider the three graphs below which portray the increase in height of water (measured to the nearest tenth of a centimeter), added 1/4 cup at a time, to three different-shaped containers.

a)

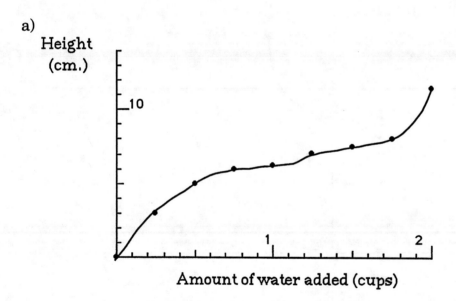

b)

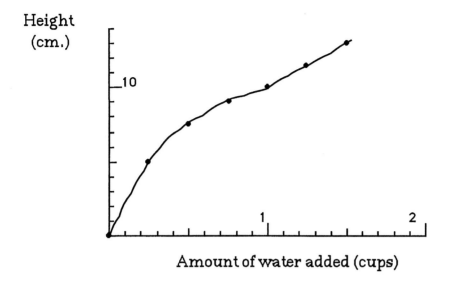

c)

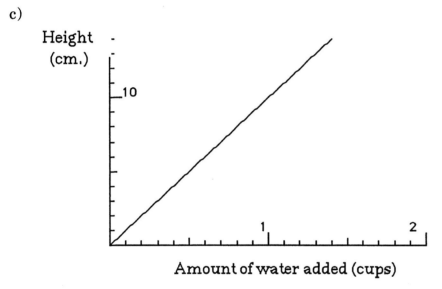

If the three containers have shapes as below, match the container with its graph. Explain your reasoning.

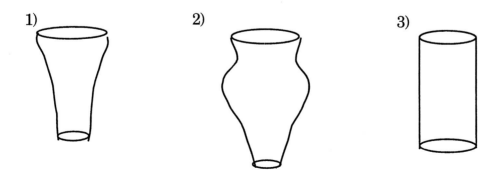

Exploration Problems

1. In order to determine how the size of an image on an overhead projector varies with the distance the projector is from the screen, the instructor will place a 3" x 5" index card on a transparency and focus the image. The first set of measurements will be to find the width and length of the card, as projected on the screen, and the distance the projector is from the screen at that point. The overhead will be moved, the card put into focus, and the same measurements taken again. This process should continue until there are at least 6 sets of measurements. Find the perimeter of the index card each time and plot these values (on the vertical axis) as a function of the corresponding distances the projector is from the screen (on the horizontal axis). On another coordinate system, plot the area of the card each time and the corresponding distances from projector to screen. What type of graphs do you have? Are the perimeter and area functions of the distance from the projector to the screen?

2. Given the function machine with input and output as shown, write at least 3 functions, using the correct function notation, that use the number 3 as the input and generate an output of 13.

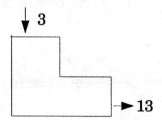

Solutions - Problem Set 7.1

1. $d = f(t) = 50t$

2. a) $t = f(s) = 3s - 2$, $D = \{0,1,2,3,...,6\}$ and $R = \{-2, -1, 0, 1, 2, ..., 8\}$

    b) $y = f(x) = x - 1$, $D = \{$ all reals$\}$ and $R = \{$all reals$\}$

3. $f(x) = x^2$

4. $x = 5/3$

5. a) $f(0) = 2$

    b) $f(-2) = -6$

    c) $f(1/2) = 4$

    d) $f(a) = 4a + 2$

6. a) 1.5   b) -1

7. Let u and v be elements in sets U and V, respectively. The domain is the set, $\{0,1,2,4\}$ and the range is $\{-1,-1/2,0,1\}$. The function is:
$$v = f(u) = \frac{u}{2} - 1$$

8. a) Is a function; for every 1 input, there is exactly 1 output.

    b) Is a function; it passes the vertical line test.

    c) Is not a function; at $x = 0$, for example, $y \approx 3.5$ and -1.5.

9. The graph is

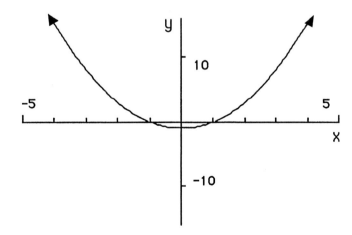

and it is a function because it passes the vertical line test.

10. a) $2.50 + $.45(5 - 3) = $3.40

   b) $2.50 + $.45(n - 3)

   c) $3.40x

   d) $15.95 + 3.40x

11. For example: {(2,3), (1,2), (2,4)}

12. Graph a) matches container 2). There is a fairly steady increase in water height at first, then the increase slows as the container widens and finally, the increase picks up as the container narrows at the top. Graph b) matches container 1). The first part indicates a steady rate of increase in the height of the water. As the container widens toward the top, the increase is steady but slower. Graph c) matches container 3). As each $\frac{1}{4}$ cup of water is added, the height increases at the same rate because the container is a cylinder.

Section 7.2 - Linear Functions

Examine the graph below of the equation, $y = \frac{1}{2}x$.

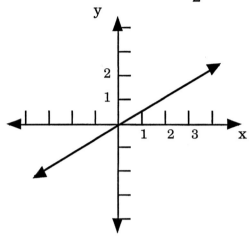

What is the y-coordinate for the point on the line with an x-value of 1? From the graph, it appears that for x = 1, y = $\frac{1}{2}$ and we can confirm this algebraically by substituting x = 1 into the equation above. However, using similar triangles, ABC and ADE, we can explore a few interesting properties of straight lines.

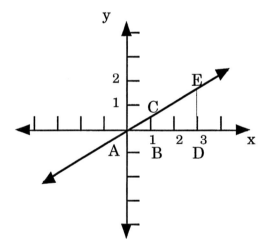

To find the exact coordinates of point E, we need to know the length, **DE**. From our work with similarity and proportions we know that the following relationship exists between sides **AB**, **AD**, **BC** and **DE**. That is,

$$\frac{AB}{AD} = \frac{BC}{DE}$$

Substituting known values

$$\frac{1}{3} = \frac{1/2}{DE}$$

and using cross-products we have

$$DE = 3(1/2) = 3/2$$

Thus, the coordinates of point E are $(3, 3/2)$. We found the coordinates of point C to be $(1, 1/2)$. Comparing the x-values and corresponding y-values of points C and E, does it seem reasonable that when the x-value tripled, the y-value also tripled? If the pattern holds, what do you think is the y-coordinate of a point on the line when x = 4?

Before we answer that question, recall that we can set up our original proportion of

as
$$\frac{AB}{AD} = \frac{BC}{DE}$$

$$\frac{DE}{AD} = \frac{BC}{AB}$$

(We have the same cross-products!)

Here **DE** represents the **directed vertical distance** $(3/2 - 0)$ from the x-axis to point E on the line, and **AD** is the **directed horizontal distance** $(3 - 0)$ from the y-axis to the same point. Similarly, **BC** is the directed vertical distance $(1/2 - 0)$ from the x-axis to point C and **AB**, the directed horizontal distance $(1 - 0)$ from the y-axis to point C. Using the diagram and our previous results, we have

$$\frac{DE}{AD} = \frac{\frac{3}{2} - 0}{3 - 0} = \frac{\frac{3}{2}}{3} = \frac{1}{2}$$

and

$$\frac{BC}{AB} = \frac{\frac{1}{2} - 0}{1 - 0} = \frac{\frac{1}{2}}{1} = \frac{1}{2}$$

In both cases, the comparison of the vertical change to the horizontal change resulted in a ratio of 1 to 2. This ratio is commonly called the **rate of change** for the line defined by the points A and E, that is, the line given by the equation $y = \frac{1}{2}x$. In other words, for *any* point on this line, when we take the ratio of the directed vertical distance from the x-axis to the directed horizontal distance from the y-axis, we get $1/2$ (or a fractional equivalent).

Returning to our question, it's reasonable to speculate that for a point x = 4 on the line, the pattern, and the equation, indicate that the y-value should be 2. To verify this geometrically, let's choose the point, F on the line with the x-coordinate of 4, use a y-coordinate of 2, and see if we generate the same rate of change.

Setting up the ratio of directed distances, that is, the vertical change compared to the horizontal change, we have

$$\frac{2}{4} = \frac{1}{2}$$

This tells us that the point (4, 2) is on the line. Also, by quadrupling the x-value, we quadrupled the y-value, which verifies our earlier supposition regarding the relationship between points on this line.

Let's now consider the Wildcat Ski Hill as shown below.

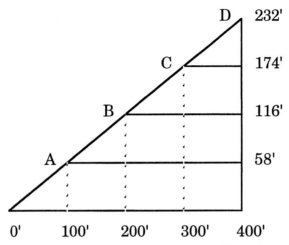

To find the rate of change of the ski hill at point A, we'll take the ratio of the vertical rise (58') to the horizontal distance (or run) (100') to that point and we have

$$\frac{58}{100} = .58$$

To find the rate of change at point C, the ratio of the vertical distance to the horizontal distance is

$$\frac{174}{300} = .58$$

Can you predict the rate of change of the hill at point D? Examining the diagram we have

$$\frac{232}{400} = .58$$

The last two examples should help us realize that the rate of change, or "steepness" of a specific line (or an incline) does not vary and is not dependent upon the point(s) on the line that are chosen.

Sample Problems

1. Find the rate of change of a line through the origin and the point (2,5).

    Solution: From (0,0) the vertical change is +5 and the horizontal change is +2, so the rate of change is

    $$\frac{5-0}{2-0} = \frac{5}{2}$$

2. On a coordinate system, draw a line with a rate of change of a) 3, and b) $-\frac{1}{2}$.

    Solution: There are many such lines but two examples are shown.

391

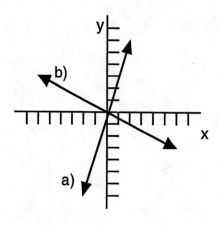

Notes: a) A rate change of 3 implies a vertical change of 3 for every horizontal change of 1.

b) We can think of $-\frac{1}{2}$ as a drop of 1 for every horizontal distance of $+2$, or, as a vertical change of $+1$ for each horizontal change of -2 (since $-\frac{1}{2} = \frac{1}{-2}$).

*It is important to note that the rate of change can be determined from **any** two points on the graph since lines do not always pass through the origin.*

To illustrate, consider another line segment, L, defined by the general points (x_1, y_1) and (x_2, y_2), on the coordinate system below.

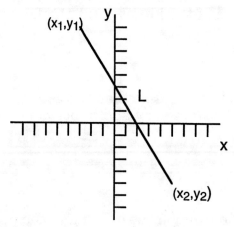

In section 6.4, in finding the length of a line segment, we sketched a right triangle on the coordinate system with the given line segment as the hypotenuse. Using this technique again and labeling the vertex at the right angle with the appropriate coordinates, we have the following triangle:

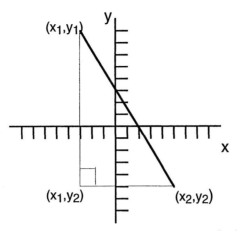

In the first example in this section, we measured the rate of change by comparing the vertical and horizontal directed distances of points from the point (0,0). For the ski hill, we again compared the vertical and horizontal distances, assuming that the very bottom of the slope was the starting point, analogous to the point (0,0). The line above, however, <u>does not</u> pass through the origin, but we can use the vertical and horizontal distances given by the legs of the right triangle to find the <u>directed</u> distances we need.

If we signify the point (x_1,y_1) as the starting point, then we get to the point (x_2,y_2) by traveling along the right triangle -- going down to the point (x_1,y_2) and then to the right. The directed length of the vertical line segment is

$y_2 - y_1$ (since the x-coordinates are the same)

and the directed length of the horizontal line segment is

$x_2 - x_1$ (since the y-coordinates are the same)

We can calculate the rate of change by forming the ratio of the directed vertical distance to the directed horizontal distance. For <u>any</u> line we now have

$$\text{rate of change} = \frac{y_2 - y_1}{x_2 - x_1} = \frac{y_1 - y_2}{x_1 - x_2} = \frac{\text{rise}}{\text{run}}$$

It is extremely important to remember that, in general, the direction of each distance CANNOT be ignored. If one point is designated as the reference for the change in the y-values, then that same point must be used as the reference point for the change in the x-values. In other words, the order of the subtractions must be consistent. Can you see why the second ratio in the box above gives the same result?

Let's reexamine the equation $y = 2x - 1$, pictured on the next coordinate system and find the rate of change for the line.

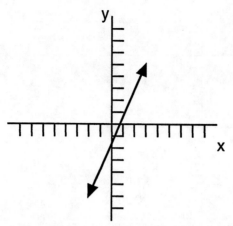

We first estimate the rate of change by selecting two points on the graph, drawing a right triangle and then determining the amount of rise compared to the amount of run between those two points. We'll select (2,3) and (0,–1) as two points that appear to be on this line and form the right triangle as shown.

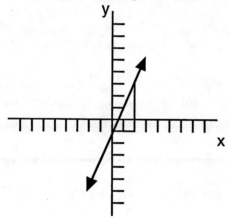

It appears that for every rise (positive) of 4 graph units there is a run (positive) of 2 graph units. Thus, we can predict the rate of change to be

$$\frac{4}{2} = 2$$

For some lines, however, it is difficult to find two points visually. We can, however, identify two points *exactly* on the line by <u>substituting</u> values for x into the equation and <u>generating</u> the corresponding values for y.

For example, again using the equation, $y = 2x - 1$,

           if $x = 1$, $y = 1$ so one point is $(1,1)$
and   if $x = -2$, then $y = -5$ so another point is $(-2,-5)$.

Using $(1,1)$ as (x_2, y_2) and $(-2,-5)$ as (x_1, y_1), the vertical change between these two points is

$$y_2 - y_1 = 1 - (-5)$$
$$= 6 \text{ [which means there's a \underline{rise} of 6 units from } (-2,-5) \text{ to } (1,1)]$$

and the *corresponding* horizontal change is

$$x_2 - x_1 = 1 - (-2) = 3 \quad \text{(which means there's a positive \underline{run} of 3 units)}$$

Thus, the
$$\text{rate of change} = \frac{6}{3} = \frac{2}{1} = 2$$
(Our estimate was very accurate in this case!)

It is interesting to note that the rate of change is 2, and 2 is also the coefficient of x in the equation, $y = 2x - 1$. Recall that in the equation, $y = \frac{1}{2}x$, we found the slope to be $\frac{1}{2}$. We'll discuss this in greater detail shortly.

It is common terminology to talk about the "slope" or "grade" of a hill or an incline. In fact, the **rate of change** of a straight line is commonly called the **slope** of the line. We will use these terms interchangeably from now on.

Sample Problems

1. Find two different points on the line given by the equation $y = 2x - 1$ and verify that the rate of change of the line segment connecting the points is equal to 2.

    Possible solution:
    If $x = 3$, then $y = 5$ so one point is $(3,5)$.
    If $x = -1$, then $y = -3$ and another point is $(-1,-3)$.

    The rate of change is

    $$\frac{5-(-3)}{3-(-1)} = \frac{8}{4} = 2 \text{ OR } \frac{-3-5}{-1-3} = \frac{-8}{-4} = 2$$

2. Approximate the slope of the line on the graph below. Then interpret the slope in terms of getting from one point to another on the graph.

    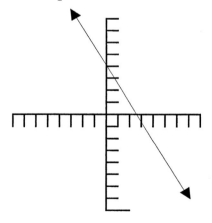

    Solution:
    First, we need to identify two points on the line. It appears that $(0,4)$ and $(4,-2)$ are both on the line.

    The vertical change is: $-2 - 4 = -6$ and
    the corresponding horizontal change is: $4 - 0 = 4$ so the

$$\text{rate of change} = \frac{-6}{4} \text{ or } \frac{-3}{2}$$

This means that to get from **(0,4)** to **(4,–2)**, we need to travel down 6 units and then to the right 4 units. (Or, since $\frac{-6}{4} = \frac{6}{-4}$, to get from the point **(4,–2)** to **(0,4)**, we'd move 6 units up and then 4 units to the left.

For any two points on the line, the rate of change or slope of the line segment will be $-3/2$ (or a fractional equivalent of $-3/2$.

3. Verify that the slope of Wildcat Ski Hill is .58 by finding the rate of change between any two points on the hill surface.

Solution:
For example, using points A and C,

$$\frac{174 - 58}{300 - 100} = \frac{116}{200} = .58$$

Now that we have examined the slope of a line, we need to consider the point on the line where x = 0. We have already determined that the line defined by the equation, y = 2x – 1, has a slope of 2. Studying the graph of the line, as duplicated below,

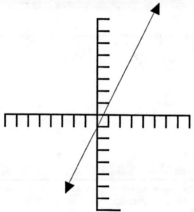

we can see that when x = 0, y = –1 (again, substituting x = 0 into the equation will verify this). This point is called the **y-intercept of** the line because it is where the line crosses the y-axis. The x-coordinate of this point will *always* be 0. Also note that –1 is the constant term in the equation y = 2x – 1.

Finally, for $y = \frac{1}{2}x$ (which we can write as $y = \frac{1}{2}x + 0$), the y-intercept is the point at which x = 0. Here, if x = 0, y = 0. The graph on the first page of this section confirms that the line crosses the y-axis at y = 0.

In summary, the slope of 2 appears in the equation, y = 2x – 1 as the coefficient of *x* and the value of the constant term actually is the y-intercept. Again in $y = \frac{1}{2}x + 0$, the slope is $\frac{1}{2}$ and the y-intercept is 0. In general, the equation of a line in **slope-intercept form** is

> **y = mx + b**
> where *m* represents the slope of the line and *b* is the y-intercept.

For the line through the points (0,4) and (4,–2) in a previous sample problem, we calculated the slope to be $-3/2$ and, from the graph, we can see that the y-intercept is 4, so the equation of the line, in slope-intercept form, is

$$y = \frac{-3}{2}x + 4$$

We can also check to make sure the points (0,4) and (4,–2) are on this line by substituting each x-value into our equation. The y-values should be 4 and –2, respectively. The vertification is shown below.

$$y = \frac{-3}{2}(0) + 4 = 4$$

and

$$y = \frac{-3}{2}(4) + 4 = -2$$

Sample Problems

1. Given the line defined by the equation y = 3x – 2, find the slope of the line and the y-intercept.

   Solution: Since the slope is the coefficient of the x-term, the slope is 3 and the y-intercept is –2 (because 3x – 2 = 3x + –2).

2. Find the equation of the line on the coordinate system below and write the equation in slope-intercept form.

   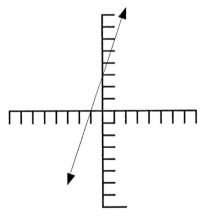

   Solution:
   Looking at two points on the graph, specifically (0,3) and (–1,0), the **y- and x-intercepts**, respectively, we can determine that the directed vertical distance between them is 3 and the corresponding directed horizontal distance is 1 so the slope = $m = \frac{3}{1} = 3$. The line crosses the y-axis at +3 (i.e., b = 3), so the equation is

   $$y = 3x + 3.$$

   (We could have also calculated the slope by using the formula and the points we have identified on the line.)

It is important to realize that the graphs we have discussed so far in this section are all linear functions. Let's examine the line defined by the equation

$$x = -2$$

On the coordinate system, the graph of $x = -2$ is

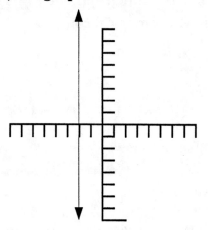

Since the only input is –2 and the output can be any real number, the equation $x = -2$ does not represent a function. (Also note that it does not pass the vertical line test since at $x = -2$, a vertical line intersects the line at all points!)

To find the rate of change for the vertical line, we need two points on the line: (–2,0) and (–2,4) will do. The slope is then

$$\frac{0 - 4}{-2 - (-2)} = \frac{-4}{0} = \text{undefined}$$

In other words, the slope of a vertical line is undefined. Do you see why the $y = mx + b$ formula does not apply here? What do you think is the slope of a horizontal line?

We must emphasize at this time that the rate of change, or slope, of a straight line can be positive, negative, zero, or undefined. If the slope is positive, the y-values increase (rise) as the x-values increase and if the slope is negative, the y-values decrease (fall) as the x-values increase. In both cases, there is a functional relationship between the x and y-values. However, when the slope of a line is undefined, that is, when a line is vertical, the relationship between x- and y-values is not a function. Does a horizontal line represent a function?

Sample Problems

1. From the graph given, identify whether the slope of the line is positive, zero, or negative and explain your reasoning. Then estimate the value of the slope.

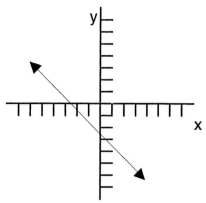

Solution: The slope of the line is negative since the values of y decrease as the values of x increase. The slope is approximately -1.

2. Find the equation of the line through the points $(4,0)$ and $(-1,2)$.

Solution:

The slope of the line is: $m = \dfrac{2-0}{-1-4} = \dfrac{2}{-5}$ or $\dfrac{-2}{5}$

We can use one point on the line and the slope to find the y-intercept. (Note: 4 is the <u>x-intercept</u> and cannot be used directly in the slope-intercept form)

Substituting the coordinates $x = 4$ and $y = 0$ into the general equation, $y = mx + b$, we have

$$0 = 4\left(\dfrac{-2}{5}\right) + b$$

and solving for b

$$b = \dfrac{8}{5}.$$

Thus, the equation of the line is

$$y = \dfrac{-2}{5}x + \dfrac{8}{5}$$

Check: When substituted into the equation, the coordinates of both points, $(4,0)$ and $(-1,2)$, make the equation true. Therefore, the points are on the line.

Problem Set 7.2

1. Given lines **a**, **b** and **c** on a coordinate system and slopes of $1/2$, -2 and 5, match each line with its corresponding rate of change. You should not have to do any calculations here.

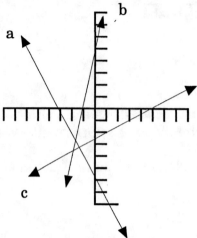

2. Calculate the rate of change for the pairs of lines given below. Then find the equation of each line. Explain the meaning of your results with respect to the graph of each pair of lines.

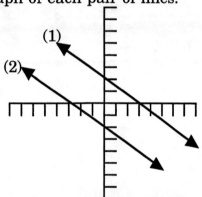   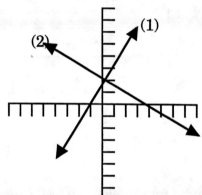

3. Find the rate of change that took triangle I and translated it to the position of triangle II.

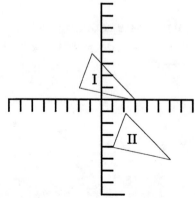

4. a) Find the slope of the roof pictured by

   b) Is the slope positive or negative? Explain your reasoning.

5. Find the slope and y-intercept of each of the following:

   a) $y = x$
   b) $y = \dfrac{3}{4}x - 1$
   c) $x - 2y = 5$
   d) $2x + y = 4$

6. Find the equation of the line through the point (0,3) with slope $1/2$.

7. Find the equation of the line through the points (1,4) and (–1,2).

8. Write the equation of the vertical line through the point (–3,2).

9. Find the equation of the line through the points (0,–3) and (3,0).

10. Find the equation of the line through the points (2,5) and (–4,2).

11. a) Find the slope of a line through the points (–2,5) and (4,5).

    b) Write the equation of this line.

12. Discuss the rate of change for each of these lines. Then find the equation of each line.

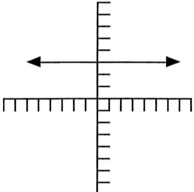 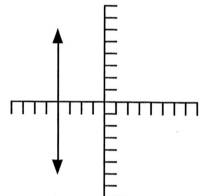

13. If a line rises to the right, what do you know about the slope? If it rises to the left?

14. A train carrying iron ore can climb a maximum grade of 2.18%. Draw a picture of this amount of grade and put appropriate numbers in for the rise and run.

15. Find the value of x such that the slope of the line passing through the points (2,4) and (x, –2) is –3.

16. Find the value of y such that the slope of the line passing through the points (2,4) and (–1,y) is $1/3$.

17. On a coordinate system, sketch two lines each with a slope of

    a) $-3/5$
    b) 0
    c) $4/3$

18. Given a function defined by the following ordered pairs: (0,–4), (1,–1), (2,2), (–1,–7). Use any method you wish to find the rule relating the x- and y-values.

19. Examine the graph below which relates the amount in an account, paying simple interest, after 10 years. No withdrawals or deposits were made in that time. Find two points on the graph and calculate the slope of the line. Interpret the meaning of this number.

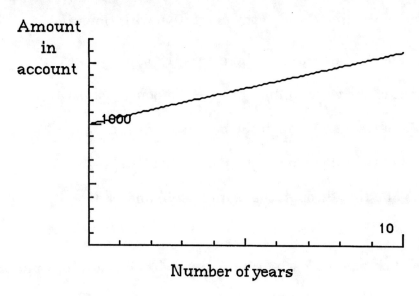

Exploration Problems

1. The *Yankee Girl* of Marquette, MI was sailing last summer on Lake Superior. A storm blew up suddenly and the skipper had to get to a safe harbor. His choices, given the wind direction, were Munising or Big Bay. His LORAN (long range navigational system) located Munising harbor at (−8,13) and Big Bay harbor at (8,3). The slope of an imaginary line between the boat and Munising harbor was −2. The slope of an imaginary line between the boat and Big Bay was $1/5$. Where was *Yankee Girl* at the time of this LORAN reading?[1]

2. Discuss a method for measuring the incline of a stairway. Then find the incline of a stairway in your building and determine whether it satisifies the building code for your state.

---

[1] Thanks to Retha Weiss.

Solutions - Problem Set 7.2

1. Line a has slope = -2, line b has slope = 5 and line c has slope = $1/2$.

2. a) Line 1 and 2 both have slope $\dfrac{-2}{3}$.

   Equations: line 1) $y = \dfrac{-2}{3}x + 2$

   line 2) $y = \dfrac{-2}{3}x - 2$

   The lines have the same slope but different y-intercepts.
   The lines are <u>parallel</u>.

   b) Line 1 has slope = 2 and Line 2 has slope = $-1/2$.
   Equations: line 1) $y = 2x + 2$

   line 2) $y = (-\dfrac{1}{2}x) + 2$

   The lines have slopes that are negative reciprocals of each other.
   These lines are <u>perpendicular</u>.

3. $m = -5/3$

4. a) m is approx. $4/12$ or $1/3$ since the run is calculated to be about 12".

   b) A negative value for the slope or pitch of a roof is <u>not</u> appropriate.

5. a) $m = 1$, y-int. = 0;   b) $m = 3/4$, y-int. = -1;   c) $m = 1/2$, y-int. = $-5/2$
   d) $m = -2$, y-int. = 4

6. $y = \dfrac{1}{2}x + 3$

7. $y = x + 3$

8. $x = -3$

9. $y = x - 3$

10. $m = 1/2$ and $b = 4$ so the equation is $y = \dfrac{1}{2}x + 4$.

11. a) $m = 0$

    b) $y = 5$

12. For the first graph the rate of change or slope = 0 and the equation is $y = 3$; for the second line the rate of change is undefined and the equation is $x = -4$.

13. It's positive. The slope is negative.

14.

    For each vertical climb of 2.18 feet, there is a horizontal run of 100 feet.

15. x = 4

16. y = 3

17. The following are examples:
    a)
    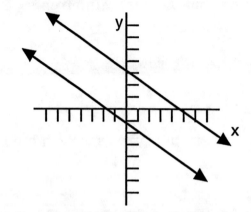

    b)

    c)
    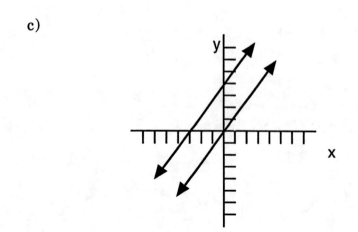

18. Using two ordered pairs of the function, namely, (0,–4) and (1,–1),

$$m = \frac{-1-(-4)}{1-0} = 3$$

and the y-intercept is –4, so the equation is f(x) = y = 3x – 4.

19. Two possible points are (3.5,1200) and (5,1300). The slope of the line through these two points is: m = 100/1.5 = 66.7. This means that the account increases by approx. $66.70 each year. This also means that the rate of change (or interest rate) is 6.67%.

Section 7.3 - Quadratic and Exponential Equations and Functions

Consider the table below, showing a relationship between x-and y-values.

x	y
0	1
1	2
2	5
3	10
−1	2
−2	5
−3	10

The graph of this relationship is

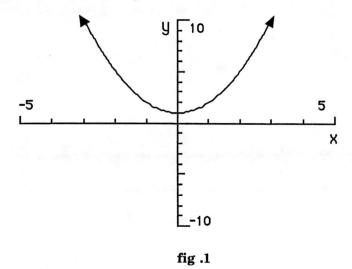

fig .1

The equation relating the x and y values in the table and on the graph is $y = x^2 + 1$. Note that this graph is very similar in shape to that of $y = x^2$ (which appeared in problem set 7.1) except that it does not contain the point (0,0). In fact, just as in $y = x^2$, there is exactly one y-value corresponding to each x-value, and the equation, $y = x^2 + 1$, also represents a function.

It is apparent that this graph does not have a constant rate of change, however. In other words, by putting your pencil on the top leftmost part of the graph and tracing the graph, you can "feel" the change in the curve, and at the *turning point*, which we have previously referred to as the **vertex**, your pencil and hand also change direction.

Recall that a graph of this shape is called a **parabola**, and it is always generated by a specific type of equation. Let's examine this type of graph and the equations generating these graphs more closely.

In Chapter One, we explored the relationship between the width and area of various rectangular enclosures, each with a perimeter of 100 feet. Generalizing this so that W = the width and L = the length of such an enclosure, we have

$$2W + 2L = 100 \quad \text{and} \quad A = LW.$$

Solving for L in terms of W in the perimeter formula results in
$$2L = 100 - 2W \quad \text{or} \quad L = \frac{100 - 2W}{2} = 50 - W$$

Now substituting (50 − W) into the Area formula for L, we have

$$A = (50 - W)W = 50W - W^2$$

It is especially important to notice that, as in $y = x^2$ and $y = x^2 + 1$, this equation also contains one variable with an exponent of 2. In general, a *quadratic equation*, or **quadratic function**, is of the form

$$\boxed{y = ax^2 + bx + c \quad \text{for } a, b, c \text{ real numbers and } a \ne 0.}$$

Let's plot the graph of
$$A = 50W - W^2$$

by substituting enough values for W so that we can generate the entire curve.

W	A
0	0
5	225
10	400
20	600
25	625
30	600
40	400
50	0

Since both width and area will be positive values, we'll use the first quadrant only. Fractional values are appropriate for both width and area so we connect the dots and generate the following graph:

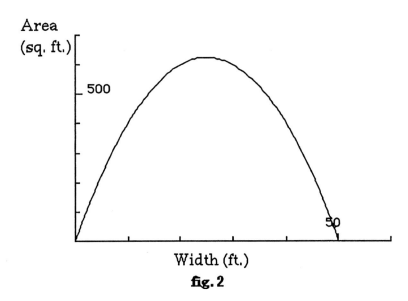

fig. 2

Note that the table of values and the shape of the graph are exactly like those we illustrated in section 1.5. Also, since each x-value corresponds to exactly one y-value, and the graph passes the vertical line test, this graph represents a function.

To verify that the rate of change in a parabola is not constant, let's use two sets of ordered pairs from the table. For example, the rate of change between the points (5,225) and (20,600) is

$$\frac{600 - 225}{20 - 5} = 25$$

and the rate of change between the points (30,600) and (25,625) is

$$\frac{600 - 625}{30 - 25} = -5.$$

Not only is the rate of change *value* different in different parts of the curve, but the *direction* is also different. This verifies our pencil "feeling" for the curve!

Sample Problems

1. List several objects or phenomena you have observed that have the shape of a parabola.

    Some possibilities:
       Shape of the surface of a satellite dish, arc of a hummingbird during courtship, flight of a kicked football, bridge cables (as shown below).

2. Write the quadratic equation that relates x and y in the table below. Sketch the graph of this equation. Explain how this graph differs from the graph of $y = x^2$.

    ```
    x  |  y
    -----------
    0    -1
    1     0
    2     3
    3     8
    -1    0
    -2    3
    -3    8
    ```

    Solution:
       Using a little trial and error, the rule, in equation form, is $y = x^2 - 1$.

The graph looks like

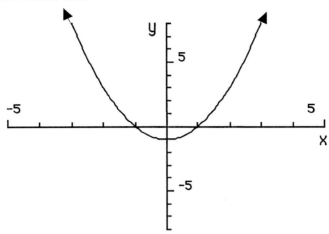

This graph has a shape similar to the graph in **fig. 1**, that is, the "cup" is about the same width across, but the vertex is at (0,–1) instead of at (0,1). Again, the graph represents a function.

At this time, it is important to compare the graphs of

a) $y = x^2$ b) $y = x^2 + 1$ c) $y = x^2 - 1$ d) $y = -x^2$ e) $y^2 = x$ and   f) $y = 2x^2 + 1$

As we have already seen, the graphs of the first three equations open upward, are symmetric with respect to the y-axis, have about the same "width," and describe functions.

The graph of $y = -x^2$ is shown below.

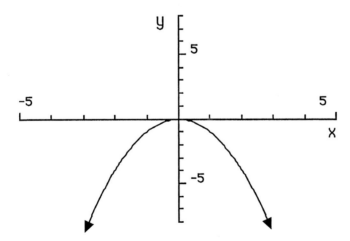

Again, the graph is symmetric with respect to the y-axis, has about the same "width" as the graph of $y = x^2$, describes a function, but the graph opens <u>downward</u>.

The graph of $y^2 = x$ is

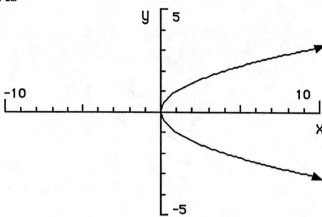

The "cup" is about the same shape as those in the previous examples, but note here the position of the parabola on the coordinate system – this graph is symmetric with respect to the *x-axis*. Also note that two y-values correspond to many of the x-values so the graph does not pass the vertical line test and does not represent a function.

Finally, a representation of the graph of $y = 2x^2 + 1$ is shown below.

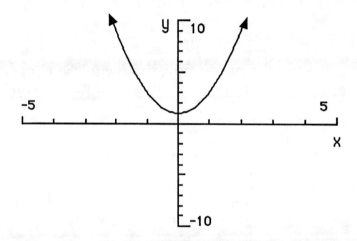

The vertex of this parabola is at the point (0,1), and the "width" of the parabola is smaller than that of the equation $y = x^2$. The graph is symmetric with respect to the y-axis and does describe a function.

We can now make a few general statements about the graphs of quadratic equations in the form of

$$y = ax^2 + c$$

1) The graphs will be symmetric with respect to the y-axis;
2) The graphs will open upward if a > 0 and downward if a < 0;
3) The vertex of the parabola is at the point (0,c);
4) As *a* increases, the width of the parabola decreases and as *a* decreases, the width increases.

Sample Problems

1. Graph the functions a) $f(x) = x^2 - 2$ and b) $f(x) = -x^2 - 2$ on the same coordinate system and compare the graphs.
   Solution:

   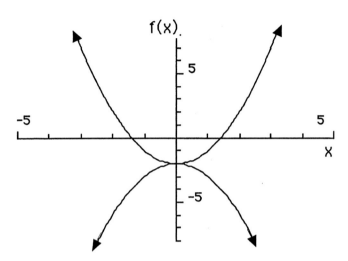

   The graphs are mirror images of each other. Each vertex is at the point $(0,-2)$. Both are symmetric with respect to the y-axis.

Exponential Functions

A certain culture of bacteria doubles every hour. If 100 bacteria are present at 2:00 P.M. one day, let's use a table to calculate the number of bacteria in the culture as time passes.

No. of hours from 2:00 P.M.	No. of bacteria in culture
0	$100 = 2^0(100)$
1	$200 = 2^1(100)$
2	$400 = 2^2(100)$
3	$800 = 2^3(100)$

From the pattern, can you predict the number of bacteria in the culture at 10 P.M. that night? It appears that the number of hours from 2:00 P.M. is the same as the number of factors of 2 needed to calculate the total, so at 10 P.M., which is 8 hours after 2 P.M., there should be

$$2^8(100) = 25{,}600 \text{ bacteria in the culture.}$$

To find the number of bacteria n hours after 2:00 P.M., we can generalize the equation to

$$\text{No. of bacteria} = 2^n(100)$$

This also gives us a method to find the number of bacteria at 2:30 P.M. that day. Since 30 minutes is $1/2$ hour, we have

No. of bacteria = $2^{1/2}(100) = 141$ (rounded to the nearest whole bacterium.)

This pattern of growth is called **exponential growth** and we have previously discussed this in the context of population growth, inflation, and compound interest.

Let's sketch the graph of the equation

$$y = f(n) = 2^n(100)$$

Substituting various values for *n*, including fractional values, and calculating the corresponding y-values, we have

n	y = f(n)
0	100
1/2	141
3/4	168
1	200
1 1/4	238
1 1/2	283
1 3/4	336
2	400
3	800
5	3200

and graphing these values

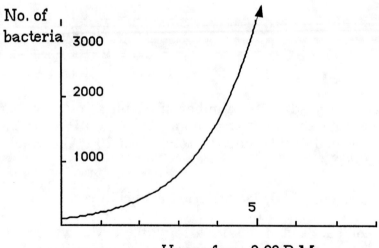

Hours from 2:00 P.M.

We can say that the number of bacteria depends upon, or is a function of, the time elapsed since 2:00 P.M., and the graph verifies that this equation is a mathematical function. The shape of this graph does not seem to be significantly different from that of one-half a parabola, but the equation

$$y = f(n) = 2^n(100)$$

is not a quadratic equation. In fact, in this equation the *exponent contains the variable.* This is a distinguishing feature of **exponential equations and functions**.

Finally, let's examine the graph of

$$y = \left(\frac{1}{2}\right)^x.$$

Substituting several values for x and using the calculator to generate the corresponding y-values, the table looks like

x	y
0	1
1	1/2
2	1/4
3	1/8
5	1/32
−1	2
−2	4
−3	8

(Using the negative exponent rule, these can be done mentally!)

and the graph is

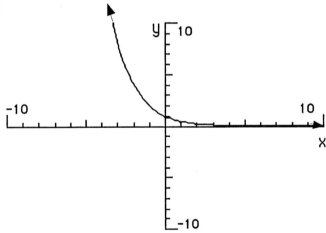

This definitely does not have a parabolic shape since the curve does not turn and change direction. As the x-values get larger and larger, the y-values get smaller and smaller, getting very close to 0 but never really reaching 0. Also, as the x-values get "more negative", the y-values keep increasing. Does this graph represent a function?

Sample Problems

1. Calculate f(x) for x = 0, −1, and 4 in the equation, $f(x) = 3^x$.

    Solution: If $x = 0$, $f(x) = 3^0 = 1$,

if $x = -1$, then $f(x) = 3^{-1} = \frac{1}{3}$

and if $x = 4$, then $f(x) = 3^4 = 81$.

2. If the population of Tinytown is now 312, and Tinytown is projected to grow at a rate of 6.25% each year over the next 10 years due to the arrival of a new herbologist, graph the population of Tinytown from now until 10 years from now.

Solution: Recalling our work in Chapter 6, the equation to find the population of Tinytown n years from now is

$$\text{Pop.} = 312(1 + .0625)^n$$

so a table of values might be

n	Pop.
0 (now)	312
.5	322
1	332
1.5	342
2	352
3	374
6	449
8	507
9	538
10	572

and the graph is

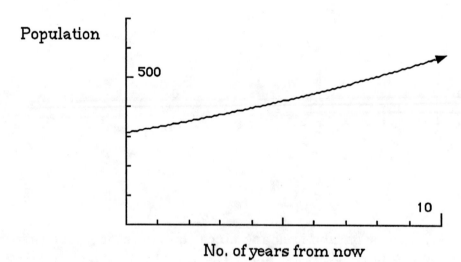

This very gradual curve is that described by an exponential function.

Problem Set 7.3

1. Draw the graph of $f(x) = (x - 1)^2$ and describe how the graph differs from the graph of $f(x) = x^2$.

2. Below are the graphs of $y = (x + 1)^2$ and $y^2 = x + 1$. Compare the graphs in shape and position on the coordinate system. Determine whether or not each is a function. If applicable, list the domain and range of the function.

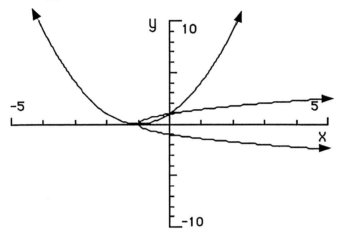

3. Without drawing the graph, compare each of the following to the graph of $y = x^2$ in terms of the vertex, shape, and position on the coordinate system. Tell whether or not each is a function.

   a) $y = \dfrac{1}{4}x^2$ b) $y = -2x^2 + 2$ c) $y^2 = \dfrac{1}{2}x$

4. A dog breeder has 300 feet of fencing and wants to enclose a large rectangular exercise yard for her golden retrievers.

   a) Write an equation for the perimeter of this rectangular yard.

   b) Write a related equation for the area of the yard as a function of L, the length.

   c) Draw a graph to show how the area of the yard depends on the length.

   d) From the graph, determine the maximum area, and corresponding length, possible with 300 feet of fencing.

5. For each of the following, write a quadratic equation that satisfies the given conditions.

   a) $a = 3$ and the vertex at the point $(0,-3)$.

   b) $a = -1$ and the vertex is at the point $(0,4)$.

   c) $a = 1/4$ and the vertex is at the point $(0,-2)$.

   d) $a = -2$ and the vertex is at the point $(0,0)$.

6. Draw the graph of $y = 2^x$ and compare it to the graph of $y = \left(\frac{1}{2}\right)^x$ in this section.

7. Given below are several graphs depicting common situations. Give at least one situation that could be depicted by each type of graph.

   a)

   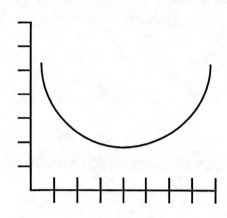

   b)

   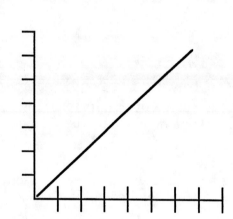

   c)

   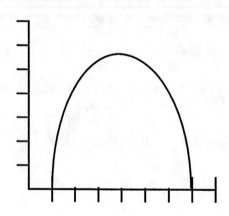

8. a) From the graph below, find the projected population of Podunk in 1993.

    b) Determine in what year the population of Podunk is about 3000 people.

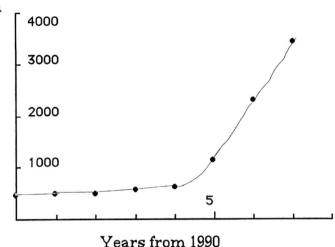
Years from 1990

9. Pick two pairs of two points on the graph below and find the rate of change for each pair. What do you conclude about an exponential function's rate of change?

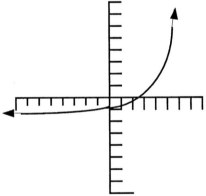

10. Graph $y = \sqrt{x}$, the positive root only. Is this graph a function? Compare this to the graph of $y^2 = x$. What do you find?

11. Draw the graph of the function: $y = 100\left(\dfrac{1}{2}\right)^x$.

12. A culture of bacteria doubles every half-hour. Starting with a culture of 50 bacteria at 8 A.M.,

    a) make a table of values to show the number of bacteria in the culture for each half-hour through 11:00 A.M.

    b) find the population of bacteria at 12:30 P.M. that day.

    c) make a graph of your results.

13. Complete the sequence: 1, 1, 2, 3, 5, 8, ... for 5 more terms. Then graph the sequence, letting the x-values be the term numbers. What kind of graph is this?

Exploration Problems

1. An amount of $2500 is put into a bank that offers interest at a rate of 5%, compounded continuously. The formula for the amount the principal grows to is

$$A = Pe^{rt}$$ where e is a constant programmed into the scientific calculator, r is the annual rate, and t is in years.

   a) Draw a graph which depicts the amount in the account during the first 8 years of the account, assuming no withdrawals or deposits are made.

   b) If the interest is compounded daily, find the amount in the account during the first 8 years, again assuming no withdrawals or deposits are made.

   c) Explain the difference in your results.

   d) Estimate the amount of time it will take for the investment to double.

Solutions - Problem Set 7.3

1. The graph of $f(x) = (x - 1)^2$ is

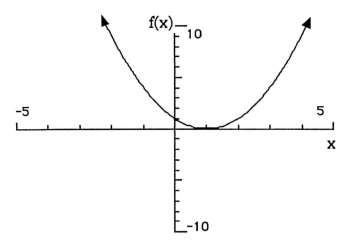

   The parabola has moved over along the x-axis so that the vertex is at (1,0). The shapes of the parabolas are the same.

2. The graph of $y = (x + 1)^2$ is a parabola with vertex at (−1,0), opening upward. The graph of $y^2 = x + 1$ is a "narrower" parabola with vertex at (−1,0), opening to the right. The first graph is a function with Domain = {all real numbers} and Range = {0 and positive real numbers}. The second graph does not represent a function.

3. a) The graph of $y = \frac{1}{4}x^2$ is "wider," although it is in the same position on the coordinate system and has a vertex at the origin. It does represent a function.

   b) The graph of $y = -2x^2 + 2$ opens downward, has a narrower shape and has its vertex at (0,2). It is symmetric with respect to the y-axis and is a function.

   c) The graph of $y^2 = \frac{1}{2}x$ is "lying down" and symmetric with respect to the x-axis. It has a vertex at (0,0) and is not a function.

4. a) $P = 2W + 2L = 300$ or $W + L = 150$

   b) $A = LW = L(150 - L) = 150L - L^2$

c) The graph is:

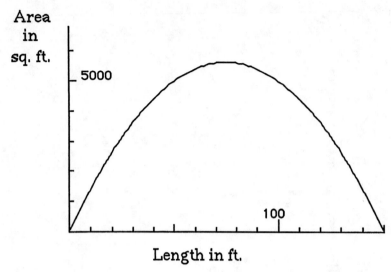

Length in ft.

d) The maximum area is approximately 5600 sq. ft. when the length is 75 ft.

5. Examples are: a) $y = 3x^2 - 3$ b) $y = -x^2 + 4$ c) $y = \frac{1}{4}x^2 - 2$ d) $y = -2x^2$

6. The graph of $y = 2^x$ is

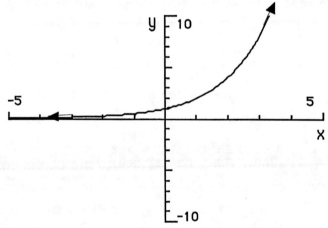

It appears to be the mirror-image of $y = \left(\frac{1}{2}\right)^x$ through the y-axis.

7. Examples:

a) The horizontal axis represents the number of weeks into the semester and the vertical axis represents the average student's study time (in hours) per week.

b) The years 1988, 1989, 1990, etc., on the horizontal axis and the constant growth of a company's staff on the vertical axis.

c) The path of an object thrown high into the air is similar to this graph.

8. a) about 600 people   b) sometime during 1996

9. Answers will vary, but the rate of change is not constant.

10. The graph is shown by

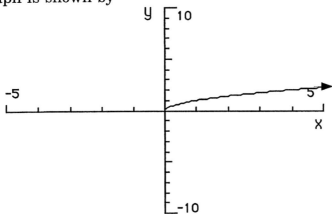

and does represent a function. It appears to be the "top half" of $y^2 = x$.

11. The graph is

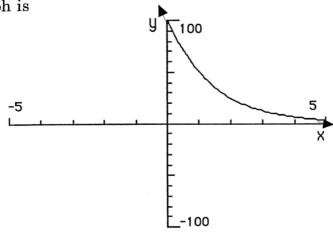

12. a)
Half-hours from 8 A.M.	No. of bacteria
0	50
1	100
2	200
3	400
4	800
5	1600
6	3200

b) Projected population at 12:30 P.M. that day is 25,600 bacteria.

c) The graph is shown below.

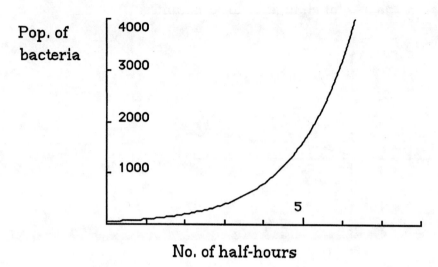

No. of half-hours

13. The next five terms are: 13, 21, 34, 55, 89.
    The graph is

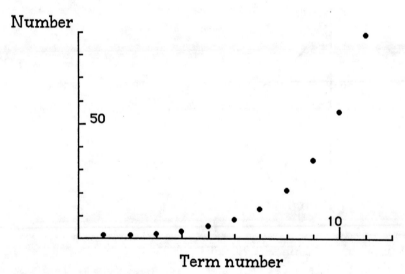

Term number

This is an exponential curve, although the dots cannot be connected since there can't be a term that is numbered $1/2$, for example.

Section 7.4 - Functional Relationships

Distance, Rate and Time

We have considered distance, rate, and time problems and graphs, but let's now examine their functional relationship. We must again keep one of the quantities constant so that we can compare how the distance is related to time or how the rate and time are related or how the distance and rate are related.

Specifically, we'll study the relationship between the average rate of travel and the time it takes to travel a distance of 500 miles. Since distance = rate • time, and we'll vary the rate of travel, it will be reasonable to solve the formula, d = rt, for t and we have

$$\frac{d}{r} = t$$

Using a table to keep track of our data, we'll find the time it takes to travel the 500 miles at various speeds.

If r = 5 (perhaps an unreasonable rate of speed by car, but for graphing purposes), then

$$t = \frac{500 \text{ miles}}{5 \text{ mph}} = 100 \text{ hrs}$$

and if r = 10 mph, then

$$t = \frac{500}{10} = 50 \text{ hrs.}$$

The times are recorded below for possible rates of travel.

Rate (mph)	5	10	20	30	40	50	55	60	65	70
Time (hrs)	100	50	25	16.7	12.5	10	9.1	8.3	7.7	7.1

Constructing a graph gives us a visual representation of how the time of travel depends on the rate of travel.

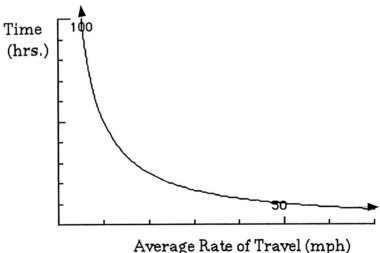

Average Rate of Travel (mph)

Several notes should be made here. If the rate is 0, $500/0$ is an undefined quantity and the graph does not generate a time-value at r = 0. Also keep in mind that using the graph for rates larger than 75 is inappropriate since common speed limits must be considered. Finally, the graph verifies that, for a given distance, we have a **functional relationship** between rate and time, and as the rate increases, the time it takes to travel the 500 miles decreases.

As another example, let's examine the path of a football kicked into the air and the relationship between the height of the ball and the time the football is in the air. The graph below shows how the height, in feet, depends on, or is a **function** of, the length of time elapsed, in seconds, after the ball is kicked.

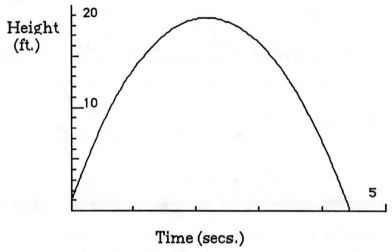

Time (secs.)

From this graph, it appears that it takes the ball approximately 4.5 seconds to reach the ground after the kick. It also shows the maximum height reached by the football. Note that at the time of the kick, when t = 0, the ball is about 1 foot off the ground. The path of the football, as well as the paths of projectiles shot into the air, baseballs hit, etc., are all parabolic in shape.

Sample Problems

1. From the graph above:

   a) Find the height of the football 1 second after it was kicked.

   b) Then find the time when the football reached a height of 10 yards.

   Solution:
   a) One second after the kick, the football reached a height of approximately 13 feet.
   b) There are two time values here corresponding to the height of 15 feet because the football achieves this height when it is climbing and then again when it is on its way back down to the ground (due to gravity). Drawing a horizontal line from the height of 15 feet so that the line intersects the entire curve visually shows that there are two values of time corresponding to that value of height. Dropping vertical lines from these points of intersection with the graph will generate the time values along the horizontal axis. These steps are shown below.

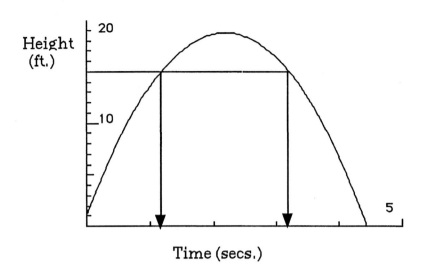

From the graph, it appears the football first reached a height of 15 feet approximately 1.2 seconds after it was kicked. It then reached a height of 15 feet approximately 3.2 seconds after it was kicked.

Business Applications

In section 6.2 we explored several types of problems in which a population grew or declined, prices inflated or depreciated, and interest earned or owed was calculated. In each of these cases we looked at changes over a period of time. In other words, our calculations were always a function of elapsed time. Now we need to consider a few specific investment and loan problems and analyze graphs of amounts earned or owed as **functions** of time. (You may wish to look back at section 6.2 to review the simple and compound interest formulas.)

If you have $10,000 to invest, you may choose a Certificate of Deposit which pays 6% simple interest per year. You could choose instead to invest $10,000 at 6% per year for 10 years, but in an account where the interest is compounded daily. **Line 1** on the next graph shows the total of your $10,000 plus interest earned over a period of 10 years. **Line 2** shows the total amount in the compounded account at various times during the 10 years.

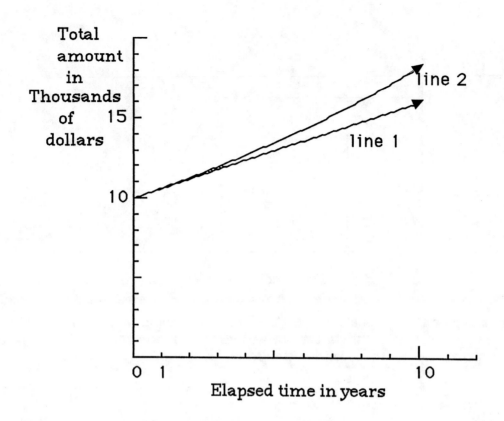

Sample Problem

From the graph, how much money will you have in each investment option after 5 years? After 10 years?

Solution:	Simple Interest:	Compound Interest:
5 years:	about $13000	about $13500
10 years:	about $16000	about $18000

Instead of investing, say you need to <u>borrow</u> $10,000 for ten years. If you kept the entire amount and paid it back in one lump sum, how much would you owe? If the interest, at 6%, was compounded annually, you would owe

$$\$10,000(1 + .06)^{10} = \$17,908.48. \quad \text{(Ouch!)}$$

Of course, if interest were compounded more frequently, you'd owe an even greater amount. However, in most cases, when money is borrowed it is paid back gradually. Because the amount of principal owed (the amount upon which interest is calculated) is gradually reduced, the portion of a month's payment allocated to interest declines and the amount allocated to principal increases. The following chart illustrates this concept. Interest is 6% per year, or .5% per month. Note that calculations for only the first 5 and last 3 payments are shown.

Payment Number	Payment Amount	Amt. as Interest	Amt. as Principal	Principal Owed
				$10000.00
1	$200	$50.00	$150.00	$9850.00
2	$200	$49.25	$150.75	$9699.25
3	$200	$48.50	$151.50	$9547.75
4	$200	$47.74	$152.26	$9395.48
5	$200	$46.98	$153.02	$9242.46
- - -	- - -	- - -	- - -	- - -
56	$200	$ 2.66	$197.34	$ 333.79
57	$200	$ 1.67	$198.33	$ 135.46
58	$136.14	$ .68	$135.46	0

Fifty-eight payments will be required to pay off the loan in this way. The final payment includes the remaining principal, $135.46, plus $.68 interest for the last month. Can you find the total amount of interest paid?

A formula, the derivation of which is beyond the scope of this course, can be used to determine the amount of a monthly payment which would pay off a loan in a set amount of time. We'll apply the formula to our $10,000 loan at 6% for 3-year, 4-year, and 10-year payoffs. The variables have the same meaning as those we've used previously in the compound interest formula, that is:

P: Principal, r: rate, n: number of compounding periods per year, t: time, in years

$$\text{Payment} = \frac{P\left(\dfrac{r}{n}\right)}{1-\left(1+\dfrac{r}{n}\right)^{-nt}}$$

For a 10-year payoff:

$$\text{Payment} = \frac{10000\left(\dfrac{.06}{12}\right)}{1-\left(1+\dfrac{.06}{12}\right)^{-12\cdot 10}}$$

Payment = $111.02

For a 4-year payoff:

$$\text{Payment} = \frac{10000\left(\dfrac{.06}{12}\right)}{1-\left(1+\dfrac{.06}{12}\right)^{-12\cdot 4}}$$

Payment = $234.85

For a 3-year payoff:

$$\text{Payment} = \frac{10000\left(\frac{.06}{12}\right)}{1 - \left(1 + \frac{.06}{12}\right)^{-12 \cdot 3}}$$

$$\text{Payment} = \$304.22$$

The monthly payment for a 10-year payoff is shown to be $111.02. For a 4-year payoff it's $234.85, and for a 3-year payoff it's $304.22.

Sample Problems

Find the total amount of interest paid in each of the cases above. What can you conclude about the relationship between the time period of a loan and the amount of interest paid?

Solutions: Interest paid = (Monthly payment • 12 • No. of years) − Principal

For   3 years: ($304.22 • 12 • 3) − $10,000 = $951.92;

4 years: ($234.85 • 12 • 4) − $10,000 = $1272.80;

10 years: ($111.02 • 12 • 10) − $10,000 = $3,322.40.

The amount of interest paid increases as the length of the time of the loan increases, and therefore the amount of interest paid can be said to be a **function** of elapsed time.

Surface Area

A cylinder can be made from a rectangular piece of paper by taping one pair of oppposite sides together. Starting with a 6 cm by 10 cm rectangle as shown below,

10 cm

6 cm

we'll take one 6 cm end and tape it to the other 6 cm edge. This forms the **lateral surface** of the cylinder. Since we started with a rectangle that is 6 cm x 10 cm, we can say that the **lateral surface area** of this cylinder is 6 x 10 or 60 square cm. To find the radius of the base of this cylinder, we need to visualize that the 10 cm base of the rectangle has now become the circumference of the base of the cylinder. We know that for a circle, the distance around, or circumference, is given by

$$C = 2\pi r.$$

In this case, C = 10 cm and so
$$10 = 2\pi r \text{ and}$$
$$10 = 2(3.14)r. \text{ Thus, } r \text{ is approx. 1.6 cm.}$$

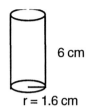

The total surface area of the cylinder is the sum of the lateral surface area and the area of the top and bottom circles. Since the areas of the circles are the same and each is given by
$$A = \pi r^2,$$
the total surface area of this cylinder is
$$2(\pi r^2) + 60 = 2(3.14 \cdot 1.6^2) + 60 = 76.1 \text{ square cm.}$$

Let's see what happens if we attach the two 10 cm sides of our original 6 cm x 10 cm rectangle together.

The 10 cm side now becomes the height of the cylinder and 6 cm is the circumference of the base.

Since again $\quad$ C = 2πr but now C = 6

we have $\quad$ 6 = 2(3.14)r and r = .96 cm.

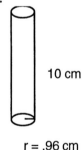

The total surface area of this cylinder is
$$2(\pi r^2) + 60 = 2(3.14)(.96)^2 + 60 = 65.8 \text{ square cm.}$$

Can you hypothesize about how, for a given lateral surface area, the shape of the cylinder affects the total surface area?

Let's examine another 60 square unit rectangle, but one that is 15 cm by 4 cm as shown.

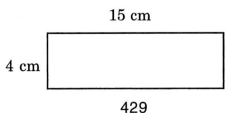

Attaching the two 4 cm sides together, we have a cylinder with height of 4 cm. Here

$$C = 15 \text{ inches}$$

so the radius of the base is

$$15 = 2(3.14)r \text{ and } r = 2.4 \text{ cm}.$$

[diagram: cylinder with height 4 cm and r = 2.4 cm]

The total surface area is then

$$2(\pi r^2) + 60 = 2(3.14)(2.4)^2 + 60 = 96.2 \text{ square cm}.$$

Comparing the first two cylinders we constructed, both with a lateral surface area of 60 square cm, it appears that the cylinder that was "tall and thin" had the smallest total surface area. Constructing the cylinder from the 15 cm x 4 cm rectangle (again with a lateral surface area of 60 sq. cm.) gave an even <u>larger</u> total surface area, and again, this cylinder is "broader" than both the first and second cylinders.

How can we find a cylinder with a lateral surface area of 60 square cm that has the smallest total surface area?

Let's construct and examine a graph that relates the radius of the base to the total surface area, again keeping the lateral surface area constant at 60 sq. cm.

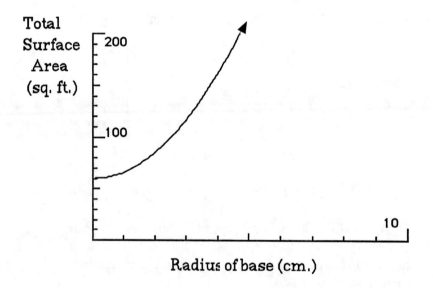

From the graph, it appears that as the radius of the base increases while the lateral surface area remains constant at 60 square cm., the total surface area increases. This seems reasonable, because if the *lateral surface area is constant*, a larger radius gives a larger base area and this increases the total surface area. Thus, the cylinder with the smallest radius (that is reasonable to construct) would generate the smallest surface area and we can say that the surface area is a **function** of the radius of the base.

Sample Problems

1. Based on the work above, can you devise a formula for calculating the total surface area of a cylinder?

   Solution: The area of each base is πr^2 so the area of the two bases together is $2\pi r^2$.

   The lateral surface area is the circumference of the base times the height of the cylinder. Thus, the lateral surface area is

   $$C \cdot h = 2\pi rh$$

   and the formula for the *total surface area* of a cylinder is

   $$S.A. = 2\pi r^2 + 2\pi rh.$$

2. Which cylinder below has the largest total surface area?

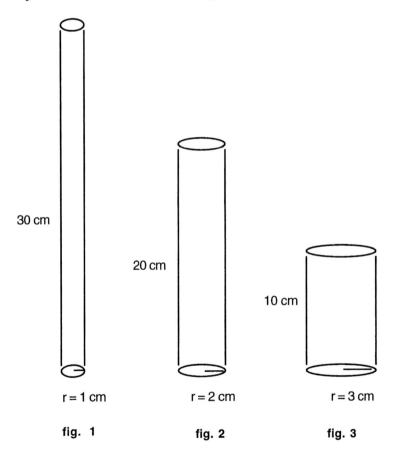

   fig. 1   fig. 2   fig. 3

   Solution: Using the pattern of results above, the cylinder that is 10 cm in height and has a base radius of 3 cm. certainly has the largest "base" area. However, it is important to realize that the lateral surface areas <u>may not</u> be the same here. The lateral surface areas are calculated below:

431

For the cylinder in **fig. 1** with r = 1 cm and h = 30 cm, the lateral surface area is

$$2\pi rh = 2(3.14)(1)(30) = 188.4 \text{ sq. cm.}$$

In **fig. 2**, for r = 2 cm and h = 20 cm, the lateral surface area is

$$2\pi rh = 2(3.14)(2)(20) = 251.2 \text{ sq. cm.}$$

and for the cylinder in **fig. 3** with r = 3 cm and h = 10 cm, the lateral surface area is

$$2\pi rh = 2(3.14)(3)(10) = 188.4 \text{ sq. cm.}$$

Of the two cylinders with the *same* lateral surface area, the one with the largest base radius, in this case, the cylinder in figure 3, has the greatest total surface area.. Thus, we only need to decide between the cylinders in **figures 2 and 3**. The larger base radius, and, thus, larger base area, is in **fig. 3**, but whether it is large enough to compensate for the larger lateral surface area of **fig. 2** is the question. Calculating the areas of the bases, the cylinder in **fig. 2** (with r = 2 cm) has combined base areas of

$$2(3.14)(2^2) = 25.1 \text{ sq. cm}$$

and the cylinder in **fig. 3** with r = 3 cm has combined base areas of

$$2(3.14)(3^2) = 56.5 \text{ sq. cm.}$$

Thus, the "taller" cylinder in **fig. 2** has a total surface area of

$$251.2 + 25.1 = 276.3 \text{ sq. cm.}$$

and the "shorter" cylinder in **fig. 3** has a total surface area of

$$188.4 + 56.5 = 244.9 \text{ sq. cm.}$$

In other words, the 10 cm increase in height more than compensated for the 1 cm reduction in base radius to make the total surface area of the cylinder in **fig. 2** larger!

Surface Area and Volume

Let's now consider the amount of sand each of the cylinders we constructed out of the 6 cm by 10 cm rectangle could hold, and then examine the relationship between the volumes of sand and the surface areas of the cylinders.

The volume of any cylinder can be found by multiplying the area of the base times the height of the cylinder. For the cylinder shown,

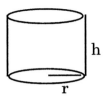

the area of the base is πr^2, and the volume is

$$V = \pi r^2 h$$

Let's compare the volumes of our first 3 cylinders, each with a lateral surface area of 60 sq. cm.

For the first cylinder with a base radius of 1.6 cm and height of 6 cm,

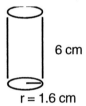

the volume is

$$V = (3.14)(1.6)^2(6) = 48.23 \text{ cubic cm.}$$
(cubic cm since we are multiplying cm • cm • cm)

The volume of the second cylinder with r = .96 cm and height 10 cm is

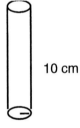

$$V = (3.14)(.96)^2(10) = 28.94 \text{ cubic cm.}$$

Finally, for the third cylinder with a base radius of 2.4 cm and height of 4 cm,

$$V = (3.14)(2.4)^2(4) = 72.35 \text{ cubic cm.}$$

Again, for a lateral surface area of 60 sq. cm, it appears that as the radius of the base increases, the volume also increases. The next graph clearly illustrates this relationship. (Note: If r = 0, we have a cylinder with V = 0.)

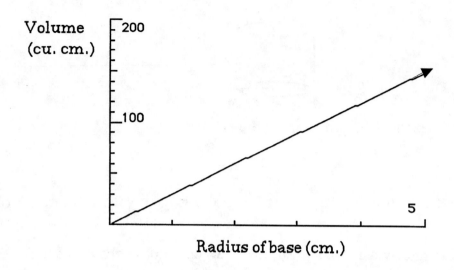

It appears that for a given lateral surface area, an increase in the radius of the base will increase the volume. For example, from the graph, if the base radius is 4 cm, the volume is approximately 120 cubic cm. We can verify this result by finding the height of a cylinder with a base radius of 4 cm and use this information to calculate the volume. Substituting r = 4 into the formula we get

$$C = 2\pi r = 2(3.14)(4) = 25.12 \text{ cm}$$

and since the lateral surface area has remained constant at 60 square cm, then

$$C \cdot h = 60 \text{ and } h = \frac{60}{25.12} = 2.4 \text{ cm.}$$

Finally, the volume can be calculated as

$$V = (3.14)(4^2)(2.4) = 120.58 \text{ cubic cm} \quad \text{(which is very close to the value we read from the graph.)}$$

It is reasonable then to conclude that for a given lateral surface area, the volume will be the greatest when the radius of the base can be made as large as possible and we can say that the volume is a **function** of the radius of the base.

[Also note that if the base radius and/or the height increases, and the lateral surface area is not held constant, then both the total surface area and volume also increase.]

Sample Problem

A closable box-shaped carton can be designed so its capacity is 1000 cm³. Out of the three possible boxes shown on the top of the next page, which uses the most cardboard? Which uses the least cardboard?[1]

---

[1] *Mathematics Resource Project*, page 741.

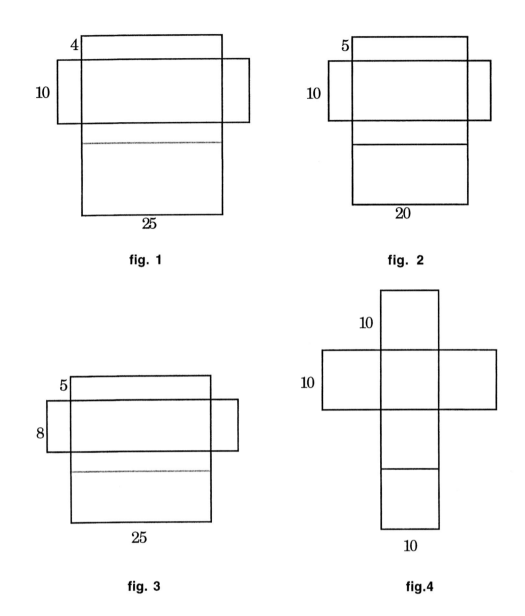

fig. 1

fig. 2

fig. 3

fig. 4

Solution: The box in **fig. 1** would look like

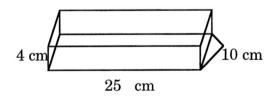

and that in **fig. 2** would look like

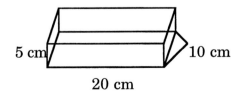

The box in **fig. 3** would look like

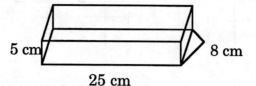

and the box in **fig. 4** would be a cube as shown below.

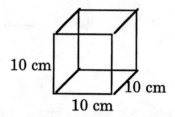

The total surface area of each box is the sum of the areas of the sides and top and bottom surfaces. The volume of each box is calculated by

$$V = \text{length of base} \cdot \text{width of base} \cdot \text{height}$$
$$\text{or} \quad \mathbf{V = l\,w\,h}$$

and, thus, you can verify that the boxes all have a volume of 1000 cm^3.

The surface area of the 4 by 25 by 10 cm box (**fig. 1**) is

$$2(4 \cdot 25) + 2(4 \cdot 10) + 2(25 \cdot 10) = 780 \text{ sq. cm.}$$

The surface area of the 5 by 20 by 10 cm box in **fig. 2** is

$$2(5 \cdot 20) + 2(5 \cdot 10) + 2(10 \cdot 20) = 700 \text{ sq. cm.}$$

The surface area of the 5 by 25 by 8 cm box (**fig. 3**) is

$$2(5 \cdot 25) + 2(5 \cdot 8) + 2(8 \cdot 25) = 730 \text{ sq. cm.}$$

Finally, the surface area of the 10 by 10 by 10 cm cube in **fig. 4.** is

$$6(10 \cdot 10) = 600 \text{ sq. cm.}$$

Thus, for the volume of 1000 cm^3, the box with largest surface area, and which would require the most cardboard to construct, is the 4 by 25 by 10 box. (Examining the boxes themselves, does this seem reasonable?) The box needing the smallest amount of cardboard for construction is the 10 x 10 x 10 cube. It should be noted that, in general, <u>for a given volume</u>, the surface area is a **function** of the shape of the box, and a cube minimizes surface area. In terms of packaging and construction costs, the cube is the "most efficient" box.

Problem Set 7.4

1. If Merle decides to travel at an average speed of 50 mph, draw a graph to show how the distance he can travel, in miles, is a function of the time he travels, in hours. (Hint: Make a table beginning with t = 0 and use incremental values of t.)

2. Given the path of a projectile thrown in to the air as shown below, answer the following questions with respect to the graph.

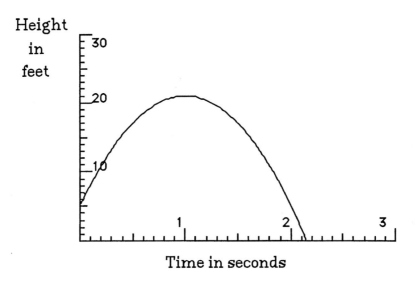

   a) How long after the object is thrown does it reach its maximum height?

   b) After 2 seconds, what is the height of the projectile? Is it on its way up or on its way back down to the earth?

   c) Find the time of travel of the projectile when it is at a height of 10 feet.

   d) How long does it take the projectile to hit the ground?

3. Make a table to find the amounts in each account for 0 to 10 years, inclusive, for $500 deposited at 2.5%

   a) if the interest is simple interest

   b) if the interest is compounded daily.

   Then graph the data from both tables on the same coordinate system to show how the amounts in the accounts are a function of the time elapsed.

4. If $12,000 is borrowed for a new car at 7% interest, calculate the amount of each payment and the total amount paid for the car if the length of the loan is

   a) 3 years   b) 4 years   c) 60 months

5. If $500 is deposited in an account paying 5.5% simple interest and another $500 is deposited in a different account paying 5.2% interest, compounded daily (and no deposits or withdrawals are made)

   a) use the table below and calculate the amount in each account at the given periods of time.

   Simple Interest

t	1	2	5	10	15
Amt.					

   Compound Interest

t	1	2	5	10	15
Amt.					

   b) Find the difference between the amounts for each of the times given and draw a graph to show how the difference in amounts is a function of the time.

6. Given the following two containers. All measurements are in centimeters.

   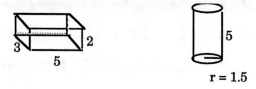

   a) Predict which would require the most light cardboard to construct.

   b) Predict which container would hold the most oatmeal.

   c) Find the surface area and volume of each.

   d) Compare your answers to your predictions.

7. Draw an arrangement of four cubes (top view) where each cube touches at least 1 edge or corner of another cube so that

   a) the surface area is minimized

   b) the surface area is maximized

8. Two dozen small boxes each measuring 3 cm by 6 cm by 8 cm will exactly fit into a large box which has a base measuring 12 cm by 12 cm. What is the height of the large box? [1]

---
[1] *Mathematics Resource Project*, page 789.

9. Which is the better buy, oranges 4 cm in radius that cost $.30 each or oranges 5 cm in radius that cost $.36 each? Justify your answer. [V of sphere = $\frac{4}{3}\pi r^3$]

10. Which will carry more water: two pipes, one with a 3-inch radius, the other with a 4-inch radius; or one pipe with a 5-inch radius? Justify your answer.

11. Three tennis balls fit exactly into this cylindrical can as shown. Which quantity is greater, the height of the can or the circumference of the base of the can? Justify your answer.

12. Compare the surface area and volume of these two figures and explain your reasoning.

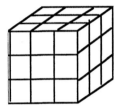 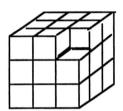

Exploration Problems

1. Use 12 plastic cubes and build as many different rectangular prisms (box-like shapes) as you can. Record the dimensions and the surface area of each shape in a table. Which 12-cube prism would be the cheapest to paint? Why?

2. A 3 in. by 3 in. by 3 in. cube is made of individual 1" cubes all placed together. If the surface only of the large cube is painted, how many of the 1" cubes have

   a) 1 face painted?   b) 2 faces painted?   c) 3 faces painted?

      d) 4 faces painted?   e) 0 faces painted?

   Think of a 4 in. by 4 in. by 4 in. cube and then a 5 in. by 5 in. by 5 in. cube, both constructed and painted in the same fashion. Make a table of the number of 1" cubes that have 0, 1, 2, 3, 4, etc., faces painted in each case. Generalize your results to an n by n by n cube.

Solutions - Problem Set 7.4

1.
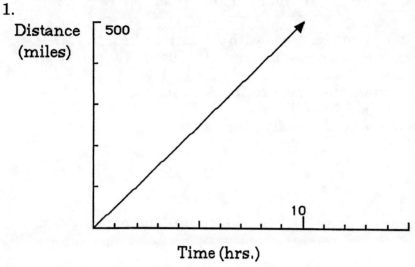

2. a) about 1 second;   b) 5 feet, on its way down
   c) about .2 second and 1.8 sec.;   d) a little more than 2.1 seconds

3.
years	simple	compound
0	$500	$500
1	$512.50	$512.50
2	$525	$525.31
3	$537.50	$538.45
4	$550	$551.91
5	$562.50	$565.70
6	$575	$579.85
7	$587.50	$594.34
8	$600	$609.20
9	$612.50	$624.43
10	$625	$640.04

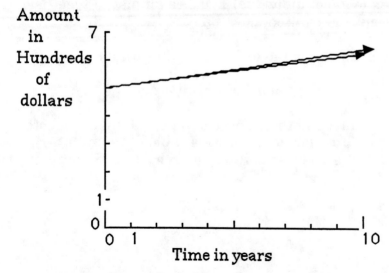

4. $t = 3$: $370.53 per month; total = $13339.08
    $t = 4$: $287.35 per month; total = $13792.80
    $t = 5$: $237.61 per month; total = $14256.60

5.
	simple	compound	difference
1 yr.	$527.50	$526.69	− $0.81
2 yrs.	$555	$554.80	− $0.20
5 "	$637.50	$648.45	$10.95
10 "	$775	$840.98	$65.98
15 "	$912.50	$1090.68	$178.18

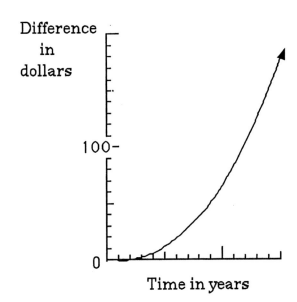

6. Answers will vary for a, b, and d.
   c) S.A. box = 62 sq. cm; S. A. cylinder = 61.23 sq. cm.
   V box = 30 cm³; V cylinder = 35.325 cm³.

7. a) [figure] b) [figure] space

8. 24 cm

9. Larger oranges cost less per cc.

10. Equal: Area of cross sections: $3^2\pi + 4^2\pi = 5^2\pi$

11. Height = 3 • diameter = 3 • 2r; circumference = 3.14 • 2r so the circumference is larger.

12. Surface areas are equal. Left-hand figure has more cubes and therefore greater volume.

Section 7.5 - Explorations With the Graphing Calculator

Linear and Quadratic Functions Revisited

The graphing calculator is an extremely useful tool for drawing and analyzing graphs. The first step is to select a range appropriate for the expected input and output values of the equation. Since graphing calculators are not all consistent in key functions, it is important that you consult your handbook for instructions. If we graph the function

$$f(x) = y = \tfrac{1}{3}x - 2,$$

The graph should appear on your display as

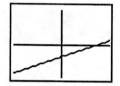

With the "trace" (or similar) key, follow along the line to verify that the y-intercept is –2, and continue along the line to determine the x-intercept. Since the *x-intercept* is the point at which y = 0, it is called the **zero** of the linear function. For $y = \tfrac{1}{3}x - 2$, the zero is 6 since the graph crosses the x-axis at the point (6,0).

On the same coordinate system, graph the equation, $y = -x - 2$.

The display will be

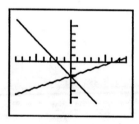

This time, use the trace function to find the point at which the two lines intersect. The coordinates of this point, (0,–2), satisfy both equations and, as we have discussed previously, this point represents the solution to the system of equations:

$$y = \tfrac{1}{3}x - 2$$
$$y = -x - 2$$

As another example, let's consider various-sized squares and compare the perimeter and area of these squares. In table form

Length of side (cm)	Perimeter (cm)	Area (sq. cm)
1	$1+1+1+1 = 4(1) = 4$	$1 \cdot 1 = 1$
2	$4(2) = 8$	$2 \cdot 2 = 4$
3	$4(3) = 12$	$3 \cdot 3 = 9$
4	$4(4) = 16$	$4 \cdot 4 = 16$
5	$4(5) = 20$	$5 \cdot 5 = 25$
6	$4(6) = 24$	$6 \cdot 6 = 36$

Again, using an appropriate range, let's first graph the perimeter, y, with respect to the length of a side, x, by graphing $y = 4x$. Then we'll graph the area, y, of the square with respect to the length, x, as the equation $y = x^2$ on the same coordinate system. The graphs should look like

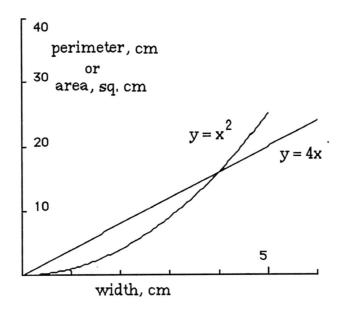

To find the square with the same perimeter and area, the trace key can be used to determine the coordinates of the point where the two graphs intersect. Algebraically, we have the system of equations

$$y = 4x$$
$$y = x^2$$

Since both equal y, we can write

$$4x = x^2.$$

Subtracting 4x from both sides we have

$$x^2 - 4x = 0$$

and using the distributive property (i.e., factoring), we can rewrite the left side as

$$x(x-4) = 0.$$

If the product of two quantities is 0, then one or both must equal zero. Therefore,

$$x = 0 \text{ and/or } x - 4 = 0 \longrightarrow x = 4.$$

Obviously, if x = 0, the area is 0 and we can consider this a trivial case. However, checking our results for x = 4, the perimeter is 4(4) or 16 and the area is 4^2 or 16. Thus, the square whose side measures 4 cm has a perimeter equal to its area. These results can be verified in the table on the previous page and by the trace function on the calculator. (Note, however, that even though the <u>values</u> of the perimeter and area are the same, the <u>units</u> are not the same since the units for area are square units.)

<u>Sample Problems</u>

1. Use the graphing calculator to find the solution to the system of equations

$$y = x - 4$$
$$y = 3x + 1$$

   Solution: The point of intersection is (–2.5,–6.5).

2. From the graph of the perimeter and area of various squares, if the perimeter of a square is 10 cm and its area is 6.25 sq. cm, what is the length of a side?

   Solution: The side is 2.5 cm.

As another example, let's graph the function

$$f(x) = y = (x - 2)(x + 1) \text{ on the domain } -8 \leq x \leq 8 \text{ and range } -10 \leq y \leq 10.$$

The graph is

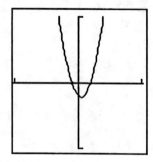

The trace function will help to determine the points at which the graph crosses the x-axis, in other words, the **zeros** of the function. In this case, it appears the zeros are at x = 2 and x = –1. To verify this algebraically, we can set y = 0 and we have

$$0 = (x - 2)(x + 1)$$

which implies $x - 2 = 0$ and/or $x + 1 = 0$.

Thus, $y = 0$ when $x = 2$ and when $x = -1$, and this agrees with our graphical solution.

Also note that
$$\begin{aligned} 0 &= (x-2)(x+1) \\ &= x(x+1) - 2(x+1) \quad \text{(using the distributive property twice.)} \\ &= x^2 + x - 2x - 2 \\ &= x^2 - x - 2 \end{aligned}$$

which is a quadratic function and our graph suggests that this a parabolic curve.

Let's try two more graphs: a) $y = f(x) = x^3$ and [Be sure to select an appropriate range in each case.]
b) $y = f(x) = (x - 4)(x + 1)(x + 2)$

For the graph below of $y = f(x) = x^3$, use the trace function to find the value(s) of x that correspond(s) to $y = 8$. (You should see that $x = 2$ is the only value).

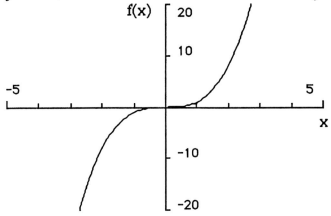

To find the zeros of that function, note that the graph crosses the x-axis only at $y = 0$. Thus, the only zero of this function is at (0,0). This is easy to verify algebraically by substituting $y = 0$ into the equation, $y = x^3$, and solving for x.

Examine the graph of $y = (x - 4)(x + 1)(x + 2)$.

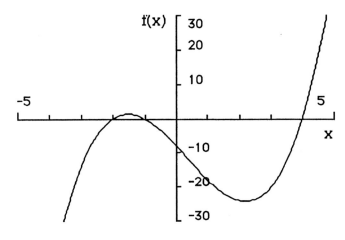

Can you find the zeros of this function? They should be at 4, −1, and −2. Letting y = 0 in the equation, we have

$$0 = (x - 4)(x + 1)(x + 2)$$

which implies that

$$x - 4 = 0, \ x + 1 = 0 \text{ and } x + 2 = 0.$$

Thus, we have algebraic verification that the zeros are

$$x = 4, \ x = -1, \text{ and } x = -2.$$

[It is important to note that the zeros of these equations and the factored forms of the equations are related. Study the results from the last several examples to understand the connection.]

Finally, the graphing calculator is a useful tool to help us determine when the function is increasing or decreasing. In the case of $y = x^3$,

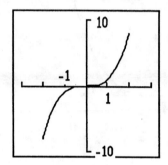

the function is constantly increasing, that is, as x increases the y-values also increase. The trace function will confirm this.

The graph of $y = (x - 4)(x + 1)(x + 2)$, however, is different. The function increases to a point between −1 and −2, changes direction and then decreases before it starts climbing again when x is about +2.3.

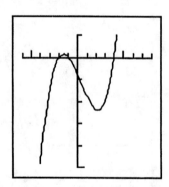

In other words, while $x < -1.5$, the function is increasing; for $-1.5 < x < 2.3$ the function is decreasing; and for $x > 2.3$ the function is again increasing. (Remember that these interval divisions are decimals approximations of the values given by the trace function.)

Sample Problem

1. Use the graphing calculator to graph the equation: $y = x^4 + 3$. Find the zeros of the function and describe whether the function is increasing or decreasing, or both.

   Solution: The graph is shown below.

   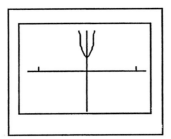

   It does not intersect the x-axis and so there are no zeros for this function. (It stands to reason that x^4 will never be negative and, thus $x^4 + 3$ will <u>always</u> be positive.) The graph is a decreasing function when x is less than zero and then it becomes an increasing function for x>0.

The Median-Median Line and Line of Best Fit

The chart below lists the study time, in hours, reported by ten individuals who were preparing for a quiz and the scores these individuals earned on the quiz.

Person	Study Time (hrs.)	Quiz Score
#1	2	65
#2	3	70
#3	3.5	75
#4	4	90
#5	5	92
#6	3.7	90
#7	4	88
#8	4.5	92
#9	7	81
#10	5.5	98

If we plot this data, letting x = study time and y = quiz score, there is an indication of a linear relationship between the two variables and it is appropriate to find a line that "fits" the data. (In other words, we'll try to find an equation that relates how the quiz score is a function of the study time.) We'll do this with the calculator later, but for now we'll use a less formal statistical method which will generate the **median median line.**

First, the data is divided into 3 sections so that there are the same number of data points in each section. If there are extra data points, they are spread accordingly, but it is important that the outer regions contain the same number of data points. Since 10 ÷ 3 = 3.3, we'll put the extra data point in the middle section. The scatterplot and regional divisions are shown below.

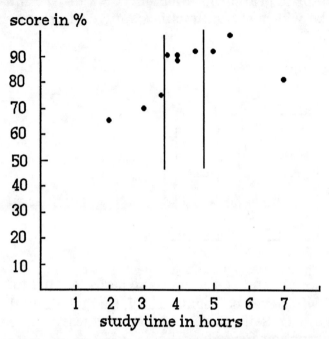

For each region, we need to find the median of the x-values and draw a vertical line through this point. Then we find the median of the y-values and draw a horizontal line through that point, making an "x" where these lines intersect.

We now have three points through which we'll draw the line. If the points do not all lie on a line, a straightedge initially must be placed to connect the two "x's" in the outside regions but then slid 1/3 of the way to the middle point. The line is drawn as shown.

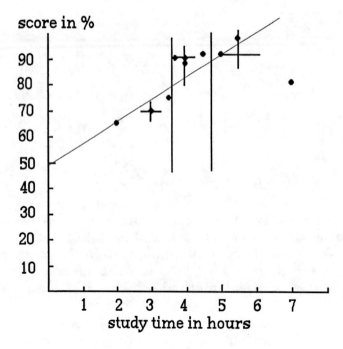

To find the *equation* of the median median line, we'll use two points on (or very close) to the line and calculate the slope of the line. The y-intercept can be read directly from the graph and, in this case, the y-intercept is about 49. We'll use the point, (0,49), and the point, (2,65) to find the slope of the line as shown below.

$$m = \frac{65 - 49}{2 - 0} = \frac{16}{2} = 8$$

Substituting that and the y-intercept into the equation

$$y = mx + b$$

gives us

$$y = 8x + 49$$

as the equation of the **median-median line** for the quiz score data.

We can now use this line to predict the quiz score for a person who studied 4.2 hours, for example. Since

$$y = 8(4.2) + 49 = 83 \text{ (rounded off)}$$

then, based on our equation, if a student studied 4.2 hours for this quiz, we expect that person's score to be about an 83. Of course, individuals studying this amount of time will not all get an 83, but the line can still be used to see that, for the most part, an increase of study time results in a better quiz score. Can you give a reason why the person who studied for 7 hours got an "81," or why any person studying for longer than 7 hours might not do well on the quiz?

The graphing calculator is programmed to draw the scatterplot and find the "line of best fit", commonly known as the **least squares regression line**, for a data set. You must select an appropriate range, enter the data with the statistical function keys, select "scatterplot," and then, if there is an apparent linear relationship, select "linear regression" to get the equation of the line that best fits the data. For the quiz score data, the graphing calculator yields an equation of

$$y = 4.52x + 65.02$$

Note that this is not at all the same equation as that of the median median line. This is because the score of 81 for a study time of 7 hours wasn't typical of the rest of the data. That value did not affect the placement of the median in the third section of the graph, but it did enter into the calculation of the least squares regression line. It is best to think of the median median line as an approximation and the least squares regression line as a more statistically accurate measure.

Sample Problems

1. From the scatterplot given, answer the following questions.

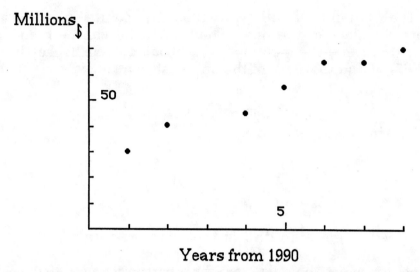

Years from 1990

a) Does there appear to be a linear relationship between the two variables?

b) Would it be appropriate to find a "line of best fit"?

c) Give a possible scenario that would generate data in this pattern.

d) Use your graphing calculator to generate the least squares regression line.

Solutions:

   a) Yes, even though the "y-values" have fluctuated some.

   b) Yes.

   c) The amount of sales each year since 1990 that a company has had, in millions of dollars.

   d) The least squares regression line is given by the equation, $y = 5.7x + 25.9$. (This does depend on how the data points are read.)

Problem Set 7.5

For each exercise, use the graphing calculator.

1. Graph $y = 3x - 1$ and $y = -\frac{1}{3}x + 2$ on the same coordinate system. What appears to be the relationship between the lines? At what point, if any, do the lines intersect?

2. Graph $y = 4x - \frac{1}{2}$ and $y = 4x + 2$ on the same coordinate system. Find the point of intersection, if it exists.

3. Graph $y = x^2$ and $y = +\sqrt{x}$ on the same axes. Find the point of intersection of the two graphs.

4. Graph $y = \frac{1}{x}$ and describe the graph, in words. Does $y = 0$ for any value of x? Why or why not?

5. Graph $y = \sin x$. Use domain values of -6.28 to $+6.28$ and range values of -2 to $+2$.

6. Graph $y = \cos x$ on the same axes as $y = \sin x$. Describe how these graphs are similar and how they are different.

7. Write a polynomial equation (in factored form) that has zeros at -3, 2, and 5.

8. Graph $y = f(x) = x^2 + 2x - 8$ and find the zeros of the function.

9. Find the point of intersection of the graphs: $y = x^2 - 2$ and $y = x$.

10. Describe the intervals on which the function $f(x) = (x + 3)(x + 1)(x - 3)$ is increasing and decreasing.

11. The Journal of Wildlife Management, October, 1970 issue, contained the following scatterplot which represents the relationship between the north/south population ratio of pintail ducks and the number of ponds on the prairies of Alberta and Saskatchewan in May. (The north/south ratio is used instead of indexes, or numbers of pintails, to eliminate yearly variation in the size of the continental population of pintails.) Find the line of best fit

    a) using the median median

    b) using the graphing calculator

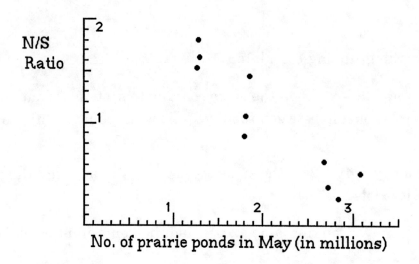

Exploration Problem

To gather biological data, each student in the class will find the circumference of his/her thumb and then of his/her wrist, both to the nearest tenth of a centimeter. (A metric tape measure will work fine, or string can be used, marked and then measured.) Gather the class data and draw a scatterplot, putting the thumb circumference as the x-value and the wrist circumference as the y-value. Does there appear to be a relationship between the thumb and wrist circumferences? If so, use the graphing calculator to enter the data and find the equation of the regression line relating these two variables. Discuss the results.

Solutions – Problem Set 7.5

Note: The graphs will not be shown here, but answers to other questions are given.

1. The lines are perpendicular and meet at the point (.9, 1.7).

2. The lines are parallel and do not intersect.

3. The graphs meet at the points (0,0) and (1,1).

4. The graph occurs only in the first and third quadrants and it does not touch either axis; y is never zero because in the equation $y = \frac{1}{x}$, for $y = 0$, the numerator would have to be zero and it is always 1.

5. and 6.
   The graphs have the same shape but the $\cos x$ is shifted to the left of the $\sin x$ graph.

7. For example, $y = (x + 3)(x - 2)(x - 5)$

8. The zeros are at $x = 2$ and $x = -4$.

9. The points of intersection are (2,2) and (−1,−1).

10. For $x < -2$, the function is increasing; for $-2 < x < 1.6$, the function is decreasing; and for $x > 1.6$, the function is again increasing.

11. The median median line will vary (one possibility is $y = -1.1x + 1.85$) but the graphing calculator line of best fit is given by the equation $y = -.73x + 2.52$.

Chapter Seven Review

1. Write the cost function of n pounds of apples purchased at $.99 per pound.

2. A furniture salesperson earns $450 per week plus a 7% commission on all sales.

    a) Write the salesperson's salary as a function of the amount of sales, x.

    b) If the salesperson sold $3588 worth of furniture one week, find that person's salary that week (to the nearest cent).

3. A function machine is drawn below. Write two different functions that would take the input value of 3 and give an output of 8.

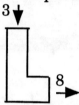

4. Using function notation, write the rule that relates the elements of the first and second sets.

    a)
    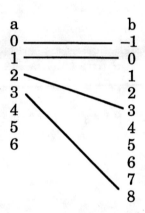

    b) {(0,2),(1,3),(2,4),(6,8)}

    c)
    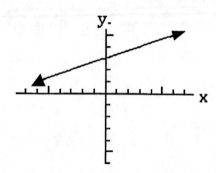

5. Explain why each of the following does or does not represent a function.

a)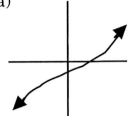

b)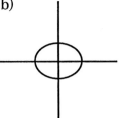

c)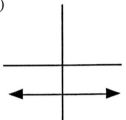

d)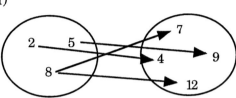

6. Find the rate of change of each of the following functions. Discuss the meaning of each.

   a)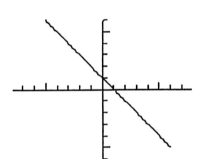

   b)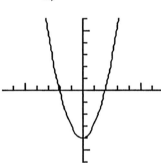

7. Find the equation of the line through the point (1,–2) with an x-intercept of –4.

8. In 1997, a small store had net sales of $195,000. In 1999, the same store had net sales of $250,000.

   a) Graph the sales as a function of the elapsed time since 1997.

   b) Find the rate of change of this function from 1997 to 1999.

   c) If the rate of change stays the same, what would the net sales be in 2001?

9. Write the equation of a horizontal line through the point (–3,–1).

10. Graph $y = 3^x$ and $y = \left(\frac{1}{3}\right)^x$ on the same coordinate system and compare the results.

11. Compare the graphs of $y^2 = 2x$ and $y = 2x^2$. Discuss how they are similar and how they are different.

12. Starting with a 10 cm by 16 cm rectangle, if we cut a 1 cm square out of each corner of the rectangle, we have:

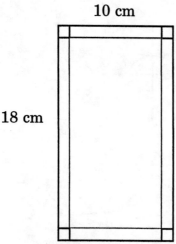

If a fold is made on the lightest lines, a box will be formed.

a) Calculate the surface area of the box (without the lid).

b) Calculate the volume of the box.

13. The population of Teaspoon Creek was 768 in 1998. If the population of Teaspoon Creek is expected to grow at an annual rate of 1.8% for the next 3 years, find the population of Teaspoon Creek in 1999, 2000, and 2001. (rounded to the nearest whole person!)

14. A box is 10 cm long, 4 cm wide and 3 cm deep. A cylinder is 10 cm high and has a base radius of 2 cm. Find the surface area and volume of each container.

15. If $5000 is deposited in an account paying 4.5% interest, compounded quarterly, find the amount of time needed (to the nearest year) for the amount in the account to double.

16. The graphs below represent two accounts paying the same interest but the interest in one account is compounded daily while the other one is compounded annually. Determine which graph corresponds to each account. Explain your reasoning.

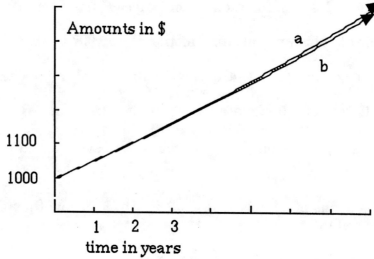

17. The following graph represents the height of a projectile thrown up into the air at a specific initial velocity. Use the graph to answer the following questions.

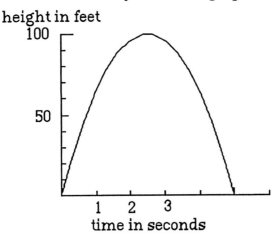

a) How long after it's thrown does the object reach its maximum height?

b) What is the height of the object 4 seconds after the initial throw?

c) Find the time the object is in the air when it is at a height of 30 feet. Explain the meaning of your answer(s).

18. Use the graphing calculator and the trace function to find:

a) the solution to each system:

i) $y = -x + 7$
$y = 2x + 4$

ii) $y = 2x^2 + x - 1$
$y = x + 5$

b) the zeros of each function:

i) $y = 3x - 5$

ii) $y = 3x^2 + 9x - 12$

iii) $y = x^2 + 3$

19. Given the graph below, answer the following questions with respect to this graph.

a) Find the zeros of the function.

b) Describe the intervals on which the function is increasing and decreasing.

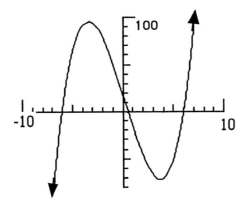

20. Given the following scatterplot of the number of sit-ups that can be done by participants after a number of weeks in a fitness program. Find the line of best fit

   a) using the median median line and

   b) using the graphing calculator

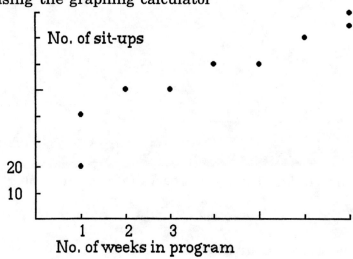

Exploration Problems

In the following 15"x 20" grid, if one square inch is cut off each corner as shown and the sides are folded along the dashed lines, we have an open-top box.

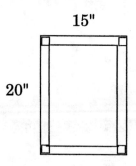

   a) Determine the surface area and the volume of this box.

   b) From the original grid, cut a 4" square from each corner. Again, fold along the dashed lines and find the surface area and volume of this open-top box.

   c) Continue in this fashion by cutting a 9" square and then a 16" square from the corners of the original grid and find the surface area and volume in each case.

   d) Describe your results.

Solutions - Chapter 7 Review

1. $C(n) = \$.99n$

2. a) $\$450 + .07x$ b) $\$701.16$

3. Possible answers: $y = 2x + 2$; $y = x^2 - 1$

4. a) $b = f(a) = a^2 - 1$ b) $y = f(x) = x + 2$ c) $y = f(x) = \frac{1}{3}x + 3$

5. a) A function, passes vertical line test
   b) Not a function
   c) A function, passes vertical line test
   d) Not a function, 8 maps to both 7 & 12

6. a) rate of change $= -1$, a constant
   b) Answers will vary, but the rate of change is not constant here

7. $y = -\frac{2}{5}x - \frac{8}{5}$

8. a)

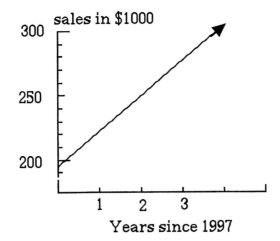

   b) rate of change $= 27500$.
   c) $s = 27500n + 195,000$ where s is the net sales and n is the number of years.
   d) In 2001 the net sales would be $\$305,000$.

9. $y = -1$

10. They are mirror images in the y-axis.

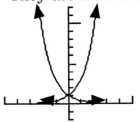

459

11. The graph of $y^2 = 2x$ is a parabola, opening to the right, and the graph of $y = 2x^2$ is a narrow parabola that opens upward. The vertex of each is at (0,0). Only $y = 2x^2$ describes a function. The domain of $y^2 = 2x$ must be $x \geq 0$.

12. a) S.A. = 176 sq. cm.   b) V = 128 cubic cm.

13. In 1999, pop. = 782 ; in 2000, 796; and in 2001, 810.

14. S.A. of box = 164 sq. cm., S.A. of cylinder = 150.72 sq. cm.
    V of box = 120 cu. cm., V of cylinder = 125.6 cu. cm.

15. 16 years

16. Graph a represents the account that is compounded daily and graph b represents the account that is compounded annually. The "compounded daily" account grows a little more rapidly, as previously illustrated.

17. a) At approximately 2.5 seconds   b) approximately 64 feet

    c) about 1.4 secs and 3.6 secs. The first time is when the object is 20 feet off the ground on its way up and the latter is when the object is on its way back to earth.

18. a) i) (1,6)   ii) at approximately (−1.8, 3.2) and (1.6, 6.6)

    b) i) $x = \frac{5}{3}$ ii) $x = -4$ and $x = 1$ iii) There are no zeros.

19. a) at −6, $\frac{1}{2}$ and 6

    b) The function is increasing when $x < -3$, decreasing when $-3 < x < 4$ and increasing when $x > 4$.

20. a) Answers will vary, one possibility: $y = 5x + 40$ b) $y = 7.0x + 27.7$

Chapter 7 Worksheet 1

1. Using the function, f(x) = –x + 1, make a t-table of values and plot these values on a coordinate system. Then, from the graph, find the x-value when f(x) = 4.

2. Given the table below. Write the equation as a function that expresses y in terms of x.

    ```
    x | y
    ---------
    0   -1
    2    0
    4    1
    6    2
    ```

3. Graph each of the following.

    a) $f(x) = (x + 1)^2$
    b) $f(x) = \sqrt{x}$ for $x \geq 0$

4. Given the two containers drawn below. If water is added to each container, $\frac{1}{2}$ cup at a time, indicate which graph best describes the increase in height of the water for each container. Justify your answer.

    1)

    2)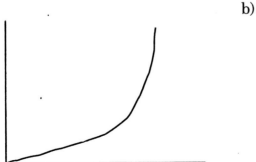

a)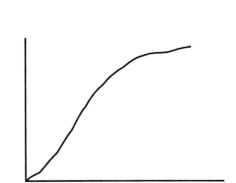

b)

5. Find the slope and y-intercept for each line.

   a) y = 6x + 5

   b) 2y = –x – 4

   c)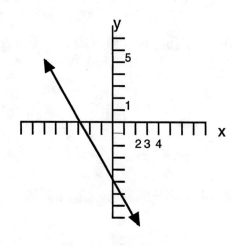

   d) x = 3

6. Graph the function, 40x – x² = f(x) for 0 ≤ x ≤ 40.

   a) For what value of x is the function a maximum?

   b) Describe a situation portrayed by this graph.

7. There are 500 bacteria in a culture. If the number of bacteria doubles each hour,

   a) make a table to show the population of bacteria for the first 8 hours;

   b) draw a graph to show how the bacteria population depends on the time elapsed.

8. Write the equation of each line:

   a) through the points (2,5) and (–1,–1).

   b) a horizontal line through (–1,3)

   c) with slope = $\frac{1}{4}$ and y-intercept = –2

   d) with slope = 0 and y-intercept of 3

   e) a vertical line through the point (6,2)

   f) through the points (0,–4) and (2,1)

9. Determine whether or not each of the following is a function. Write YES or NO.

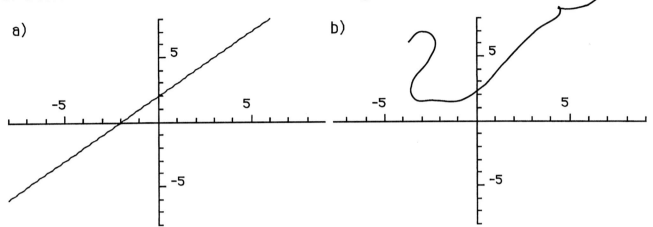

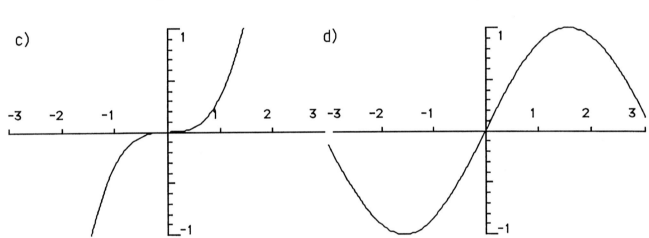

10. Determine the slope of the lines given by $y = \frac{1}{3}x + 1$ and $3y = x - 6$. Without drawing the graphs, what do you know about these lines?

Chapter 7 Worksheet 2

1. In words, compare the graphs of $y = \frac{1}{3}x^2 + 4$ and $y = -3x^2 - 5$.

2. The graph of $f(x) = (x - 3)(x + 3)$ is given below. Answer the following questions from the graph.

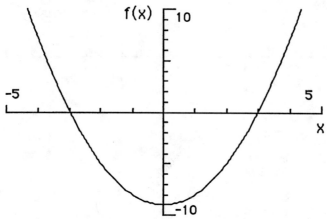

   a) What are the coordinates of the vertex of the parabola?

   b) What are the zeros of the function? How do they relate to the function as written above?

   c) Simplify $(x - 3)(x + 3)$ and describe, in words, the connection between the new expression and the graph.

3. Rominati's Pizza has two sizes, medium (12") and super-large (16"). If you can get two mediums for $10.99 or two super-large pizzas for $14.99, which size is the better buy?

4. If $5200 was deposited in a savings account, paying 5.1% interest, compounded daily, find the amount in the account after 10 years (assuming no withdrawals or deposits are made).

5. Graph the functions, $f(x) = 3x + 1$ and $f(x) = -x - 3$ on the same coordinate system. Find and interpret the point of intersection.

6. Use a graphing calculator to graph the function, $f(x) = (x + 1)(x - 3)$. Then find the zeros of the function and describe the intervals in which the function is increasing and decreasing.

7. A light filter placed a given distance from a light source allows only $1/2$ of the light to come through. Another filter placed directly behind the first will let $1/2$ of $1/2$, or $1/4$, of the light through. Find the amount of light that will pass through if 5 of these filters are used consecutively. Graph the relationship between the number of filters and the amount of light filtering through, based on an original intensity of 1.

8. Use the graphing calculator to find the solution to this system of equations:

$$y = x + 1$$
$$y - 2x = 4$$

9. Given the two cylinders below. Estimate which has the smaller surface area and which has the smaller volume. Then calculate the surface area and volume of each to check your results.

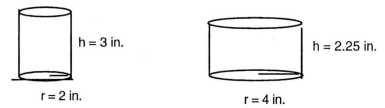

10. Given two types of cartons for popcorn as shown.

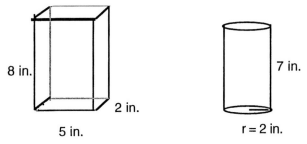

   a) Which holds the most popcorn?

   b) If the rectangular box sells for $2 and the cylindrical carton sells for $2.35, which is the better buy?

11. Given the data plotted below relating the number of sit-ups different people can do after a period of weeks in a fitness program. Approximate the line of best fit. Then use the graphing calculator to find the regression line and compare your results.

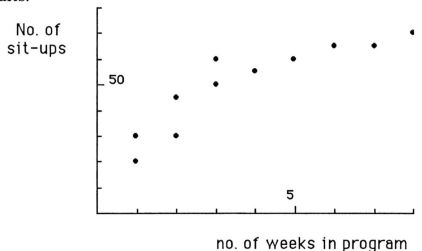

12. From the graph below, determine the approximate distance between Car 1 and Car 2 after 3 hours.

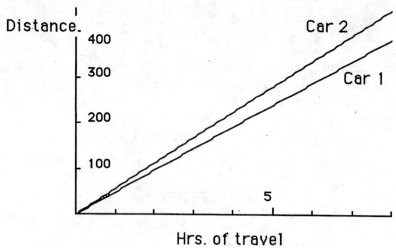